American Romanian Academy of Arts and Sciences (ARA), California, USA

http://www.AmericanRomanianAcademy.org

I0038059

Nanostructured Oxide Thin Films Synthesized by Spray Pyrolysis:

Characterizations and Applications

Edited by

Ruxandra Vidu
Najoua Kamoun-Turki

Title: Nanostructured Oxide Thin Films Synthesis by Spray Pyrolysis: Characterizations and Applications

Editors: Ruxandra Vidu and Najoua Kamoun-Turki

Contributors: Olfa Kamoun, Nasreddine Beji, Wafa Nafoutti, Refka Mimouni, Tarek Ben Nasr, Mehdi Souli, Mosbah Amlouk, and Najoua Kamoun-Turki.

Cover illustrator: Iulian Gherstoaga

Published by
ARA Publisher, a Publishing house of the American Romanian Academy of Arts and Sciences, University of California Davis, http://www.AmericanRomanianAcademy.org
Address: P.O. Box 2761
Citrus Heights, CA 95611-2761

Table of Contents

About the Editors

Ruxandra Vidu is currently an Assoc. Adj. Prof. of the University of California Davis, USA. She graduated from the University Politehnica Bucharest, Romania, and completed her PhD in 2000 at the School of Engineering, Osaka University, Japan. Dr. Vidu's research focuses on advanced materials and processes, characterization and nano-device integration. Her research interests include materials for energy and sustainable developments (iron oxide nanoparticles, superparamagnetic iron oxide nanoparticles (SPIONs), thermoelectrics, solar cells, water treatments); fabrication of thin films and nanowire arrays using electrochemistry and template synthesis; characterization of nanostructures and advanced materials; relationship between nanostructured materials and properties (physical, electrical, chemical, optical, magnetic and electrochemical properties). She is an invited speaker at scientific conferences, authored more than 100 papers in scientific international journals, book chapters and patents, and contributed to hundreds of communications at international conferences. Since 1996, Dr. Vidu has worked with leading universities in Japan, United States, Tunisia and Romania. She is a co-inventor of Bloo Solar's Solar Brush™ core technologies that developed the unique 3rd generation three dimensional architecture, thin films, and solar module. She is an elected member of the American Romanian Academy of Arts and Sciences since 2003.

Address: University of California Davis, Faculty of Engineering, Davis, California
Email: rvidu@ucdavis.edu, info@AmericanRomanianAcademy.org

Najoua Kamoun-Turki is a full Professor at the Faculty of Sciences of Tunis (FST) University of Tunis El Manar Tunisia since 2007. She obtained her PhD thesis in 1992 and the Habilitation (HDR) in Physics in Tunisia (FST) in 2000. Her academic research focuses on transparent conductive oxides (TCO: ZnO, SnO_2, In_2O_3, TiO_2, MoO_3, Fe_2O_3 and Fe_3O_4), binary semi-conductors (In_2S_3, SnS, CdS, Cu_2S, ZnS, PbS and MgS), ternary (P3HT and P3OT, $CuInS_2$, $In_{2-x}Ga_xS_2$,) and quaternary ($CuIn_{1-x}Ga_xS_2$:CIGS, Cu_2ZnSnS_4:CZTS) compounds for optoelectronic applications such as photocatalysis, gas sensors, solar cells, UV and IR detectors. Nanomaterials and thin films are grown by various low cost techniques (e.g. spray pyrolysis, chemical bath deposition (CBD), spin coating, electro-deposition and sputtering). She published about 100 papers in International Journals with impact factor and supervised more than 20 PhD students. Since 1989 she was a researcher in Physics Condensed Matter Laboratory (LPMC) where she was a head (2011-2015). During 2013-2014 she was the General Director of Physico-Chemical Analysis Institute (INRAP) in the Technopole of Sidi Thabet. She was a Vice President of AUF COMARES (for MAGHREB: 2013-2015) and she is, since 2013, a representative of the Ministry of Higher Education and Scientific Research on the Board of Directors of the National Metrology Agency (ANM). Since 2001 she is the Director of Synthesis of Nanomaterials and Thin Film Semiconductors for Optoelectronic Applications, Physics Condensed Matter Laboratory. She is an elected ARA and TWAS member since 2015 and 2016, respectively.

Address: University de Tunis El Manar, Faculty des Sciences de Tunis, Tunisia
Email: n.najouakamoun@gmail.com, Tel:00216 98 347470, fax: 0021671871666

Synthesis and Characterization of Europium-doped ZnO, In₂O₃ and MoO₃ Thin Films for Water Purification

Olfa Kamoun and Mosbah Amlouk

Université Tunis El Manar, Faculté des Sciences de Tunis, Département de Physique, LR99ES13 Laboratoire de Physique de la Matière Condensée (LPMC), 2092 Tunis Tunisie, Tunisia

Abstract: This chapter highlights emerging doped binary oxides based on zinc, molybdenum and indium in photocatalysis purpose which is currently a focus of research at laboratories around the world. First, the synthesis of europium doped ZnO, MoO₃ and In₂O₃ thin films by the spray pyrolysis method has been described. Moreover, the physical characterizations by means of XRD, MEB microscopy and optical spectroscopy techniques of these films are made in terms of Eu content. Second, the photoluminescence measurements of such doped films have been carried out under a large wavelength spectrum. Finally, the films prepared using appropriate content of etch doping element have been tested for photocatalysis purpose towards photodegradation of methylene blue (MB) dye.

Table of Contents

Introduction

Metals oxides semiconductors such as ZnO [1], In_2O_3 [2] and MoO_3 [3] have been widely studied because they can be obtained in several structures and nanostructures, and because of their high chemical and thermal stability [4]. Most of these oxides have wide band-gaps which make them great candidates for various applications such as solar cells [5, 6] and transparent conducting oxide thin film electrodes [7, 8]. These thin films metals oxides are also used in others interesting applications such as gas sensors [9, 10], acoustic and optical devices [11] and selective catalytists [12]. To date, the group of the most important nanomaterials includes simple metal oxides such as titanium oxide (TiO_2), zinc oxide (ZnO), copper oxide (CuO), magnesium oxide (MgO), indium oxide (In_2O_3), molybdenum oxide (MoO_3) and iron oxide (Fe_3O_4, Fe_2O_3). Metal oxides are finding an increasing number of applications in a wide range of fields and represent about one-third of the consumer products in the nanotechnology market. These materials are used as pigments in paints, sunscreens and cosmetics, antimicrobial agents in industrial and medical applications, etc.

The binary oxides based on zinc, indium and molybdenum are of great interest in the field of optoelectronics. Doping them with europium, increases their use in applications such as gas sensors, photocatalysis and photovoltaic solar cells. However, there are only a small number of studies on Eu-doping of these oxides. In the literature, europium element is frequently used as a doping element to improve the photocatalytic performances of the oxide thin films, while only a few research teams were interested in the influence of Eu doping on the optoelectronic properties of oxides. In this study, europium, which is a rare element in the earth-crust, was preferred as doping element because it would generate more intrinsic defects when incorporated in matrix, and would impact directly the optical and electrical properties of the oxide. Consequently, other properties related to the modification of the surface chemistry would be affected. On the other hand, in the field of gas sensors and waste water (including photocatalytic degradation and the decomposing of organic pollutants, contaminants and harmful dyes), the characteristic response of a polycrystalline semiconductor oxides can be controlled by several factors as follows: the size of particles, pore structure, grain density. Therefore, a comprehensive understanding of the influence of these parameters on the final product is essential.

Nowadays, semiconductor materials and oxides are more and more used in various industrial applications. Transparent conductive oxides (TCO) are materials that have good transparency in the visible range with relatively high electrical conductivity, making them attractive for optoelectronic applications. Morcover, the properties of nanomaterials depend on the size and morphology of nanoparticles. As a result, various nanostructured morphologies have been synthesized for various applications. A decrease of the crystallite size results in an increase in the surface-to-volume ratio. This allows that the surface defects play an important role in all surface phenomena such as gas adsorption, photocatalytic activity, wettability etc.

In this work, a comprehensive study of the effects of europium doping on the physico-chemical properties of thin layers of metal oxides deposited by chemical spraying. Among them, we chose those oxides that can be used in optoelectronic applications, as follows: zinc oxide (ZnO), a n-type semiconductor with a valence of 2 for Zn; indium oxide (In$_2$O$_3$), n-type semiconductor with a valence of 3 for In; molybdenum oxide (MoO$_3$), n-type semiconductor with a valence of 6 for Mo.

This chapter has several parts that will be briefly described. The first part describes in detail the basic properties of ZnO, In$_2$O$_3$ and MoO$_3$ thin films along with a review on the deposition methods to prepare thin films. Efficiency of these oxides in optoelectronic devices, photocalysis and gas sensing applications is also reviewed. Second part presents the spray pyrolysis technique setup that was used to deposit europium doped ZnO, In$_2$O$_3$ and MoO$_3$ thin films on glass substrate at 460°C. The crystal structures were investigated by XRD and Raman spectroscopy techniques. Then, the lattice parameters, crystallite size, the microstrain and Raman modes of the samples were calculated and correlated with the incorporation of metal ions. The chemical composition, growth rates of the thin films and verification of the Eu incorporation in the MoO$_3$ lattice were achieved using the XPS analytical technique. In the third part, the optical and the emission properties of the Eu-doped ZnO, In$_2$O$_3$ and MoO$_3$ thin films are presented. The reflectance and transmittance measurements were carried out by spectrophotometry, and then the absorption coefficient α, the band gap energy E$_g$ as well as the Urbach energy EU values have been estimated. The fourth part deals with the rate of electron-hole recombination and the amount of intrinsic defects in the doped oxide films obtained using photoluminescence (PL) spectroscopy. Investigation of the surface morphology has been performed using scanning electron microscopy (SEM). Finally, photocatalytic tests have been carried out using Eu doped ZnO, In$_2$O$_3$ and MoO$_3$ thin films against MB dye under UV-Vis radiations. These results have been discussed in terms of PL data as well as the morphology of thin film surface as revealed by SEM observations.

1. Properties

1.1 Physico-chemical properties ZnO, In$_2$O$_3$ and MoO$_3$ basic elements

Zinc (Zn) is a chemical element with the atomic number **30**, its most common oxidation state is **+2**. Zinc is a blue-gray color metal, medium reactive, which combines with oxygen and other non-metals. It reacts with dilute acids to produce hydrogen. Zinc crystallizes in hexagonal system. Above the boiling temperature (1180 K), zinc distillation occurs, but it starts to emit vapors well below this temperature. Zinc oxide is widely used as a catalyst in the manufacture of rubber. It is also used as a heat disperser for the rubber and acts to protect its polymers from ultraviolet radiation (the same UV protection is conferred to plastics containing zinc oxide) [13].

Indium (In) is a chemical element with the atomic number **49**. Its most common oxidation state is **+3**. It is a silvery-white, shiny, metal and highly ductile and one of the softest metals known. It crystallizes in a tetragonal structure. Indium metal dissolves in acids, but does not react with oxygen at room temperature. At relatively higher temperatures, it combines with oxygen to form indium oxide (In$_2$O$_3$). Indium is of considerable industrial importance, most notably is in the production of the transparent conductive coatings. Indium oxide (In$_2$O$_3$) and indium tin oxide (ITO) are used as a transparent conductive coating applied to glass substrates in the making of electroluminescent panels. [14] Indium is used in photovoltaics for the synthesis of the semiconductor copper indium gallium selenide (CIGS), which is used in the manufacturing of CIGS solar cells, a type of second generation thin film solar cell [15]. It is also used in light-emitting diodes (LEDs) and laser diodes based on compound semiconductors such as InGaN, InGaP.

Molybdenum (Mo) is a transition metal in group VI B of the periodic table, located between chromium (Cr) and tungsten (W). It has an atomic number of **42.** The name is from Neo-Latin molybdenum, from Ancient Greek molybdos, meaning lead. This metal is silvery white, strong, malleable. It dissolves in diluted nitric acid and aqua regia (mixture of nitric acid and hydrochloric acid) and it can be found in a variety of oxidation states (II to VI).

Oxygen (O) is a chemical element of the chalcogen group, with the atomic number **8.** Its most common oxidation states are **+2, -1**. This is a non-metal that forms very easy compounds including oxides, with almost all the other elements, it is colorless, odorless and slightly magnetic. Also, it crystallizes in a cubic structure. The main physic-chemical properties of these elements are listed in table I.1.

Table I.1 Physico-chemical parameters of zinc (Zn), Indium (In), Molybdenum (Mo) and Oxygen (O) elements.

Properties	Zinc (Zn)	Indium (In)	Molybdenum (Mo)	Oxygen (O)
Electronic configuration	$[Ar]$ $3d^{10}4s^2$	$[Kr]\ 4d^{10}\ 5s^2$ $5p^1$	$[Kr]\ 4d^5\ 5s^1$	$[He]\ 2s^2$ $2p^4$
Atomic number	30	49	42	8
Atomic mass (g.mol^{-1})	65.39	114.82	95.94	15.99
Density (g.cm^{-3})	7.13	7.31	10.2.	1.43
Atomic radius (pm)	142	156	190	48
Ionic radius (nm)	0.074	$0.080\ 10^{-3}$	0.06	0.14
Ionic charge (e)	(+2)	(+3)	(+6)	(-2)
Boiling Point (°C)	907	2072	4 639	-182,95
Electronegativity	1.65	1.79	2.16	3.44
Oxidation state	+2	+3	-2, 0, 1, 2, 3, 4, 5, 6	-2

1.2. Physico-chemical properties of the dopant: Europium

Europium (Eu) is the most reactive rare earth element. It is a chemical element with atomic number **63**. Europium usually assumes the oxidation state +3. It is a ductile metal with hardness similar to that of lead. It crystallizes in a body-centered cubic lattice [16]. Most applications of europium exploit its phosphorescence property. It is a dopant in some types of glass in lasers and other optoelectronic devices. Europium oxide (Eu_2O_3) is widely used as a red phosphorin television sets and fluorescent lamps, and as an activator for yttrium-based phosphors [17]. A recent (2015) application of europium is in quantum memory chips which can reliably store information for days at a time; these could allow sensitive quantum data to be stored to a hard disk-like device (Figure I.1) [18]. The most important physico-chimical properties of europium element are listed in Table I.2.

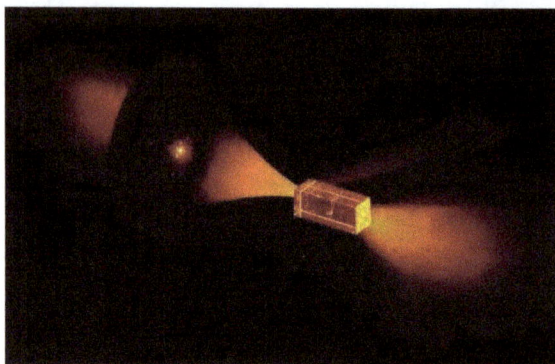

Figure I.1 Writing quantum information onto a europium ion embedded
in a crystal. Image Solid State Spectroscopy Group, ANU [18].

Table I.2. Physico-chemical parameters of Europium element.

Properties	Electronic configuration	Atomic number	Atomic mass, g/mol	Density, g.cm^{-3}	Atomic radius, pm	Ionic radius, pm
Europium (Eu)	[Xe] $4f^7 6s^2$	63	151.964	5.24	233	240

Boiling Point, °C	Oxidation state
1529	+2, +3

1.2. Crystal Structure
- **Zinc Oxide (ZnO)**

ZnO is a II-VI semiconductor binary compound. It can adapt to three different crystallographic phases (Figure I.2): (a) cubic phase (Rocksalt), (b) Blende phase and (c) hexagonal phase (wurtzite). At ambient conditions, the thermodynamically stable phase is hexagonal wurtzite. The blende structure can be stabilized only by growth on cubic substrates, and the rocksalt structure (NaCl) may be obtained at relatively high pressures (10-15 GPa).

Zinc oxide hexagonal wurtzite structure has a polar hexagonal c-axis which is chosen to be parallel to z. This structure binary belongs to space group of P63mc (C_{6v}^4). Each cation (zinc ion) is surrounded by four anions (oxygen ions) at the corners of a tetrahedron, and vice versa. This tetrahedral coordination is characteristic of sp^3 covalent bonding nature. The primitive unit cell contains two units of ZnO. Also, this binary oxide can be described by a hexagonal Bravais lattice with constants **a=b** =3.2496Å and **c** =5.2042Å; $\alpha = \beta = 90°$ and γ =120°[19].

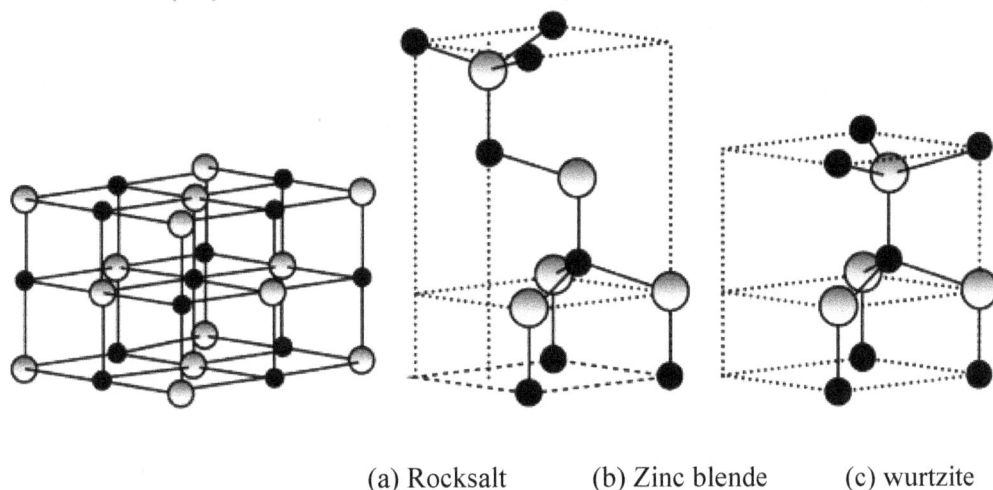

(a) Rocksalt   (b) Zinc blende   (c) wurtzite

Figure I.2. Representation of the crystal structures of ZnO: (a) cubic rocksalt, (b) cubic zinc blende and (c) hexagonal wurtzite.

In this compact hexagonal lattice, 6 atoms are in the same compact plan, 3 in the top plan and 3 in the bottom plan which makes 12 neighboring atoms. These atoms are located in the following positions:

Zn: (0, 0, 0); (1/3, 2/3, 1/2)
O: (0, 0, 3/8); (1/3, 2/3, 7/8)

- **Indium Oxide (In$_2$O$_3$)**

The crystalline form of In_2O_3 exist in two phases, rhombohedral and the cubic, Figure I.3. Both phases have a band gap of about 3 eV [20]. The rhombohedral phase is produced at high pressures and temperatures. It has a space group R3c No. 167, a = b = 0.5487 nm, c = 0.57818 nm, Z = 6 and calculated density 7.31 g/cm^3. However, the cubic phase has a lattice parameter a=1.011 nm.

- **MoO$_3$**

The basic crystalline structure of molybdenum oxide is an octahedron. The Mo atom is at the center of the octahedron, while O atoms are at the corners. Molybdenum trioxide can exist in different crystalline structures (Figure I.4). The orthorhombic phase α-MoO$_3$ is a thermodynamically stable phase. The two metastable phases are: (a) monoclinic β-MoO$_3$ and (b) hexagonal h-MoO$_3$. The fundamental structural properties of hexagonal ZnO, In$_2$O$_3$ and MoO$_3$ are listed in **Table I.3.**

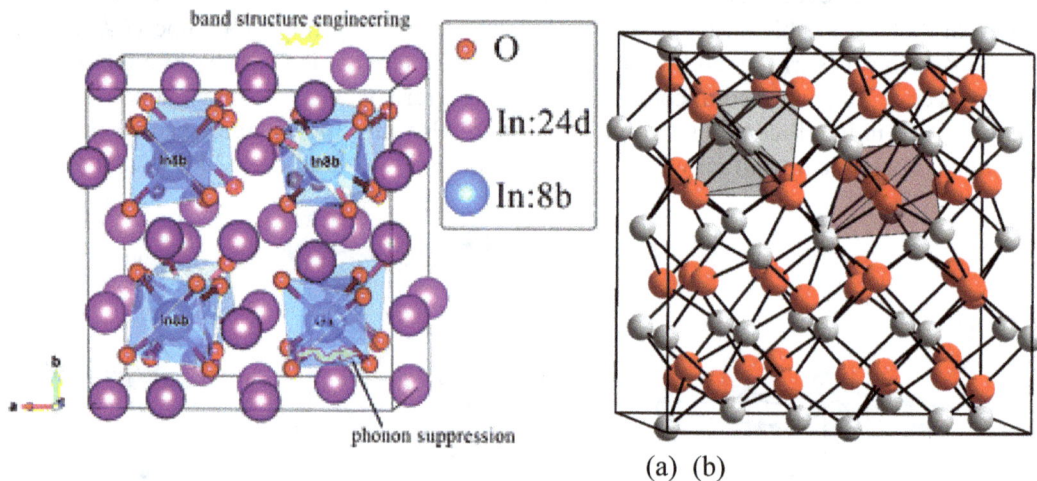

(a) (b)

Figure I.3. The crystal structures of In_2O_3: rhombohedral (a) and cubic (b) structure.

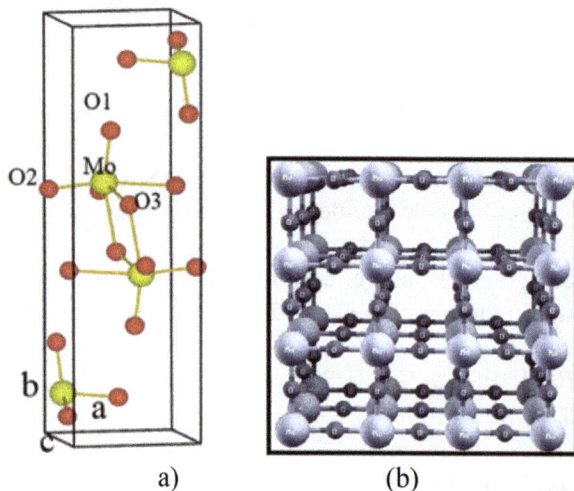

a) (b)

Figure I.4. Representation of the crystal structures of MoO_3: (a) α-MoO_3 an orthorhombic structure (b) monoclinic β-MoO_3 phase.

At room temperature, the orthorhombic structure is the most stable and favored phase of molybdenum oxide. It consists of corner-sharing chains of MoO_6 octahedra with two similar chains to form the layers of MoO_3 stoichiometric (Figure I.4). In each octahedron of MoO_6 there is only an oxygen that is not shared (O_1), two oxygen atoms that are common to two octahedra (O_2) and three oxygen atoms that are shared between three octahedra (O_3). Monoclinic structure of MoO_3 is metastable and it is similar to that of WO_3 [21] represented by ReO_3 distorted perovskite (pseudo cubic), which consists of a network of octahedrons MoO_6 interlinked by oxygen atoms. There are many oxidation states of molybdenum such as: MoO_2, Mo_4O_{11}, Mo.

Table I.3. Structural properties of zinc oxide In_2O_3 and MoO_3.

	ZnO	In_2O_3	MoO_3
Lattice parameter a_0 (nm) at 300K	0.32495	1.01	0.3962
Lattice parameter c_0 (nm) at 300K	0.52069	-	0.3697
c_0/a_0	1.602	-	0.933
Coordination (Z)	2	16	4
Stable phase at 300 K	Wurtzite	Cubic	Orthorhombic
Space group	P63mc (C_{6V})		Pbnm
Melting point	1975 °C	1910 °C	795 °C

1.3. Electronic Structure

- **Electronic Band Structure of ZnO**

The band structure of zinc oxide is critical in determining its potential utility. It is essentially due to its crystallographic structure and to electronic configurations. The electronic configurations structures of oxygen band and zinc are:

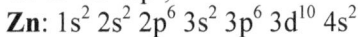

$\quad$ **O**: $1s^2\, 2s^2\, 2p^4$;
$\quad$ **Zn**: $1s^2\, 2s^2\, 2p^6\, 3s^2\, 3p^6\, 3d^{10}\, 4s^2$

The 2p oxygen states form the valence band and the 4s zinc states constitute the conduction area of the ZnO semiconductor. Therefore, each of zinc atom releases two electrons from the orbital 4s to the orbital 2p of an oxygen atom to form the ionic bond.

The reaction of ZnO formation is as follows:

$\quad$ $Zn^{2+} + 2\,e^- + 1/2\,O_2\; ->\; ZnO$ $\quad$ (II)

Since ZnO is a direct gap semiconductor with the global extrema of the uppermost valence (valence band) and the lowest conduction bands (conduction band) at the same point Γ in the Brillouin zone, we are interested in the region where $k = 0$ (at the Γ-point). The lowest conduction band is formed by the empty 4s states of Zn^{2+} or the anti-bonding sp^3 hybrid states. The ZnO band gap energy is about 3.3 eV. This band gap corresponds to the energy required by an electron to move from the valence band to the conduction band. A typical representation of the band structure is illustrated in Figure I.5.

- **Electronic Band Structure of In_2O_3**

The band structure of indium oxide is essentially due to its crystallographic structure and to electronic configurations. The electronic configurations structures of oxygen band and indium are:

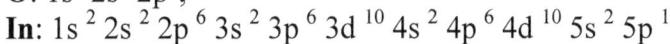

$\quad$ **O**: $1s^2\, 2s^2\, 2p^4$;
$\quad$ **In**: $1s^2\, 2s^2\, 2p^6\, 3s^2\, 3p^6\, 3d^{10}\, 4s^2\, 4p^6\, 4d^{10}\, 5s^2\, 5p^1$

The 2p oxygen states form the valence band, 5s the 5p indium states constitute the conduction area of the In_2O_3 semiconductor. Therefore, each of indium atom releases three electrons from the orbital 5s and 5p to the orbital 2p of an oxygen atom to form the ionic bond. The reaction of In_2O_3 formation is as follows: $In^{3+} + 3\,e^- + 1/2\,O_2\; ->\; In_2O_3$ $\quad$ (II)

- **Electronic Band Structure of MoO$_3$**

The band structure of a Molybdenum Oxide is essentially due to its crystallographic structure and to electronic configurations. The electronic configurations structures of oxygen band and molybdenum are:

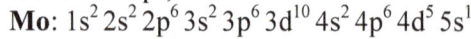

O: $1s^2 2s^2 2p^4$;

Mo: $1s^2 2s^2 2p^6 3s^2 3p^6 3d^{10} 4s^2 4p^6 4d^5 5s^1$

The 2p oxygen states form the valence band and the 4d and 5s molybdenum states constitute the conduction area of the MoO$_3$ semiconductor. Therefore, each of molybdenum atom releases six electrons from the orbitals 4d and 5s to the orbital 2p of an oxygen atom to form the ionic bond. The reaction of MoO$_3$ formation is as follows:

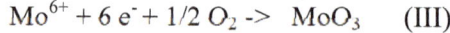

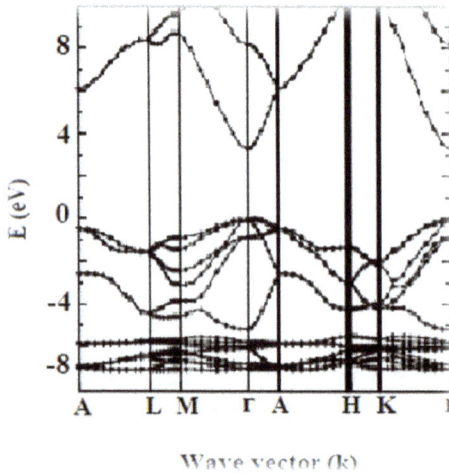

$$Mo^{6+} + 6\ e^- + 1/2\ O_2 \rightarrow\ MoO_3 \quad \text{(III)}$$

Figure I.5. Band diagram of ZnO [22].

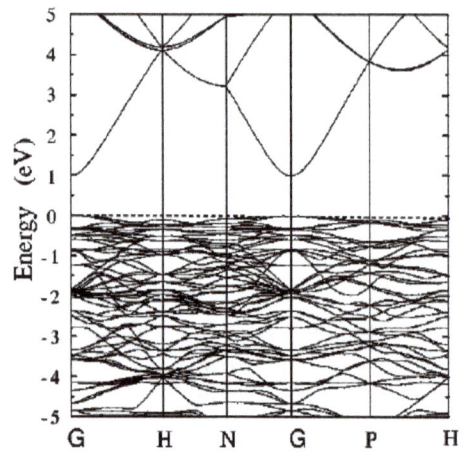

Figure I.6 Electronic band structure of In$_2$O$_3$ [23]

Since MoO$_3$ is a direct gap semiconductor with the global extrema of the uppermost valence (valence band) and the lowest conduction bands (conduction band) at the same point in the Brillouin zone, we are interested in the region where k = 0 (at the Γ-point). The lowest conduction band is formed by the empty 5s states of Mo^{6+} or the anti-bonding sp^3 hybrid states. The MoO$_3$ band gap energy is about 3.5 eV. This band gap correspond to the energy must be provided to an electron to move it from the valence band to the conduction one. A typical representation of the band structure is illustrated in Figure I.7:

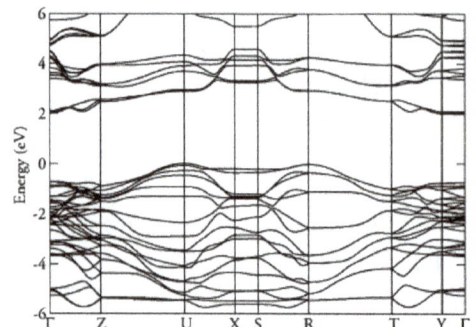

Figure I.7
Electronic band structure of MoO$_3$ [24]

1.4. Optical properties

The interaction of light (electromagnetic wave) with a material occurs when the electromagnetic wave interacts with the matter and it will be completely absorbed by it if the energy associated with the electromagnetic wave is capable of transferring electrons from the valence band to the conduction one, that is to say if the energy is at least equal to the width of the forbidden band. When going from the solid to thin film state, the optical properties differ considerably. These properties depend on the preparation method, the quality of thin films, the heat treatment applied, the type and concentration of the dopant. In Table I.4, some of optical properties of zinc, indium and molybdenum oxides are summarized.

Table I.4. Optical properties of zinc, indium and molybdenum oxides [26-28].

	ZnO	In_2O_3	MoO_3
Dielectric constant	$\varepsilon// = 8.7$ $\varepsilon^\perp = 7.8$		$\varepsilon_\infty = 5.31$
Absorption coefficient	104 cm-1	2500 cm-1	-
Refractive index	2.006 – 2.029	1.958	2.3
Band gap energy	3.3 eV (Direct)	3.7 eV (Direct)	3.05
Exciton binding energy	60 meV	130 meV	-
Transmittance	> 90%	~90%	~90%

1.5. Electrical properties

ZnO is an intrinsic n-type semiconductor II-VI. This behavior has been reported in the literature and it is assigned to the presence of native defects in its crystal structure like oxygen vacancies and zinc interstitials [29].

The electron mobility values of undoped ZnO nanostructures are estimated to vary from 120 to 440 cm^2Vs^{-1} at room temperature, depending on the fabrication method. Moreover, it is possible to modify the zinc oxide properties by doping with metal ions. It has been reported that after ZnO doping, the highest carrier concentration for holes and electrons was ~ $10^{19} cm^{-3}$ and $10^{20} cm^{-3}$, respectively [30].

Until now, producing a stable p-type ZnO was challenging because it is unstable Doping with p-type materials (N, P, Sb) is difficult due to the low solubility of the dopants and the resistance to p-type doping, which is due to the self-compensation of shallow acceptors resulting from various generated donor defects such as oxygen vacancies or interstitial zinc. In table I.5, some electrical properties of ZnO, In_2O_3 and MoO_3 are presented.

1.6. Native defects in ZnO, In₂O₃ and MoO₃

Native defects are imperfections in the crystal lattice that involve only the constituent elements [31]. They strongly influence the electrical and optical properties of a semiconductor. Examples of defects include interstitials, vacancies as well as anti-sites. Generally, native defects are related to the compensation of the donor or acceptor dopants. Acceptor defects are easier to form p-type while donor defects form n-type semiconductors, as follows:

- Oxygen vacancies are deep donors and cannot explain the observed n-type conductivity in ZnO. However, they can compensate p-type doping.
- Oxygen interstitials are electrically inactive, they can be divided as deep acceptors or interstitials at the octahedral site in n-type samples and they have high formation energies.
- Oxygen anti-sites are deep acceptors and they have the highest formation energies among the acceptor-type native point defects.
- Zinc vacancies are deep acceptors and they have low formation energies below n-type conditions, they consequently take place as compensating defects in n-type samples.
- Zinc interstitials are shallow donors and they have high formation energies below n-type conditions.
- Zinc antisites are also shallow donors and they have high formation energies in n-type samples.

Table I.5. Electrical properties of ZnO, In₂O₃ and MoO₃

	ZnO	In₂O₃	MoO₃
Effective mass of the electron	$0.24\ m_e$	$0.3\ m_e$	$0.85\ m_e$
Hole effective mass	$0.59\ m_h$	$0.78\ m_h$	-
Electron Hall mobility at 300 K	$200\ cm^2V^{-1}s^{-1}$	$15\ cm^2V^{-1}s^{-1}$	-
Intrinsic carrier concentration	$< 10^6 cm^3$	$3.3 \times 10^{20}\ cm^3$	-
Thermal conductivity	$1\text{-}1.2\ Wm^{-1}K^{-1}$ at 300K	$10\ Wm^{-1}K^{-1}$ at 300K	-

1.7. Strategies for ZnO, In₂O₃ and MoO₃ performance

Generally, the doping of zinc oxide, indium oxide and molybdenum oxide with appropriate elements are one of the most effective ways in the search for applications. Therefore, understanding these interactions will facilitate the fundamental research and technical applications. Interaction efficiency depends on the difference in ionic radius and electronegativity between the dopant and host element. Doping ZnO, In₂O₃ and MoO₃ with transitional metals have been successfully employed. Also, they improve the optical properties of the oxide material. Certain results and applications of various research works on doped ZnO, In₂O₃ and MoO₃ are listed in Table I.6, I.7 and I.8, respectively.

1.8. Certain applications of doped zinc oxide, molybdenum oxide and indium oxide

Because of their unique properties, ZnO, In_2O_3 and MoO_3 have attracted considerable attention, which make it suitable for a broad variety of applications. During the last decade, thousands of research papers on ultraviolet absorbance, wide chemistry, piezoelectricity and luminescence at high temperatures, have featured ZnO, In_2O_3 and MoO_3 and introduced them to industry. The ability of a material to be a catalyst in a specific process depends on its chemical nature and surface properties. ZnO, In_2O_3 and MoO_3 efficiency are essentially due to their crystal lattice, semiconductor properties (vacancies, interstitials...), which act on energy surface and semi-conductor doping [57].

Table I.6 A summary table of various elements used for doping ZnO.

Doping element	* Ionic radius * electro-negativity	Technique growth	Band Gap energy	applications	Ref
Al	0.054 nm 1.61	pulsed laser	3.08	Light emitting devices	[32]
Co	0.072 nm 1.88	spin coating	3.05 eV	MB photocatalyst	[33]
Cr	0.063nm 1.66	solvothermal route	-	MO photocatalyst	[34]
F	0.133nm 3.98	spray pyrolysis	-	p-type ZnO layer	[35]
In	0.08 nm 1.78	pulsed laser	3.07 eV	Transparent conducting oxide	[36]
Li	0.076 nm 0.98	Dc reactive magnetron sputtering	3.36 eV	p-type ZnO layer	[37]
N	0.171 nm 3.04	molecular beam epitaxy	3.32 eV	p-type ZnO layer	[38]
P	0.038 nm 2.19	molecular beam epitaxy	-	p-type ZnO layer	[39]
Sb	0,076 nm 2.05	molecular beam epitaxy	-	p-type ZnO layer	[40]

Table I.7. A summary table of various elements used for doping In_2O_3.

Doping element	* Ionic radius * electro-negativity	Technique Growth	Band Gap energy	applications	Ref
Fe	75 pm 1.83	spray pyrolysis	3.29–3.45	Transparent conducting oxide	[41]
Zn	74 pm 1.65	spray pyrolysis	3.29	Transparent conducting oxide	[42]
Sn	80 pm 1.96	Electron beam evaporation	3.61-3.89	Transparent conducting oxide	[43]

Mo	73 pm 2.16	Spray	3.69-3.82	optoelectronic	[44]
W	72 pm 2.36	Reactive plasma	3.60-3.83	solar cells	[45]
Sb	76 pm 2.05	Spray	3.90	acetone sensing	[46]
Ga	62 pm 1.81	Sol gel	-	transistors	[47]
Zr	86 pm 1.33	Spray	3-3.5	Transparent conducting oxide	[48]
Cr	80 pm 1.66	Sol gel	3.76	Electrical transport	[49]
Eu	117 pm 1.2	Spray	3.43–3.51	optical window	[50]

Table I.8. A summary table of various elements used for doping MoO_3

Doping element	* Ionic radius * electro-negativity	Technique growth	Band Gap energy	applications	Ref
Eu	240 pm 1.2	hydrothermal method	-	MB Photocatalyst	[51]
Cd	109 pm 1.69	hydrothermal method	2.88	H_2S sensing	[52]
Sn	80 pm 1.96	spray pyrolysis	4.004	TCO	[53]
Si	54 pm 1.9	Flame spray pyrolysis	3.63	sensing of NH_3	[54]
Ce	115 pm 1.12	impregnation	-	MB Photocatalyst	[55]
Fe	75 pm 1.83	hydrothermal method	-	H_2S sensing	[56]

Electrochromic study

Transition metal oxides are studied with respect to their electrochromic properties for application as smart windows and display devices. Among them, tungsten oxide and molybdenum oxide are the most investigated materials as promising electrochromic layers for smart windows application [58, 59].

In the literatures, electrochromic is a phenomenon of color change in a persistent but reversible manner produced the reversible intercalation-deintercalation mechanism of ions and electron. The insertion of ions and electrons in amorphous metal oxide thin films can switch from their bleached state to a colored state according to the reaction given bellow:

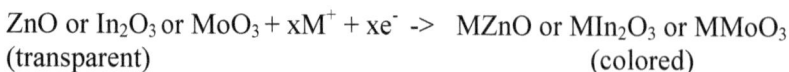

$$ZnO \text{ or } In_2O_3 \text{ or } MoO_3 + xM^+ + xe^- \rightarrow MZnO \text{ or } MIn_2O_3 \text{ or } MMoO_3$$
(transparent)   (colored)

where M^+ is H^+, Li^+, Na^+ or K^+ ions.

The optical properties, i.e. transmittance, absorbance and reflectance, are modulated when ions are inserted into or extracted from the electrochromic film under negative applied potential, which occurs by electrically controlling (EC) the oxidation state. Smart windows are able to control these properties throughput of visible light and solar radiation into buildings and can impart energy efficiency as well as human comfort by having different transmittance levels depending on dynamic needs showed in Figure I.8.

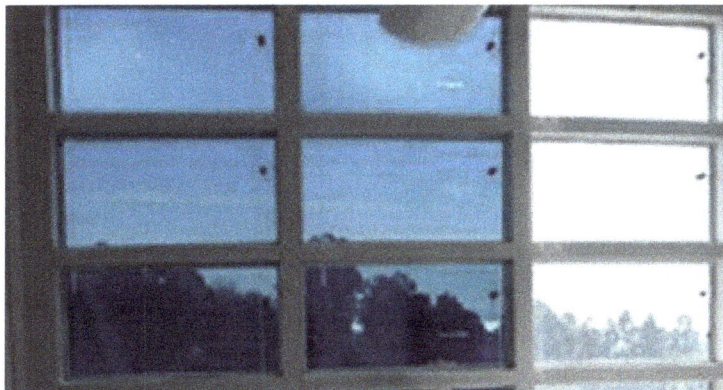

Figure I.8. Test installations of the electrically controlling EC smart windows.

Studies done on these materials confirmed that atmospheric CVD deposited MoO_3 and ZnO or In_2O_3 films possess optical and electrochromic characteristics suitable for application in smart windows. Deepa *et al.* **[60]** have successfully obtained mixed oxide films based on Mo and In and this study led to conclusion that In_2O_3-MoO_3 combines the excellent electrochromic performance of In_2O_3 and the high growth rate of MoO_3 when fabricated by spray process. Hsu *et al.* **[51]** studied the effect of annealing temperature ranging from 100 °C to 500 °C on the electrochromic properties of MoO_3 thin films prepared by sol-gel technique. The electrochromic properties of these films were investigated by performing cyclic voltammetry in propylene carbonate solutions containing 1 M of $LiClO_4$. The heat treated MoO_3 thin films at 100 °C was light yellow, the layers treated at 200, 300 and 350 °C were colorless and transparent but the films treated at 400 and 500 °C presented white and opaque layers. During the repetitive voltammetry sweeps, the cyclic voltammograms and the optical transmittance showed a good reversibility between colored and bleached states. The transparent films which heat treated at 350 °C exhibited the best and the excellent electrochromic properties in this study. When the temperature increases above 350 °C, the colored states changed slightly while the bleached states degraded significantly and the decrease in grain boundaries caused an increase in grain size and degraded the coloration performance. Also, the layer structure of VO_2/ZnO/glass is considered to be applicable in smart windows of high total energy efficiency in buildings or automobiles [52].

Gas sensors

Metal oxide semiconductor, such as tungsten oxide and molybdenum oxide which are wide band gap n-type semiconductors, are the most promising sensing material for detection of toxic and dangerous gases like NO_2, H_2, O_3, NH_3, H_2S. For this application, metal oxide semiconducting gas sensors are very attractive due to their sensitivity, stability, small size and low cost. The gas sensor technology is based on the change in resistance of a metal oxide layer due to the chemical adsorption and desorption of the gas on the sensor surface. For an n-type semiconductor exposed to a reducing gas, the resistance decreases, whereas, upon exposure to oxidizing gas, the resistance increases. The resistance variation of a sensor with time of exposure to gas is depicted by a typical response curve as shown in Figure I.9.

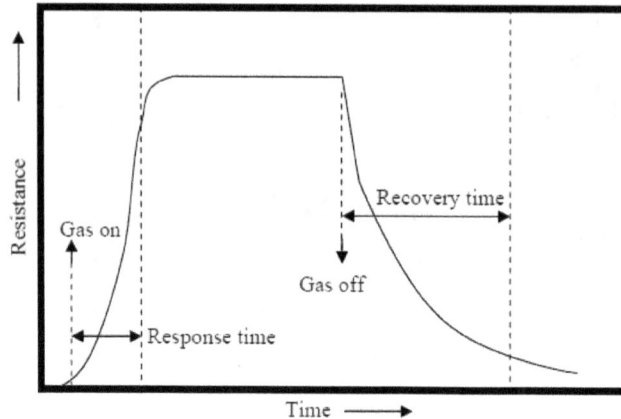

Figure I.9. A diagram representing the response and recovery time in gas sensors.

The response of a gas sensor is characterized by various parameters such as:
• Sensitivity: defined as the ratio of resistance change of a sensor upon exposure to gas and for the n-type materials;

$$S = \frac{\Delta R}{R_{gas}}$$

Sensors of high S value are desirable in order to sense low concentration of gases.
• Selectivity: expressed in terms of a dimension that compares the concentration of the corresponding interfering gas that produces the same sensor signal and it expressed as:

$$Selectivity = \frac{sensitivity\ of\ the\ sensor\ for\ interfering\ gas}{sensitivity\ towords\ the\ desired\ gas}$$

• Response time: is the time interval over which the resistance of the sensor material attains a fixed percentage of final value when the sensor is exposed to full scale concentration of the gas.
• Recovery time: is the time interval over which sensor resistance reduces to 10% of the saturation value when the sensor is exposed to full scale concentration of the gas and then placed in the clean air.

When the metal oxide semiconducting gas sensor exposed to air, a mass of oxygen molecules will adsorb on the surface of thin films and trap its electrons in the conduction band and transform them to O_2^-, O^{2-}, O^-. H_2O due to the formation of thick electron depletion layer [53]. As exposing the sensor to a reducing gas such as the ethanol, O^{2-} on the materials surface will oxidize ethanol molecules into CO_2 and H_2O seen in Figure I.10. The electron depletion layer will become thinner, which will reduce the resistance of the sensor.

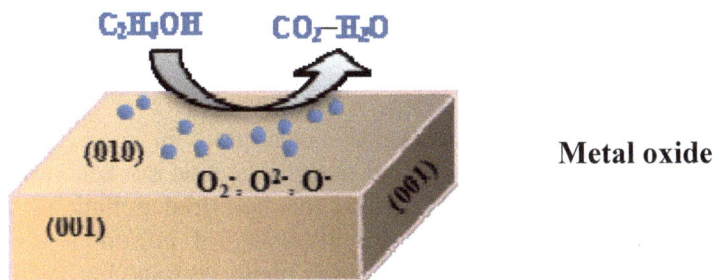

C_2H_5OH CO_2–H_2O

(010)

O_2^-, O^{2-}, O^-

(001)

(001)

Metal oxide

Figure I. 10 Schematic of the sensing reaction process.

For the n-type sensor material, there will be pre-adsorbed oxygen ions on the surface of the sensor. The gas will then react with oxygen ions to form molecules, and the electron transfer to the sensor material will result in a decrease of the resistance. When sensors were to O_2, O_3, CO, NO_2 gases and ethanol vapor, molybdenum trioxide based sensors showed promising O_3 gas sensing characteristics and the resistivity was reduced by mixing it with TiO_2. In cases of O_3, CO and NO_2 gas and ethanol vapor sensing, its performance is enhanced. The MoO_3 also showed a high response to ethanol vapor outperforming the single metal oxides [54].

ZnO based gas sensors have been the topic of research works over the last few decades. However, researchers have been exploring different possibilities for lowering the operating temperatures and optimizing the sensor response in terms of sensitivity, selectivity, shorter response times and faster recovery times. The surface sensitivity is essentially attributed to the presence of adsorbed species. The conductivity of ZnO thin films is very sensitive to the exposure of the surface to various gases such as (CO, CO_2, H_2, C_2H_5OH...) [27, 41]. The reaction on the surface between the adsorbent and the adsorbate causes the change in current which is recorded in order to obtain the sensitivity of the material. The current change depends on the number of reaction sites available for a given material.

Inyawilert *et al.* [55]. synthesized In_2O_3 sensing film by a sparking process which can be used to detect ethanol and acetone. In_2O_3 is sensitive to various reducing and oxidizing gases and is thus promising for environmental gas-sensing applications [55-59].

Transparent conductive oxides (TCO)

Transparent conducting thin films achieve large values of electrical conductivity, whereas maintaining a high transmission in the visible range. In last decades, TCOs have undergone significant development and have attracted much interest. They are used as transparent electrodes in flat panel displays [44,45], capable of transporting electrical charges and transmitting visible photons, promising materials for UV-emitting devices and electroluminescent equipment [46]. ZnO is one of the most important transparent conductive oxides in the classification of wide band gap materials because of his high binding energy (60 meV) at room temperature and his wide band gap (3.3 eV) [47,48]. Moreover, TCO such as doped zinc oxide has attracted much interest in the application of optoelectronic devices such as solar cells due to their high conductivity and high transparency in the visible region [28, 29, 32]. Among the n-type TCO materials, indium tin oxide (ITO) doped with Sn, has been widely commercialized. This is because the ITO film has high performance of both good high transmittances of ~90% and electrical conductivity of ~10-4 $\Omega \cdot cm$ [60]

Photocatalysis

Intensive research work has taken place in recent years on photocatalysis. The continuous development of global industrialization has become a major issue that has been addressed by researchers to reduce the industrial pollutions. As the demands for solar energy, pure air and water, and disposing of poisonous and hazardous pollutant are increasing, the use of semiconductor photocatalysts for environmental purification has attracted considerable attention. The removal of a wide variety of pollutants is due to the small size, low cost, simple operation and good reversibility.

In photocatalysis, an electron-hole pair is produced below the intensity of light through intermediary oxidation or reduction reactions that take place on the surface of the catalyst. An organic pollutant can be oxidized directly by means of a photogenerated hole or indirectly via a reaction with characteristic reactive groups, for example the hydroxyl radical $OH^{\cdot}$, produced in solution [38], in the presence of a photocatalyst. Zinc oxide, molybdenum trioxide and indium oxide have been extensively studied because of their particular physical properties, good chemical stability and efficient absorption of most of the UV-light and part of the visible light. Zinc oxide provides similar or higher activity than TiO_2 below the UV range [39]. Also, it has higher stability as nanomaterials, i.e. better crystallinity and smaller defects [40]. The photocatalytic activity of ZnO can be improved by doping [31, 34, 35, 40]

1.9. Statistics

We can see in Figure I.10 the number of articles on M:Eu, where M is ZnO or In_2O_3 or MoO_3. There is a small number of papers regarding the Eu doping, and among the earth rare constituent, it is the best candidate as photocatalytic material.

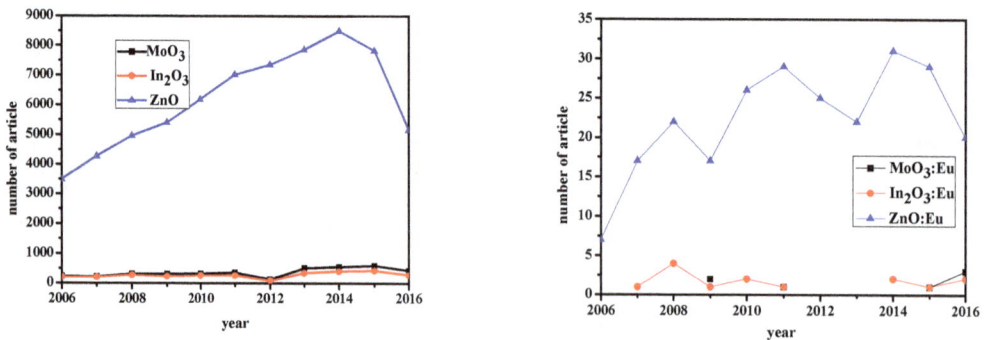

Figure I.10 Number or articles published on this subject during 2006-2016.

1.10 References

[1] K. AL-Abdullah, Elaboration of ZnO Based Varistors and the Effect of the Rare-Earths on their Electrical Behaviour, Energy Procedia, 19 (2012) 116-127

[2] P.C. Lansåker, P. Petersson, G.A. Niklasson, C.G. Granqvist, Thin sputter deposited gold films on In_2O_3:Sn, SnO_2:In, TiO_2 and glass: Optical, electrical and structural effects ,Solar Energy Materials and Solar Cells, 117 (2013) 462-470

[3] L.Boudaoud, N.Benramdane, A.Bouzidi, A.Nekerala, R.Desfeux, $(MoO_3)_{1-x}(V_2O_5)_x$ thin films: Elaboration and characterization, Optik - International Journal for Light and Electron Optics, 127 (2016) 852-854

[4] M. Peuckert, XPS study on surface and bulk palladium oxide, its thermal stability, and a comparison with other noble metal oxides, The Journal of Physical Chemistry, 89 (1985) 2481-2486

[5] M.A.Martfnez, J.Herrero, M.T. Gutièrrez, Deposition of transparent and conductive Al-doped ZnO thin films for photovoltaic solar cells, Sol. En. Materials and Solar Cells, 45 (1997) 75-86

[6] J.V.Lagemaat, T.M.Barnes, G.Rumbles, S.E.Shaheen, T.J.Coutts, C.Weeks, I.Levitsky, J.Peltola, P.Glatkowski, Organic solar cells with carbon nanotubes replacing In_2O_3:Sn as the transparent electrode APPLIED PHYSICS LETTERS 88 (2006) 23350- 233503

[7] T.Minami, Transparent Conducting Oxide Semiconductors for Transparent Electrodes, Semiconductor Science and Technology, 20 (2005) S35-S44

[8] T.Minami, Substitution of transparent conducting oxide thin films for indium tin oxide transparent electrode applications, Thin Solid Films, 516 (2008) 1314-1321

[9] Q.Wan, Q.H.Li, Y.J.Chen, T.H.Wang, X.L.He, J.P.Li, C.L.Lin, Fabrication and ethanol sensing characteristics of ZnO nanowire gas sensors, Applied Physics Letters, 84 (2004) 3654-3656

[10] N.Barsan, U.Weimar, Conduction Model of Metal Oxide Gas Sensors, Journal of Electroceramics, 7 (2001) 143-167

[11] C. J. Lee, T. J. Lee, S. C. Lyu, Y. Zhang, H. Ruh, H. J. Lee, Field emission from well-aligned zinc oxide nanowires grown at low temperature, Applied Physics Letters, 81 (2002) 3648-3650

[12] M. L. Cubeiro, J. L. G. Fierro, Selective Production of Hydrogen by Partial Oxidation of Methanol over ZnO-Supported Palladium Catalysts, Journal of Catalysis, 179 (1998) 150-162

[13] J. Emsley, Nature's Building Blocks: An A-Z Guide to the Elements, (2001) 503

[14] Azonano, The Electroluminescent Light Sabre, Nanotechnology News Archive, 2005

[15] M.Powalla, B.Dimmler, Scaling up issues of CIGS solar cells, Thin Solid Films, 2000, 361

[16] A. F. Holleman, E. Wiberg, Inorganic Chemistry, Academic Press: San Diego, 2001

[17] Caro, Paul, Rare earths in luminescence, Rare earths. (1998) 323–325

[18] http://phys.org/news/2015-01-quantum-hard-breakthrough.html

[19] Ü. Özgür, Y.I. Alivov, C. Liu, A. Teke, M. Reshchikov, S. Doğan, V. Avrutin, S.-J. Cho, H. Morkoc, A comprehensive review of ZnO materials and devices, Journal of Applied Physics 98 (2005) 041301.

[20] A. Walsh, Juarez L. F. Da Silva, Su-Huai Wei, C. Körber, A. Klein, L. F. J. Piper, A. DeMasi, K. E. Smith, G. Panaccione, P. Torelli, D. J. Payne, A. Bourlange, and R. G. Egdell, Nature of the Band Gap of In_2O_3 Revealed by First-Principles Calculations and X-Ray Spectroscopy, Physical Review Letters. 100 (2008) 167402

[21] K. Khojier, H. Savaloni and S. Zolghadr, On the dependence of structural and sensing properties of sputtered MoO_3 thin films on argon gas flow, Applied Surface Science, 320 (2014) 315–321

[22] A. Janotti, C.G. Van de Walle, Fundamentals of zinc oxide as a semiconductor, Reports on Progress in Physics 72 (2009) 126501

[23] B. Zhang, Electronic band structure of indium-based transparent and conductive multicomponent oxide, Int. J. Microstructure and Materials Properties, 9 (2014) 79-87

[24] D. O. Scanlon, G. W. Watson, D. J. Payne, G. R. Atkinson, R. G. Egdell and D. S. L. Law, Theoretical and Experimental Study of the Electronic Structures of MoO_3 and MoO_2, The Journal of Physical Chemistry C , 114 (2010) 4636–4645

[25] N. Preissler and O. Bierwagen, Electrical transport, electrothermal transport, and effective electron mass in single-crystalline In_2O_3 films, Physical Review B, 88 (2013) 085305

[26] D.P. Norton, Y. Heo, M. Ivill, K. Ip, S. Pearton, M.F. Chisholm, T. Steiner, ZnO: growth, doping & processing, Materials today 7 (2004) 34-40

[27] K. Irmscher, M. Naumann, M. Pietsch, Z. Galazka, R. Uecker, T. Schulz, R. Schewski, M. Albrecht, and R. Fornari, On the nature and temperature dependence of the fundamental band gap of In2O3 Physica status solidi, 211 (2014) 54–58

[28] S. K. Deb and J. A. Chopoorian, Optical Properties and Color-Center Formation in Thin Films of Molybdenum Trioxide, Applied physics, 37 (1966) 4818.

[29] C.G. Van de Walle, Hydrogen as a cause of doping in zinc oxide, Physical Review Letters 85 (2000) 1012.

[30] D. Seghier, H. Gislason, Shallow and deep donors in n-type ZnO characterized by admittance spectroscopy, Journal of Materials Science: Materials in Electronics 19 (2008) 687-691.

[31] J. Bourgoin, Point Defects in Semiconductors: Theoretical Aspects, Springer1981.

[32] G. Kaur, A. Mitra, K. Yadav, Pulsed laser deposited Al-doped ZnO thin films for optical applications, Progress in Natural Science: Materials International 25 (2015) 12-21.

[33] G. Poongodi, P. Anandan, R.M. Kumar, R. Jayavel, Studies on visible light photocatalytic and antibacterial activities of nanostructured cobalt doped ZnO thin films prepared by sol–gel spin coating method, Spectrochimica Acta Part A: Molecular and Biomolecular Spectroscopy 148 (2015) 237-243.

[34] C. Wu, L. Shen, Y.-C. Zhang, Q. Huang, Solvothermal synthesis of Cr-doped ZnO nanowires with visible light-driven photocatalytic activity, Materials Letters 65 (2011) 1794-1796.

[35] A. Sanchez-Juarez, A. Tiburcio-Silver, A. Ortiz, E. Zironi, J. Rickards, Electrical and optical properties of fluorine-doped ZnO thin films prepared by spray pyrolysis, Thin Solid Films 333 (1998) 196-202.

[36] S. Chirakkara, K. Nanda, S. Krupanidhi, Pulsed laser deposited ZnO: In as transparent conducting oxide, Thin Solid Films 519 (2011) 3647-3652.

[38] D.C. Look, D. Reynolds, C. Litton, R. Jones, D. Eason, G. Cantwell, Characterization of homoepitaxial p-type ZnO grown by molecular beam epitaxy, Applied physics letters 81 (2002) 1830-1832.

[39] F. Xiu, Z. Yang, L. Mandalapu, J. Liu, W. Beyermann, p-type ZnO films with solid-source phosphorus doping by molecular-beam epitaxy, Applied physics letters 88 (2006) 052106.

[40] F. Xiu, Z. Yang, L. Mandalapu, D. Zhao, J. Liu, W. Beyermann, High-mobility Sb-doped p-type ZnO by molecular-beam epitaxy, Applied Physics Letters 87 (2005) 152101-152101.

[41] N. Beji, M. Souli, M. Ajili, S. Azzaza, S. Alleg and N. Kamoun Turki, Effect of iron doping on structural, optical and electrical properties of sprayed In_2O_3 thin films, Superlattices and Microstructures, 81(2015) 114–128

[42] N. Beji, M. Souli, S. Azzaza, S. Alleg and N. Kamoun Turki, Study on the zinc doping and annealing effects of sprayed In_2O_3 thin films, Journal of Materials Science: Materials in Electronics, 27 (2016) 4849–4860

[43] V. Senthilkumar, P. Vickraman, M. Jayachandran, C. Sanjeeviraja, Structural and optical properties of indium tin oxide (ITO) thin films with different compositions prepared by electron beam evaporation, Vacuum, 84 (2010) 864–869

[44] P. Prathap, N. Revathi, K.T. Ramakrishna Reddy and R.W. Miles, Thickness dependence of structure and optoelectronic properties of In_2O_3:Mo films prepared by spray pyrolysis, Thin Solid Films, 51 (2009) 1271–1274

[45] F. Meng, J. Shi, Z. Liu, Y. Cui, Z. Lu and Z. Feng, High mobility transparent conductive W-doped In_2O_3 thin films prepared at low substrate temperature and its application to solar cells, Solar Energy Materials and Solar Cells, 122 (2014) 70–74

[46] N.G. Pramod and S.N. Pandey, Influence of Sb doping on the structural, optical, electrical and acetone sensing properties of In_2O_3 thin films, Ceramics International, 40 (2014) 3461–3468

[47] S. Jeong, J. Y. Lee, M. H. Ham, K. Song, J. Moon, Y. H. Seo, B. H. Ryu and Y. Choi, Bendable thin-film transistors based on sol–gel derived amorphous Ga-doped In_2O_3, semiconductors, Superlattices and Microstructures, 59 (2013) 21–28

[48] C. Manoharan, M. Jothibas, S. Johnson Jeyakumar and S. Dhanapandian, Structural, optical and electrical properties of Zr-doped In_2O_3 thin films, Spectrochimica Acta Part A: Molecular and Biomolecular Spectroscopy, 145 (2015) 47–53

[51] A.Phuruangrata, U.Cheed-Ima, T.Thongtemb, S.Thongtemc, High visible light photocatalytic activity of Eu-doped MoO_3 nanobelts synthesized by hydrothermal method, Materials Letters 172 (2016) 166–170

[52] S. Baia, C. Chena, D. Zhangb, R. Luoa, D. Li, A. Chen, C. C. Liu, Intrinsic characteristic and mechanism in enhancing H_2S sensing of Cd-doped α-MoO_3 nanobelts, Sensors and Actuators B: Chemical, 204 (2014) 754–762

[53] A. Boukhachem, O. Kamoun, C. Mrabet, C. Mannai, N. Zouaghi, A. Yumak, K. Boubaker, M. Amlouk, Structural, optical, vibrational and photoluminescence studies of Sn-doped MoO_3 sprayed thin films,Materials Research Bulletin, 72 (2015) 252

[54] A. T. Güntner, M. Righettoni, S. E. Pratsinis, Selective sensing of NH_3 by Si-doped α-MoO_3 for breath analysis, Sensors and Actuators B: Chemical, 223 (2016) 266

[55] Z. Liu, Y. Jin, F. Teng, X. Hua, M. Chen, An efficient Ce-doped MoO_3 catalyst and its photo-thermal catalytic synergetic degradation performance for dye pollutant, Catalysis Communications, 66 (2015) 42

[56] Q. Y. Ouyang, L. Li, Q. S. Wang, Y. Zhang, T. S. Wang, F. N. Meng, Y. J. Chen, P. Gao, Facile synthesis and enhanced H_2S sensing performances of Fe-doped α-MoO_3 micro-structures, Sensors and Actuators B: Chemical, 169 (2012) 17

[57] Y. Zeng, Z.-z. Ye, J. Lu, W. Xu, L. Zhu, B. Zhao, S. Limpijumnong, Identification of acceptor states in Li-doped p-type ZnO thin films, Applied Physics Letters 89 (2006) 1-3.

[58] K.A.Gesheva, T.M.Ivanova, G.Bodurov, Transition metal oxide films:Technology and Smart Windows electrochromic device performance, Progress in Organic Coatings, 74 (2012) 635–639.

[59] I. Hamberg and C. G. Granqvist, Evaporated Sn-doped In_2O_3 films: Basic optical properties and applications to energy-efficient windows, Journal Applied Physics, 60 (1986) R123- R159

[50] P.Deepa, V.Sivaranjani, P.Philominathan, MoO_3- In_2O_3 Binary Oxide Thin Film Deposition for TCO Application in Solar Cell, International Journal of Advance Research In Science And Engineering, 4 (2015) 419-426

[51] C. S. Hsu, C. C. Chan, H. T. Huang, C. H. Peng and W. C. Hsu, Electrochromic properties of nanocrystalline MoO_3 thin films, Thin Solid Films 516 (2008) 4839–4844

[52] J. W. Ma, Y. R. Song, G. Xu, L. Miao, Vanadium Dioxide Thin Films Deposited on ZnO Buffer Layer for Smart Thermochromic Glazing of Windows, Applied Mechanics and Materials, 361 (2013) 370-373

[53] K. Galatsis, Y. X. Li, W. Wlodarski, E. Comini, G. Sberveglieri, C. Cantalini, S. Santucci and M. Passacantando, Comparison of single and binary oxide MoO_3, TiO_2 and WO_3 sol-gel gas sensors, Sensors and Actuators B, 83 (2002) 276–280.

[54] M. Ferroni, V. Guidi, G. Martinelli, P. Nelli, M. Sacerdoti, G. Sberveglieri, Characterization of a molybdenum oxide sputtered thin film as a gas sensor, Thin Solid Films, 307 (1997) 148-151

[55] K. Inyawilert, A. Wisitsora-at, A. Tuantranont, P. Singjai, S. Phanichphant, C. Liewhiran, Ultra-rapid VOCs sensors based on sparked-In_2O_3 sensing films, Sensors Actuat. B-Chem, 192 (2014) 745–754

[56] M. Seetha, S. Bharathi, D. Mangalaraj, D. Nataraj, Reducing gas response kinetics of nanostructured indium oxide thin films, Thin Solid Films 518 (2010) 125–128.

[57] A. Qurashia, E.M. El-Maghrabyb, T. Yamazakia, Y. Shena, T. Kikuta, A generic approach for controlled synthesis of In_2O_3 nanostructures for gas sensing applications, J. Alloys Compd. 481 (2009) 35–39.

[58] J. Xu, Y. Chen, J. Shen, Ethanol sensor based on hexagonal indium oxide nanorods prepared by solvothermal methods, Mater. Lett. 62 (2008) 1363–1365.

[59] L.Liu, T.Zhang, S.Li, L.Wang, Y.Tian, Preparation, characterization and gas-sensing properties of Pd-doped In2O3 nanofibers, Mater. Lett. 63 (2009) 1975–1977.

[60] S.Sohn and H.M.Kim, Transparent Conductive Oxide (TCO) Films for Organic Light Emissive Devices (OLEDs), Organic Light Emitting Diode – Material, Process and Devices, 233-275.

2. Synthesis and characterization of the Eu doped ZnO and In_2O_3 thin films

The crystal structures of all prepared thin films were analyzed by X-ray diffraction (Philips PW 1729 system) using Cukα monochromatic radiation (λ = 1.54059 Å). Raman spectroscopy measurements were recorded at room temperature using Jobin Yvon HR LabRAM in backscattering co-focal configuration with a spatial resolution of 1μm and spectral resolution less than 0.35 cm^{-1}. The light excitation is an Ar^+ laser at the wavelength of 488 nm. Finally, all these structural results are discussed in terms of Eu incorporation in each binary oxide.

2.1. Films preparation by spray pyrolysis

Spray pyrolysis deposition is a cost-effective technique that does not require complicated equipment and can be easily transferred to large scale industry fabrication processes that are presently used to obtain good quality films. The spray pyrolysis setup used to prepare the ZnO, In_2O_3 and MoO_3 un-doped and doped films is presented schematically in Figure II.1 [1].

2.2 Precursors

All thin films oxides have been deposited on glass substrates at 460 °C. Based on previous studies, the temperature to prepare binary oxides was selected [1, 3]. During the spray pyrolysis process, the carrier gas was nitrogen (pressure $\approx$ 0.35 bar) through a 0.5 mm-diameter nozzle. The nozzle-to-substrate distance was fixed at the optimal value of 27 cm as demonstrated by Boubaker *et al.* [2]. During the whole deposition process, precursor mixture flow rate was approximately 4 ml/min^{-1}.

ZnO thin films were prepared using propanol and zinc acetate, $Zn(CH_3CO_2)_2$:10^{-1} M [1]. The precursor mixture was acidified using acetic acid (pH=5). Under similar experimental conditions, europium-doped ZnO:Eu thin films solutions have been fabricated by adding europium oxide to the precursor solution while maintaining the acidity value. In this study, the europium-to-zinc molar ratios y= $[Eu^{3+}]$ / $[Zn^{2+}]$ were 0%, 0.4%, 0.7% and 1%. After deposition, the films were cooled to room temperature. These values of y ratio did not exceed 1% because beyond this value, the solution was not clear and the presence of impurities that might have clogged the nozzle.

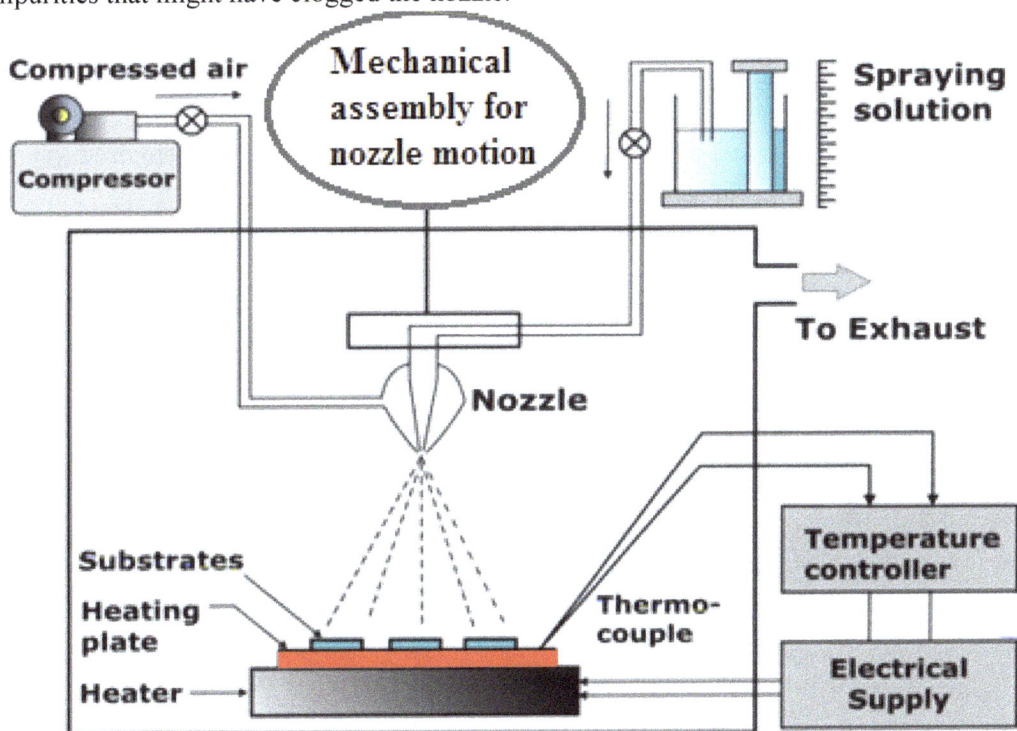

Figure II.1. Spray pyrolysis technique

In_2O_3 thin films were obtained using 0.01 M aqueous solution of indium (III) chloride tetrahydrate [$InCl_3$ ·4H_2O]. Nitrogen was used as the carrier gas (pressure at 0.35

bar) through a 0.5 mm diameter nozzle. During the deposition process, the precursor mixture flow rate was taken constantly at 4 ml min−1. Under similar experimental conditions, europium-doped In_2O_3: Eu thin film solutions have been prepared by adding europium oxide to precursor solutions, while maintaining acidity level. In the elaborated samples, the europium-to-indium molar ratios $y = [Eu^{3+}] / [In^{3+}]$ were 0.5, 1, 1.5 and 2%.

MoO_3 thin films were deposited using 0.01 M aqueous solution of ammonium molybdate tetrahydrate $[(NH_4)_6Mo_7O_{24}, 4H_2O]$. Europium oxide Eu_2O_3 dissolved in acetic medium was used as a source of europium and mixed to the previous solution at various europium-to-molybdenum molar ratios to produce the Eu-doped MoO_3 films. In practice the molar Eu/Mo ratio was fixed to 0.5, 1.0, 1.5 and 2%.

All growth parameters of undoped and Eu doped ZnO, In_2O_3 and MoO_3 are summarized in Table II.1.

Table II.1. Growth parameters of un-doped and Eu doped ZnO, In_2O_3 and MoO_3

	ZnO	In_2O_3	MoO_3
Precursors	zinc acetate dehydrate $(Zn(CH_3-COO)_2 2H_2O)$	indium (III) chloride tetrahydrate $[InCl_3 \cdot 4H_2O]$	ammonium molybdate tetrahydrate $[(NH_4)_6Mo_7O_{24}, 4H_2O]$
Solvent	propanol-2 (C_3H_8O) and deionized water (3:1)	deionized water	deionized water
Deposition temperature	460 °C	460 °C	460 °C
Concentration	$10^{-1}M$	$10^{-2}M$	$10^{-2}M$
Doping percentage	$y= [Eu^{3+}] / [Zn^{2+}]$ were 0%, 0.4%, 0.7% and 1%	$y = [Eu^{3+}] / [In^{3+}]$ were 0.5, 1, 1.5 and 2%.	$y = [Eu^{3+}] / [Mo^{6+}]$ were 0.5, 1, 1.5 and 2%.

2.3. X-ray diffraction analysis

X-ray diffraction analysis provides information regarding the crystalline structure. In the following, we present X-Ray patterns of ZnO, In_2O_3 and MoO_3 and Eu doped films. Also, some structural constants have been calculated and compared to those found in JCPDF cards.

2.3.1. ZnO thin film

Figure II.2, is the X-ray diffraction patterns of the doped and undoped zinc oxide thin films, showing the effect of Eu concentrations in the spray solution $(y= [Eu^{3+}]/[Zn^{2+}])$ on the crystallinity of ZnO.

X-ray diffraction spectra show well-defined peaks of (100), (002), (101), (102) and (103), corresponding to hexagonal würtzite phase according to JCPDS 036-1451 card. Also, it is noted that XRD peaks are very narrow showing a good crystallinity of the prepared thin films. The interplanar spacing d_{hkl} values of ZnO thin films were also calculated by using Bragg equation [4 –6]:

$$2d_{hkl}sin\theta = n\lambda \qquad (II.1)$$

where n is the order of diffraction (usually n = 1) and λ is the X-ray wavelength. In the ZnO hexagonal structure, the plane spacing is related to the lattice parameters a and c as well as the Miller indices by the following relation [4]:

$$\frac{1}{d_{hkl}^2} = \frac{4}{3}\left(\frac{h^2 + hk + k^2}{a^2}\right) + \frac{l^2}{c^2} \qquad (II.2)$$

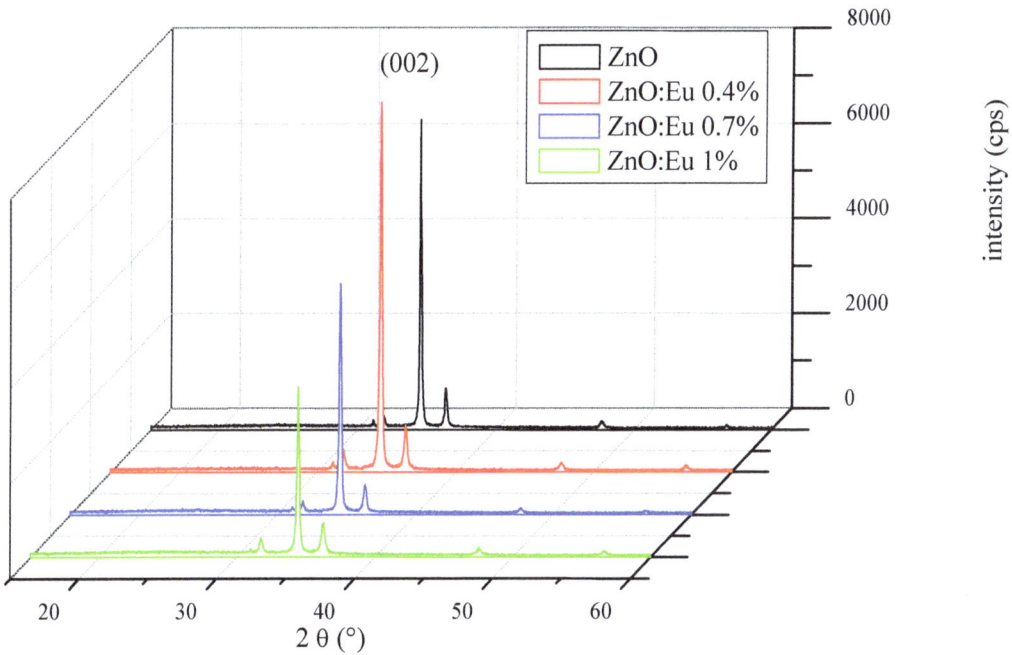

Figure II.2. XRD spectra for the ZnO:Eu thin film grown by spray pyrolysis for different Eu content.

Calculated values of the lattices parameters and their (c/a) ratios of ZnO:Eu sprayed thin films as well as the volume unite cells are presented in Table II.2. It is found that these parameters reach high values for 1% Eu doping level and are minimum for 0.7% of Eu content. The later doping level can be considered particular for improving some physical properties of such doping II–VI binary thin films. Indeed, the compactness of the host material has a maximum value at this special [Eu]/[Zn] ratio, this can pave the way for high interaction of this 0.7% Eu-doped ZnO oxide against photon light and electric response as well as sensitivity applications. Regarding the X-ray spectra of the ZnO:Eu thin films, it is found the intensities of the X-Ray diffraction decrease with Eu concentration for y=[Eu]/[Zn] higher than 0.4 at%, which is related to the influence of various factors such as spray temperature, flow rate of the sprayed solution and spontaneous cooling. In this case, it may happen that the crystallites are oriented, not completely random, but preferably in one or more particular directions, i.e. the material has a texture. This effect is studied through texture coefficient using the following relation [5–7]:

$$TC(hkl) = \frac{I(hkl)/I_0(hkl)}{N^{-1}\sum_n I(hkl)/I_0(hkl)} \quad \text{(II.3)}$$

where I(hkl) is the measured relative intensity of (hkl), I_0(hkl) is the standard intensity of the same plane (hkl) taken from JCPDS 036-1451 card and N is the reflection number.

The calculated texture coefficients are listed in Table II.3. These results reveal the existence of a preferred orientation within (002) plan for different concentration of europium indicating that the crystallites are parallel to the $\vec{c}$ axis. Also, the TC value of (002) orientation decreases with the europium content. This result may be related to the difference in ionic radius: Zn^{2+} radius (74 pm) is smaller than Eu^{3+} one (94.7 pm) [8].

Table II.2. Lattice parameters of hexagonal structure in ZnO:Eu thin films

	a(Å)	c(Å)	Volume (Å^3)	c/a
ZnO	3.2413	5.1915	47.2346	1.6017
ZnO:Eu 0.4%	3.2418	5.1887	47.2239	1.6006
ZnO:Eu 0.7%	3.2397	5.1887	47.1610	1.6016
ZnO:Eu 1%	3.2474	5.2058	47.5423	1.6031

Table II.3. Texture coefficient TC(hkl) values of ZnO:Eu grown for different Eu content

	TC			
	ZnO	**ZnO :Eu 0.4%**	**ZnO :Eu 0.7%**	**ZnO :Eu 1%**
(100)	0.294	0.383	0.32373	0.429
(002)	6.097	6.094	5.532	3.769
(102)	0.171	0.155	0.716	0.224
(110)	--	0.106	0.159	0.137
(103)	0.236	0.162	0.176	0.205
(112)	0.095	--	--	0.122
(004)	0.104	0.097	0.090	0.089

XRD spectra were also used to study the effect of incorporation of Eu element in ZnO matrix. Indeed, some structural constants have been calculated such as: the crystallite size (D) using the formula Deby–Scherrer [3, 8], the stress ξ [9] and dislocation density δ_{dis} [3, 9]:

$$D = \frac{k\lambda}{\beta_{1/2}\,cos\theta} \quad (II.4)$$

$$\xi = \frac{\beta\,cos\,\theta}{4} \quad (II.5)$$

$$\delta_{dis} = \frac{1}{D^2} \quad (II.6)$$

where k=0.90 is the Scherrer constant, $\beta_{1/2}$=corrected half width of the peak and λ=1.54Å is the wavelength of CuKα radiation. These parameters are calculated along the preferred (002) orientation their values are presented in Table II.4.

Table II.4. Dislocation density of ZnO:Eu thin films

	D(nm)	$\xi(10^{-4})$	δ_{dis} (10^{13} lines/m^2)
ZnO	60.8	19.2	27
ZnO:Eu 0.4%	74.1	15.7	18
ZnO:Eu 0.7%	63.7	18.3	24
ZnO:Eu 1%	44.4	26.3	51

From this structural study, it was found a particular effect for 1% $[Eu^{3+}]/[Zn^{2+}]$ ratio. For low doping, levels, a high value of lattice parameter c of the hexagonal structure was obtained (see Table II.2). The same doping level gives the less crystallite size and a high value of both stress and dislocation density parameters. Moreover, it can be seen that the highest value of crystallite size (D) 74.1nm and the lowest values of the stress ξ=15.7.10^{-4} and dislocation density δ_{dis}=1.8 10^{14} lines/m^2 are obtained for Eu doping level of 0.4 at% with the highest intensity of (002) peak. This may be due to the coalescence of the crystallites of ZnO caused by lower Eu doping.

2.3.2. In$_2$O$_3$ thin film

X-ray diffraction spectra of europium-doped indium oxide thin films are displayed in Figure II.3, which shows the effect of concentrations ratios of europium and indium ions in the spray solution (y=$[Eu^{3+}]/[In^{3+}]$, $0 \leq y \leq 2$ at%) on the crystallinity of indium oxide. XRD patterns are found to match the JCPDS card 06-0416. With further increase of europium concentration until 1.5 at%, a sharp and intense (004) peak is obtained, which corresponds to an enhancement of crystal quality. When europium content is higher than 1.5 at%, the intensity of (004) peak decreases to the detriment of the increase of (222) one. For an europium content of 1.5 at%, we note the presence of new peak at 2θ equal to 31.7° corresponding to Eu$_3$O$_4$ orthorhombic phase oriented on to (211) direction. The texture coefficient (TC) which indicates the maximum preferred orientation of the films along the diffraction plane means that the increase in the preferred orientation is associated with the increase in the number of grains along that plane. The TC(hkl) calculated values of Eu doped In$_2$O$_3$ thin films are given in Table II.5.

Table II.5. Texture coefficient TC(hkl) values of Eu-doped In_2O_3 spayed thin films.

y= $[Eu^{3+}]/[In^{3+}]$ at%	TC				
	0	0.5	1	1.5	2
TC(112)	0.685	1.011	0.927	0.573	1.118
TC(222)	0.516	0.801	0.852	0.288	0.996
TC(004)	3.269	2.726	2.557	3.755	2.451
TC(044)	0.252	0.181	0.287	0.131	0.198
TC(226)	0.275	0.278	0.374	0.251	0.235

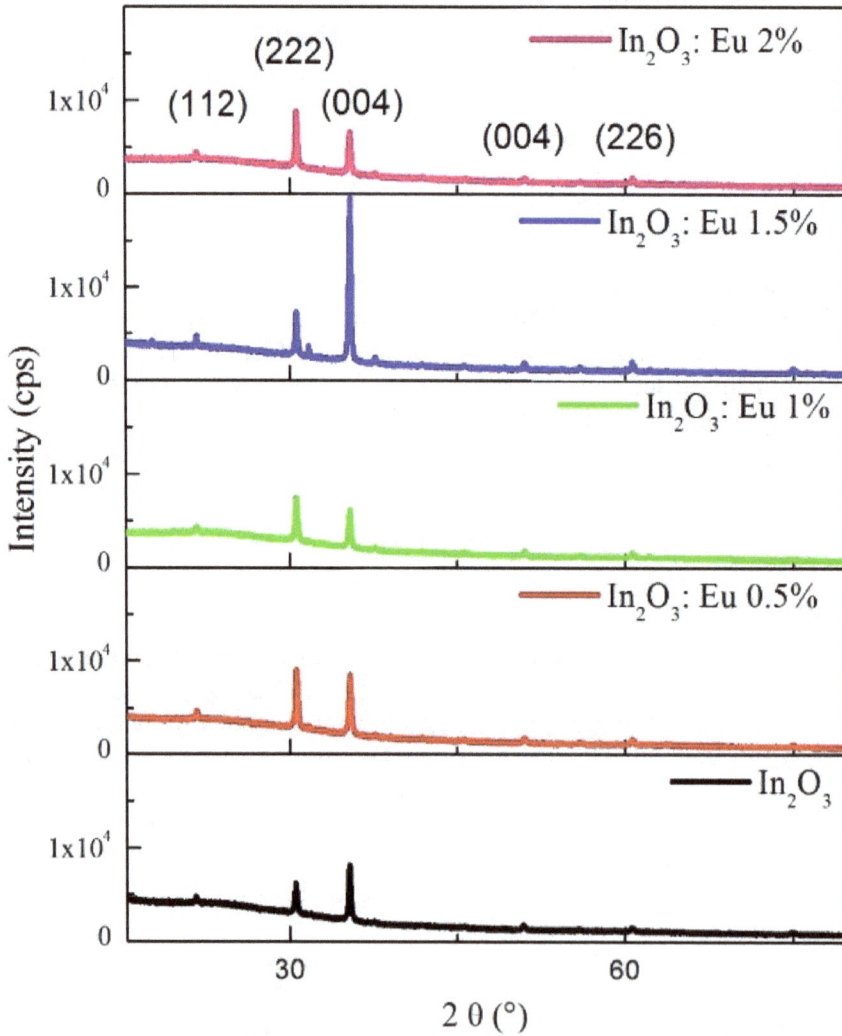

Figure II.3. X-ray diffraction spectra of Eu-doped In_2O_3 thin films grown by spray for different Eu content.

The XRD reflection peaks can be indexed only to cubic In_2O_3 and the plane spacing is related to the lattice constant a and the Miller indices (hkl) by the following relation [10]:

$$d_{hkl} = \frac{a}{\sqrt{h^2 + k^2 + l^2}} \quad (II.7)$$

The lattice parameter values a of In_2O_3:Eu are given in Table II.6. From This parameter reaches a lowest value for 1% and 1.5% Eu contents. Correlating these results with those from XRD spectra, we can deduce that the best crystallinity is obtained for Eu content equal to 1.5%. This value seems to be an appropriate value for possible incorporation of this doping inside the indium oxide thin film. Also, when Eu content increases, the (a) parameter also increases, showing a possible escape of this doping element to form oxides with high index such as Eu_2O_3. The latest has a crystalline phase different to cubic structure and goes with the structural and optical properties described above. Indeed, the compactness of the host material increases at this special doping and induces a high photon interaction of the 1.5% Eu-doped In_2O_3 oxide.

Table II.6. Lattice parameters of In_2O_3: Eu thin films.

y= $[Eu^{3+}]/[In^{3+}]$at%	0	0.5	1	1.5	2
a(Å)	10.1423	10.1423	10.1277	10.1277	10.1314

The average grain size D of indium oxide calculated from (112), (222), (004), (044) and (226) diffraction peaks using Sherrer's formula are listed in Table II.7.

Table II. 7. Grain size of Eu-doped In_2O_3 thin films using different XRD peaks.

y= $[Eu^{3+}]/[In^{3+}]$ at%	D (nm)				
	0	0.5	1	1.5	2
D(112)	33.2	42.8	42.6	53.4	46.7
D(222)	34.9	38.9	36.0	51.3	44.3
D(004)	36.3	35.6	3.6	49.5	39.2
D(044)	41.6	28.1	37.4	41.9	24.5
D(226)	36.9	33.6	39.0	46.8	29.8

The grain size was found to range from 24.4 to 53.4 nm when the concentration of Eu varies from 0 to 2 at%. We note that the larger size of the crystallite is obtained for europium rate equal to 1.5 at%. Then, we can conclude that the best crystallinity of In_2O_3: Eu is obtained for Europium content equal to 1.5 at%.

The microstrain ξ, which is an interesting structural parameter of In_2O_3: Eu sprayed-thin films is calculated using relation (II.5). All results are listed in Table II.8. From these results, it is noted that ξ decreases uniformly when y is from 0 to 1.5 at%. While for (004), direction decreases. ξ decreases from 7.49 (y = 0%) to 5.08 (y = 1.5%) and from 6.21 (y = 0%) to 4.55 (y = 1.5%), respectively, for (222) and (004) planes. For all (hkl) planes, the smallest values of ξ are obtained for Eu content close to 1.5 at%.

Table II.8. Microstrain of Eu-doped In_2O_3 thin films for different Eu content

y= $[Eu^{3+}]/[In^{3+}]$ at%	ξ (10^{-3})				
	0	0.5	1	1.5	2
ξ (112)	11.19	8.65	8.70	6.94	7.93
ξ (222)	7.49	6.70	7.24	5.08	5.89
ξ (004)	6.21	6.32	5.83	4.55	5.75
ξ (044)	3.78	5.59	4.20	3.75	6.45
ξ (226)	3.60	3.94	3.40	2.83	4.44

Using grain size values listed in Table II.7, the dislocation density δ_{dis} is defined as the imperfection in crystal. The values of this structural parameter are listed in Table II.9. They confirm the effect of europium incorporation in In_2O_3 matrix. Indeed, a maximum value of this parameter is obtained at 1% Eu doping level. This may be due to the nanosize character of the crystallites (Table II.7) of such doped film. However, the lowest values of δ_{dis} for all (hkl) directions are obtained for Eu content equal to 1.5% (Table II.9). We can conclude therefore that Eu- In_2O_3 grown using $[Eu^{3+}]/[In^{3+}]$=1.5% has the best order with well crystallinity.

Table II.9. Dislocation density of Eu-doped In_2O_3 thin films.

y= $[Eu^{3+}]/[In^{3+}]$ at%	δ_{dis} $(10^{12}$ line m$^{-2})$				
	0	0.5	1	1.5	2
δ (112)	907	546	551	351	907
δ (222)	821	661	772	380	821
δ (004)	759	789	7716	408	759
δ (044)	578	126	715	570	578
δ (226)	734	886	657	457	734

The probability made for the texture coefficient is for a given orientation; it can be used to calculate the average of crystallite size D and microstrain ξ and can thus outperform the Williamson and Hall relationship [11,12] using:

$$D_{moy} = \frac{\sum TC(hkl).D(hkl)}{\sum TC(hkl)} \quad (II.8)$$

$$\xi_{moy} = \frac{\sum TC(hkl).\xi(hkl)}{\sum TC(hkl)} \quad (II.9)$$

Likewise, the average value of dislocation density can be estimated using a similar formula:

$$\delta_{moy} = \frac{\sum TC(hkl).\delta(hkl)}{\sum TC(hkl)} \quad (10)$$

Average values of crystallite size D_{moy}, stress and dislocation density δ_{dismoy} are given in Table II. 10, which shows that the biggest crystal size (D_{moy} = 49.7 nm) and the smallest values of microstrain (ξ_{moy} = 5.6 × 10^{-3}) and dislocation density (δ_{dismoy} = 5.4×10^{14} lines m^{-2}) are obtained for Eu of 1.5 at%. From this structural analysis, it is found that a clear improvement of In_2O_3 crystallinity is obtained for a ratio between europium and indium concentrations in the spray solution equal to 1.5 %.

Table II.10. Average values of grain size D_{moy}, microstrain ξ_{moy} and dislocation density δ_{dismoy} of In_2O_3 : Eu thin films grown for $(0 \leq y=[Eu^{3+}]/[In^{3+}] \leq 2\%)$

	D_{moy}(nm)	ξ_{moy} (10^{-3})	δ_{dismoy} (10^{13}lines m^{-2})
In_2O_3	36.0	6.76	76
In_2O_3 :Eu 0.5%	37.2	6.69	74
In_2O_3 :Eu 1%	38.8	6.33	67
In_2O_3 :Eu 1.5%	49.7	5.60	54
In_2O_3 :Eu 2%	40.8	6.22	64

2.3.3 MoO₃ thin film

XRD patterns of undoped and Eu-doped MoO_3 sprayed thin films deposited on glass substrates at 460 °C are shown in Figure II.4. These spectra show the presence of (020), (040) and (060) peaks characteristic of α-MoO_3 orthorhombic structure according to [JCPDS n_: 76-1003] card. It is obvious that the crystallites have a preferential orientation along (020) direction. It is also worth noting that no peaks related to Europium oxide was found in these spectra. This may be due to the low Eu content (< 2 at %).

The calculated values of plane spacing d_{hkl} of MoO_3: Eu thin films are summarized in Table II.11. From these results, we can deduce that all values of d_{hkl} are not affected by Eu content. It is assumed that Eu ions are placed in the vacant sites or substitutional sites and not in the interstitial ones.

Table II.11. Interplanar spacing $d_{(hkl)}$ of MoO_3: Eu thin film grown for different ratios

d(hkl)	d(020)	d(110)	d(040)	d(021)	d(060)	d(0100)
MoO$_3$	6.895	-	3.461	-	2.307	1.385
MoO$_3$: Eu 0.5%	6.895	3.799	3.461	3.257	2.307	1.385
MoO3: Eu 1%	6.899	3.802	3.463	3.257	2.312	1.387
MoO3: Eu 1.5%	6.897	3.802	3.462	3.257	2.311	1.386
MoO3: Eu 2%	6.893	3.802	3.459	3.257	2.307	1.385

The values of d_{hkl} were used to calculate the lattice parameters a, b and c related to orthorhombic structure according to the following relation [13]:

$$\frac{1}{d_{hkl}^2} = \frac{h^2}{a^2} + \frac{k^2}{b^2} + \frac{l^2}{c^2} \quad (11)$$

Calculated lattice parameters are equal to: a=3.96 Å, b=13.86 Å and c=3.70Å

$$y = [Eu^{3+}]/[Mo^{6+}] \ (0{\leq}y{\leq}2 \ at \ \%)$$

On the other hand, the TC (hkl) calculated values of MoO_3: Eu thin films are presented in Table II.12. From these results, it is noted that TC (020) and TC (040) values are relatively high, showing that the crystallites are mainly parallel to substrate plane.

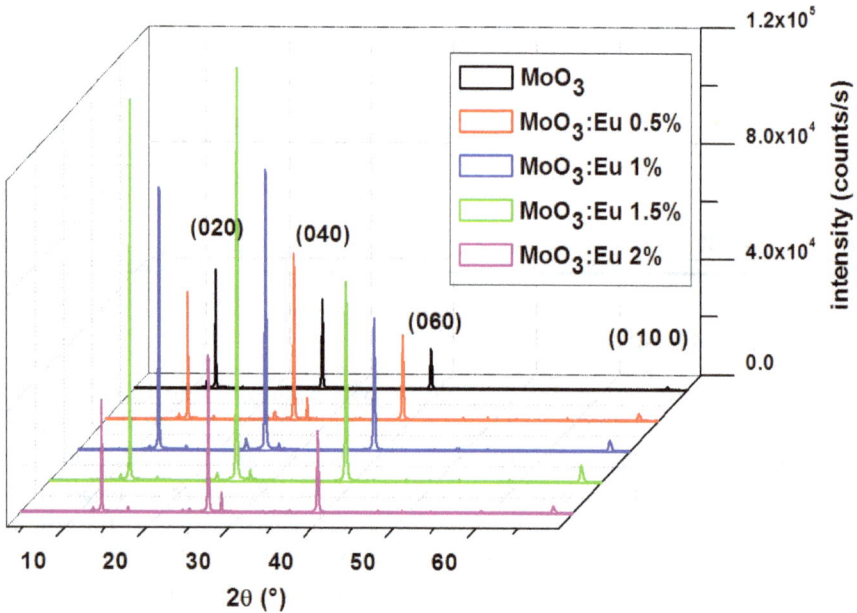

Figure II.4. X-ray diffraction spectra of Eu-doped MoO_3 thin films grown for different ratios in the spray solution, $y = [Eu^{3+}]/[Mo^{6+}]$ $(0{\leq}y{\leq}2 \ at \ \%)$

Table II.12. Different texture coefficient TC (hkl) values of Eu-doped MoO_3 spayed thin films

	MoO_3	MoO_3: Eu 0.5%	MoO_3: Eu 1%	MoO_3: Eu 1.5%	MoO_3: Eu 2%
TC (020)	2.85	1.86	2.28	2.26	1.79
TC (110)	-	0.11	0.08	0.04	0.04
TC (040)	2.13	2.43	2.39	2.41	2.47
TC (021)	-	0.3	0.05	0.06	0.31
TC (060)	0.95	1.21	1.1	1.14	1.27
TC (0100)	0.07	0.09	0.09	0.09	0.11

In the same way, from XRD spectra, the effect of incorporation of the europium in the matrix on the crystallite size (D) has been studied by using Debye-Scherrer formula. The crystallite size (D) values varied from 74 to 85 nm with Eu content in Table II.13, which confirms the presence of nanocrystallites in the MoO_3 thin films. The values of the micro strain (ξ) and dislocation density (δ_{dis}) which vary with doping ratio are listed in Table II.13.

Table II.13. Crystallite size D, micro strain ξ and dislocation density δ_{dis} for MoO_3 thin films grown for different content of europium in the spray solution

	D (nm)	$\xi(10^{-4})$	δ_{dis} (10^{14} lines /m^2)
MoO_3	84.4	12.9	1.55
MoO_3 : Eu 0.5%	85.1	11.3	1.76
MoO_3 : Eu 1%	74.0	13.6	2.36
MoO_3 : Eu 1.5%	75.1	13.4	2.12
MoO_3 : Eu 2%	76.9	12.17	1.90

2.4. Raman spectroscopy study

Raman spectroscopy is used to reach vibrational and rotational modes in a crystal material. For a radiation of a sample with a monochromatic light, different phenomena can be induced like absorption, reflection or scattering. Two kinds of scattering could occur: Rayleigh scattering and Raman scattering. Rayleigh scattering resulted from scattering process without a change in the wavelength and accounts for the blue color of the sky. A change in the frequency (wavelength) of the light is called Raman scattering. Raman shifted photons of light can be either of higher or lower energy, depending upon the vibrational state of the molecule.

- Stokes Raman scattering: the material, initially in the ground state, absorbs energy and the emitted photon has a lower energy than the absorbed photon resulting in a creation of a phonon.

- Anti-Stokes Raman scattering: the material, already in an excited state, absorbs energy and the emitted photon has a higher energy than the absorbed photon resulting in an annihilation of a phonon.

Raman spectroscopy measurements of all Eu doped oxides were recorded at room temperature using Jobin Yvon HR LabRAM in backscattering co-focal configuration with a spatial resolution of 1μm and spectral resolution less than 0.35 cm^{-1}. The light excitation is an Ar$^+$ laser at the wavelength of 488 nm.

2.4.1 Eu doped ZnO thin films

The Group theory shows that $\Gamma = A_1 + 2B_1 + E_1 + 2E_2$, where B_1 corresponds to the silence mode; A_1, E_1 and E_2 are Raman active modes. A_1 and E_1 are polar and infrared active modes; they correspond to transverse-optical (TO) and longitudinal-optical (LO) phonons respectively. E_2 mode is rather referred to two modes of low and high frequency phonons [14]. Figure II.5 shows the Raman spectra performed at room temperature of the undoped ZnO thin films, followed by those of gradually Eu:doped ones (Figs. II. 6–8). First, the observable Raman peak at 109 cm^{-1} corresponds to non-polar optical phonon modes E_{2LOW}. It has the highest intensity for ZnO:Eu 1 at% thin film.

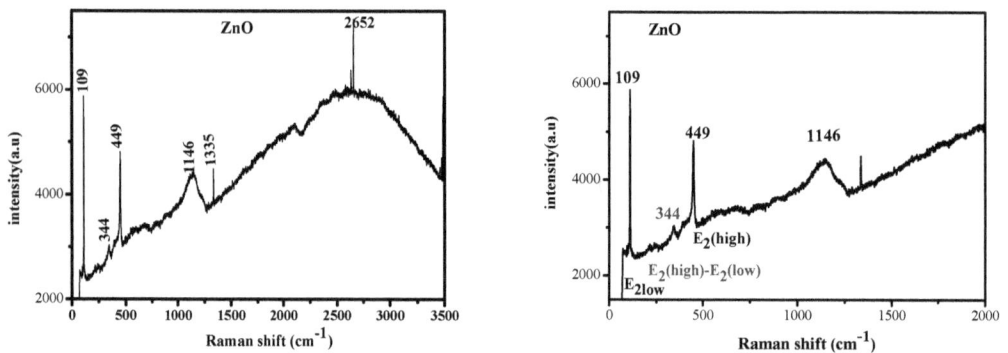

Figure II.5. Raman spectra performed at room temperature for undoped ZnO thin films

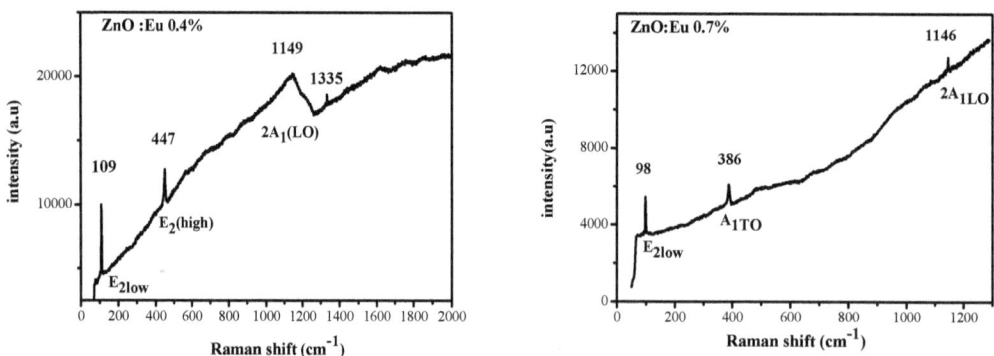

Figure II.6. Raman spectra performed at room temperature for (0.4%) Eu-doped ZnO thin films

Figure II.7. Raman spectra performed at room temperature for (0.7%) Eu-doped ZnO thin films

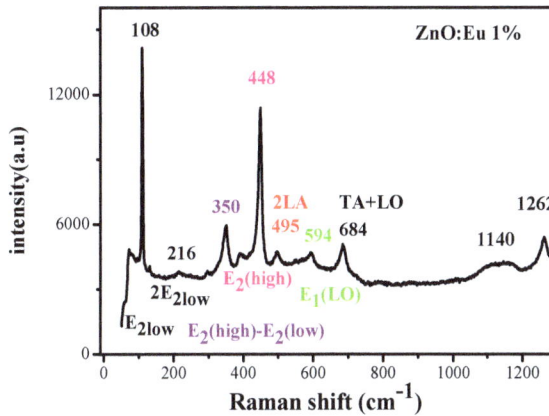

Figure II.8. Raman spectra performed at room temperature for
1% Eu-doped ZnO thin films.

These results are similar to that obtained previously by Rajeh *et al.* [15]. Second, peak located at 350 cm^{-1} can be attributed to A$_{1TO}$ mode. Also, peak situated at 448 cm^{-1} corresponds to ZnO nonpolar optical phonons of E$_{2high}$ mode [16-18]. In the same line, small Raman peaks assigned at 495 and 594 cm^{-1} for ZnO:Eu 1 at%, are related to 2LA and E$_{1LO}$ modes respectively [18,19]. The E$_{1LO}$ peak may be due to O-vacancy defects and also to Zn interstitial states and free carriers in ZnO:Eu as reported elsewhere [19,20]. Finally, The peak located around 1150 cm^{-1} for ZnO and ZnO:Eu 0.4% shifts toward low wavelength by increasing the Eu content. This peak can be attributed to 2A$_{1LO}$ which is similar to one obtained by Ya-Ping *et al.* [14] and discussed by Yang *et al.* [21]. The molar mass of europium is higher than that of zinc and since that energy is proportional to the square root of the reduced mass, we can find the mass per unit volume increase, a simple and plausible explanation to the recorded Raman peak shift at low frequencies (Figures II.6–8).

2.4.2 Eu doped In$_2$O$_3$ thin films

Additional information on the structure of the sample was obtained by Raman spectroscopy. Figure II.9 depicts Raman spectra of as deposited Eu doped In$_2$O$_3$ thin films. For such structure, the vibrations with symmetry A$_g$, E$_g$ and T$_g$ represent Raman active and T$_u$ vibrations represent infrared-active modes; in fact, 22 Raman-active and 16 infra-red active modes are expected but only 6 Raman modes and 11 IR modes were previously detected as truly In$_2$O$_3$ cubic modes [22,23].

$$\Gamma = 4A_g + 4E_g + 14T_g + 5A_u + 5E_u + 16T_u$$

In$_2$O$_3$ spectrum displayed in Figure II.9 shows the expected vibrational modes at 106, 131, 305 and 366 cm^{-1} (E$_g$), which are the signatures of cubic In$_2$O$_3$ structure [24]. Also, the same figure shows that the raman peak intensities vary with Eu content but the highest intensity of Raman peaks are obtained for In$_2$O$_3$: Eu 1.5 at% thin films.

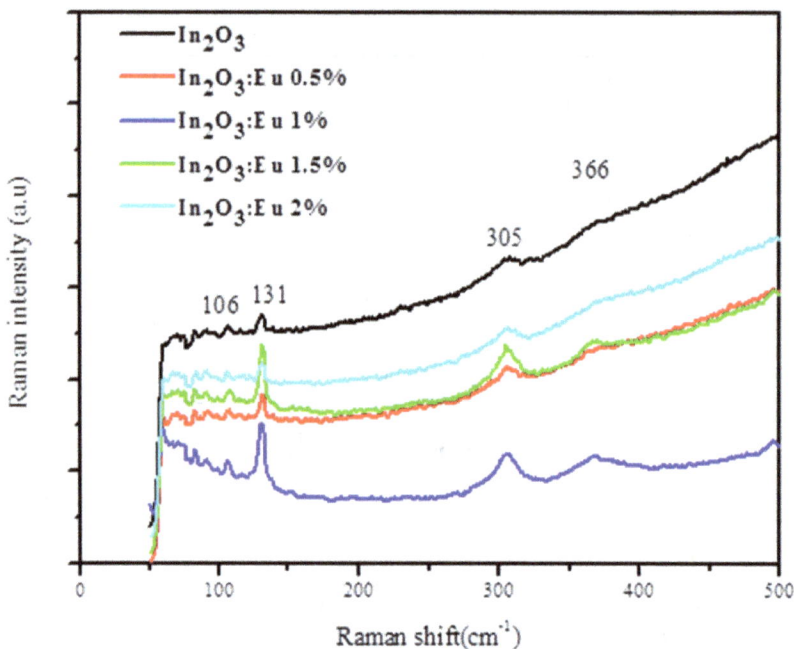

Figure II.9. Raman spectra of Eu-doped In_2O_3 thin films.

2.4.3. Eu doped MoO_3 thin films

It is known that MoO_3 has Pbnm (62) space group. Figure II.10 presents Raman spectra of undoped and Eu doped MoO_3 thin films. First, the obtained Raman peaks located at 162, 200, 221, 247, 290, 340, 379, 474, 668, 819 and 994cm^{-1} correspond to orthorhombic α-MoO_3 phase [25, 26]. The vibrational analysis of Raman spectra reveals that the bands in 900–600 and 400–200 cm^{-1} regions are mainly owing to Mo-O stretching and bending modes respectively. Moreover, Raman peaks assigned at 994, 819, and 668 cm^{-1} for the undoped film confirms the structure of α-MoO_3 variety.

First, Raman peak situated at 994 cm^{-1} is assigned to the terminal oxygen (Mo^{6+}=O) stretching mode along a- and b- axes which are the cause of the layered structure of α-MoO_3 [27]. The generation of oxygen vacancies should therefore lead to anion vacancies along c-axis. Therefore, a move of Mo atom towards the terminal oxygen in the b-direction can be anticipated with the loss of bridging oxygen, thus weakening the terminal bond along a-axis [28]. Also, the sharp intense peak situated at 819 cm^{-1} can be is attributed to the doubly coordinated oxygen (Mo-O-Mo) in stretching mode as shown by Ramans *et al.* [29]. Besides, the peak located at 668 cm^{-1} assigned to the triply coordinated oxygen (Mo_3-O) stretching mode which results from edge-shared oxygen in common with three octahedra [29]. Finally, low intense Raman peaks located at 247, 290, 340, and 379 cm^{-1} can be attributed to O-Mo-O scissoring and O=Mo=O wagging modes respectively [30]. All Raman shift and observed modes in MoO_3:Eu are summarized in Table II.14.

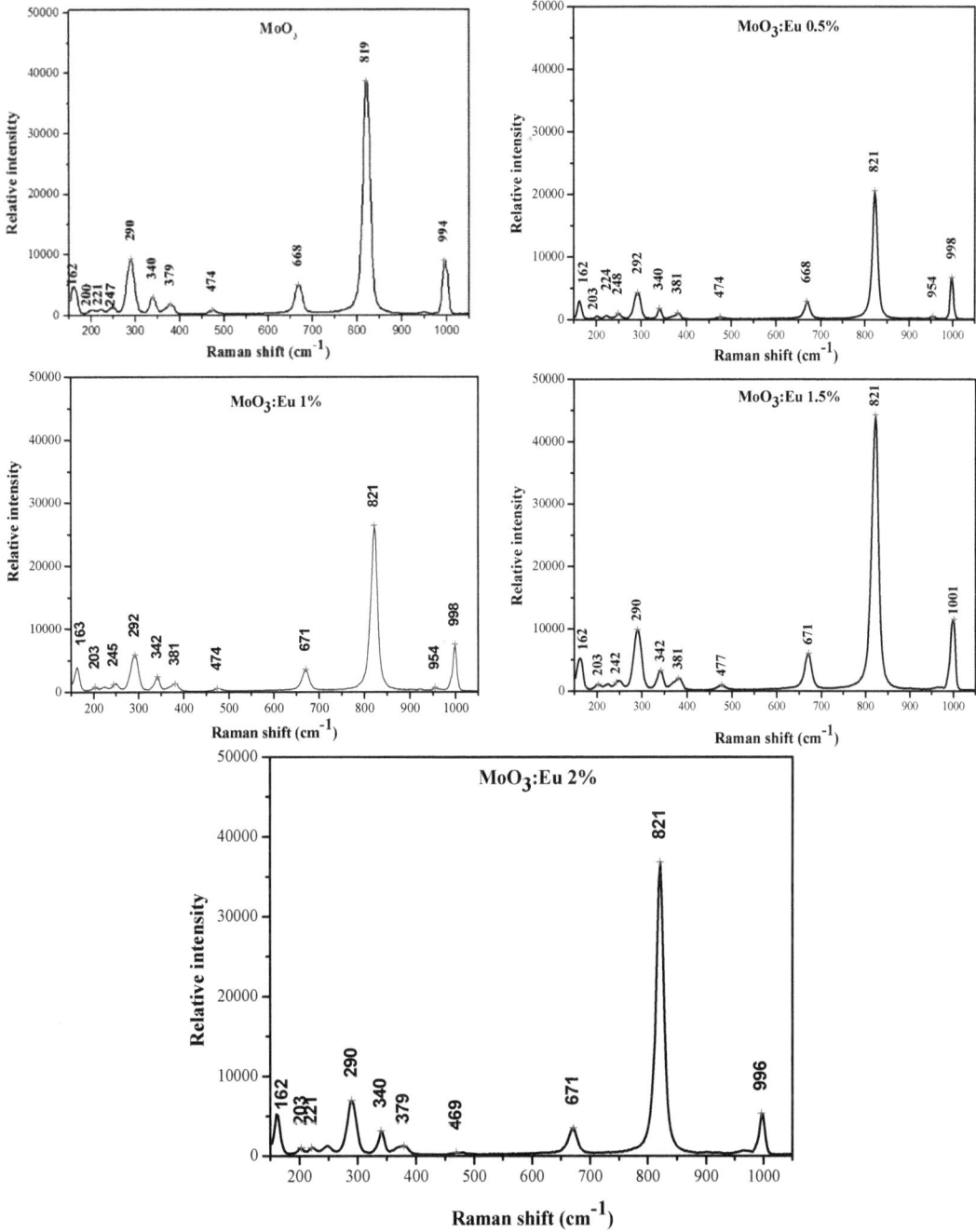

Figure II.10. Raman spectra of MoO$_3$: Eu thin films grown for different ratios y in the spray solution

y = $[Eu^{3+}]/[Mo^{6+}]$ (0≤y≤2 at %)

Table II.14. Raman shift and observed modes in MoO_3:Eu sprayed thin films.

MoO_3	MoO_3: Eu 0.5%	MoO_3: Eu 1%	MoO_3: Eu 1.5%	MoO_3: Eu 2%	Mode	Assignment
162	162	163	162	162	Ag	Translational rigid MoO^{4-} chain mod. Tb
200	203	203	203	203	B_{2g}	τ O=Mo=O twist
221	224	223	226	221		
247	248	245	242	249	B3g	τ O=Mo=O twist
290	292	292	290	290		
340	340	342	342	340	Ag. B1g	δ Mo-O-Mo bend
379	381	381	381	379	B_{1g}	δ O=Mo=O scissors
474	474	474	477	469		
668	668	671	671	671	B2g	ν Mo-O-Mo stretch
819	821	821	821	821	Ag	vs O=Mo=O stretch
948	954	954	-	-		
994	998	998	1001	996		

2.5. XPS investigations

In the literature, there are extensive XPS studies performed on ZnO and In_2O_3 undoped and doped thin films. However, few papers report on XPS of MoO_3 binary oxide. In this section we present the XPS results on Eu doped MoO_3 thin films prepared by spray pyrolysis. X-ray photoelectron spectroscopy (XPS) technique has been used in parallel with the XRD study.

The XPS results are presented in Figure II.11 and Figure II.12, and listed in Table II.15. First of all, it can be observed that the europium percentage in the film is lower than the one in the starting solution. Also, XPS measurements are carried out on the surface of the film without etching. From the XPS data, we can conclude that the surface structure is given by molybdenum. Based on NIST Database on XPS for such oxide, it was found that the oxidation state is related only to MoO_3 as shown in Figure II.11. The Mo 3d spectrum exhibits the characteristic of 3d5/2 and 3d3/2 doublet caused by the spin–orbit coupling. Figure II.11 shows XPS spectrum of undoped MoO_3. A peak related to Mo 3d5/2 located at 232.8 eV could be due to Mo (VI) reduction into a lower oxidation state component [31-34]. The binding energies (BE) of the different oxidation states of Mo are: 235.85 and 232.65 eV for MoO_3 [32], 234.9 and 231.7 eV for Mo_2O_5 and 232.3, 229.1 eV for MoO_2 as reported previously [31].

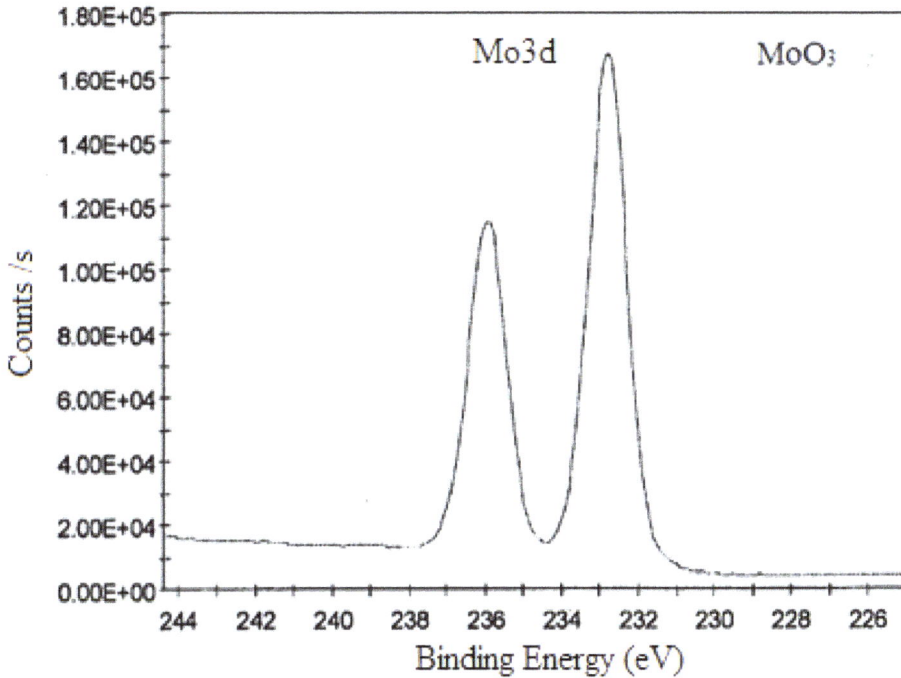

Figure II.11. XPS spectrum of MoO_3

A second observation is that the oxidation of Eu element inside the film is in the form of Eu_2O_3 binary oxide. In fact, the peak intensity related to Eu_2O_3 increases with Eu doping. It is worth noting that the presence of Eu_2O_3 was not revealed by X-Ray diffraction analysis of the Eu doped MoO_3 thin films. However, XPS measurements reveal the presence of Eu_2O_3 and prove that a little content of this doping element is not incorporate in the MoO_3 matrix but it is present along with MoO_3 in the film. On the other hand, this technique is used to characterize different ratios of Eu doping MoO_3 and to evaluate and determine the nature of the chemical species present on the surface as well as in bulk by an appropriate etching procedure. Moreover, it was found that XPS electron binding energy (BE) values shows that MoO_3:Eu thin films contain a mixed-oxide phases of the highest oxidation states of Mo (VI) and Eu (III).

Finally, the XPS analysis shows a band corresponding to MoO_3 oxide phase, which may exist on the surface. For samples Eu doped ones, the spectra exhibit the presence of peak related to Eu_2O_3 phase. This study reveals that Eu doesn't incorporated into the MoO_3 matrix, but remains as a secondary phase dispersed as nanocrystallites in the doped films. All XPS results corresponding to undoped and Eu-doped MoO_3 for different content of europium in the spray solution are summarized in Table II.15.

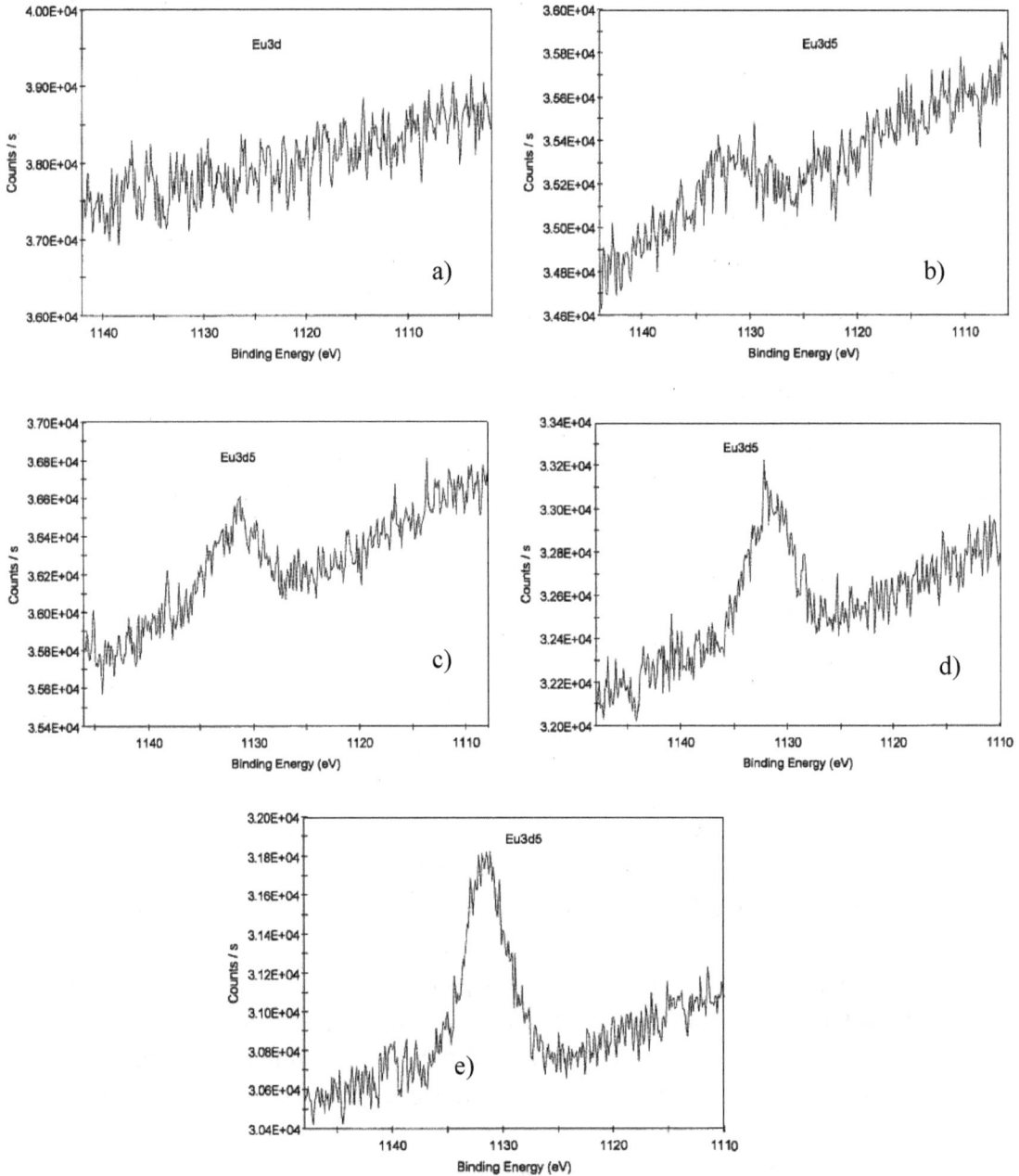

Figure II.12. XPS identification of Eu_2O_3 in MoO_3: Eu thin films (a) undoped thin film
(b) MoO_3:Eu0.5%, (c) MoO_3:Eu1 %, (d) MoO_3:Eu 1.5% and (e) MoO_3:Eu 2%

Table II.15. XPS measurements of MoO_3:Eu thin films grown for different content of europium in the spray solution (y = $[Eu^{3+}]/[Mo^{6+}]$ (0≤y≤2 at %). BE is the binding energy. FWHM is full at half maximum of the peak and the calculated area of the peak.

(a) MoO$_3$

Name	Peak BE	FWHM eV	Area (P) CPS.eV	At. %
Mo3d	232.80	1.14	330913.62	72.08
Eu3d	1141.62	0.00	-19.15	0.00
C1s	284.95	1.55	52961.82	11.65
O1s	530.84	1.56	204428.41	16.27

(b) MoO$_3$: Eu 0.5%

Name	Peak BE	FWHM eV	Area (P) CPS.eV	At. %
Mo3d	232.77	1.13	286758.04	72.07
Eu3d5	1137.24	0.00	1456.91	0.01
C1s	285.07	1.64	47084.13	11.95
O1s	530.91	1.61	173773.51	15.96

(c) MoO$_3$:Eu 1%

Name	Peak BE	FWHM eV	Area (P) CPS.eV	At. %
Mo3d	232.68	1.17	273679.99	70.46
Eu3d5	1135.74	1.77	3241.07	0.03
C1s	284.96	1.71	53026.48	13.79
O1s	530.8	1.64	167170.96	15.73

(d) MoO$_3$:Eu 1.5%

Name	Peak BE	FWHM eV	Area (P) CPS.eV	At. %
Mo3d	232.82	1.17	268132.14	72.3
Eu3d5	1135.86	4.52	4190.53	0.04
C1s	285.01	1.6	41365.73	11.26
O1s	530.89	1.57	166431.07	16.4

(e) MoO$_3$:Eu 2%

Name	Peak BE	FWHM eV	Area (P) CPS.eV	At. %
Mo2p	232.82	1.15	262104.09	73.08
Eu3d5	1135.9	3.81	5703.72	0.06
C1s	284.98	1.64	37709.95	10.62
O1s	530.9	1.54	159452.36	16.25

2.6. Conclusion

It has been shown that ZnO, In_2O_3 and MoO_3 thin films are highly dependent on the processing conditions. Zinc oxide, indium oxide and molybdenum oxide compounds crystallize generally into hexagonal structure, centered cubic structure, and orthorhombic structure, respectively. Moreover, some structural, morphological study of ZnO, In_2O_3 and MoO_3 thin films were discussed in terms of Eu content. XRD has defined certain structural parameters including grain size, lattice parameters, micro-strain and phases amounts. Also, the description of the synthetized technique "Chemical Spray Pyrolysis" of zinc oxide, indium oxide and molybdenum oxide thin films and its deposition process were investigated.

2.7 References

[1] A. Boukhachem, B. Ouni, M. Karyaoui, A. Madani, R. Chtourou and M. Amlouk, "Structural, opto-thermal and electrical properties of ZnO:Mo sprayed thin films", Materials *Science* in *Semiconductor Processing, 15 (2012) 282–292.*

[2] K. Boubaker, A. Chaouachi, M. Amlouk, and H. Bouzouita, "Enhancement of pyrolysis spray disposal performance using thermal time-response to precursor uniform deposition", The European Physical Journal Applied Physics. 37 (2007) 105–109.

[3] A. Chakraborty, T. Mondal, S.K. Bera, S.K. Senc, R. Ghosh and G.K. Paul, "Effects of aluminum and indium incorporation on the structural and optical properties of ZnO thin films synthesized by spray pyrolysis technique, Materials Chemistry and Physics", 112 (2008) 162–166.

[4] A. Bouzidi, N. Benramdane, H. Tabet-Derraz, C. Mathie, B. Khelifa and R. Desfeux, "Effect of substrate temperature on the structural and optical properties of MoO3 thin films prepared by spray pyrolysis technique", Materials Science and Engineering: B, 97 (2003) 5–8.

[5] M. Yilmaz and Ş. Aydoğan, "The effect of Pb doping on the characteristic properties of spin coated ZnO thin films: Wrinkle structures", Materials Science in Semiconductor Processing, 40 (2015) 162–170.

[6] O. Lupan, T. Pauporté, L. Chow, B. Viana, F. Pellé, L.K. Ono, B. Roldan Cuenya and H. Heinrich, "Effects of annealing on properties of ZnO thin films prepared by electrochemical deposition in chloride medium", Applied Surface Science, 256 (2010) 1895–1907.

[7] L.P. Peng, L. Fang, X.F. Yang, Y.J. Li, Q.L. Huang and F. Wu, "Effect of annealing temperature on the structure and optical properties of In-doped ZnO thin films", Journal of Alloys and Compounds, 484 (2009) 575-579.

[8] R.D. Shannon, "Revised effective ionic radii and systematic studies of interatomic distances in halides and chalcogenides", Acta Crystallographica. Section A, 32 (1976) 751-767.

[9] A. Boukhachem, C. Bouzidi, R. Boughalmi, R. Ouerteni, M. Kahlaoui, B. Ouni, H. Elhouichet and M. Amlouk, "Physical investigations on MoO_3 sprayed thin film for selective sensitivity application", Ceramics International, 40 (2014) 13427-13435.

[10] M.B Amor, A. Boukhachem, K. Boubaker and M. Amlouk, "Structural, optical and electrical studies on Mg-doped NiO thin films for sensitivity applications", Materials Science in Semiconductor Processing, 27 (2014) 99-1006.

[11] K. Arshak, E. Moore, G. M. Lyons, J. Harris and S. Clifford, "A review of gas sensors employed in electronic nose applications", Sensor Review, 24 (2004) 181-198.

[12] J. Flueckiger, K. Ko and K. C. Cheung, "Microfabricated Formaldehyde Gas Sensors", Sensors, 9 (2009) 9196-9215.

[13] H. M. Martínez, J. Torres, L. D. López Carreño and M. E. R. García, "Effect of the substrate temperature on the physical properties of molybdenum tri-oxide thin films obtained through the spray pyrolysis technique", Materials Characterization, 75 (2013) 184-193.

[14] Y. P. Du, Y. W. Zhang, L. D. Sun and C. H. Yan, "Efficient Energy Transfer in Monodisperse Eu-Doped ZnO Nanocrystals Synthesized from Metal Acetylacetonates in High-Boiling Solvents", Journal of Physical Chemistry C, 112 (2008) 12234–12241.

[15] S. Rajeh, A. Mhamdi, K. Khirouni, M. Amlouk and S. Guermazi, "Experiments on ZnO:Ni thin films with under 1% nickel content", 69 (2015) 113–121.

[16] T.C. Damen, S.P.S. Porto and B. Tell, "Raman Effect in Zinc Oxide", Physical Review, 142 (1966) 570–580.

[17] A. Kaschner, U. Haboeck, M. Strassburg, G. Kaczmarczyk, A. Hoffmann, C. Thomsen, A. Zeuner, H.R. Alves, D.M. Hofmann and B.K. Meyer, "Nitrogen-related local vibrational modes in ZnO:N", Applied Physics Letters, 80 (2002) 1909–1911.

[18] R. Cusco, E. Alarcon-Llado, J. Ibanez, L. Artus, J. Jimenez, B. Wang and M.J. Callahan, "Temperature dependence of Raman scattering in ZnO", Physical Review B, 75 (2007) 165202.

[19] J. Zuo, C.Y. Xu, L.H. Zhang, B.K. Xu and R. Wu, "Sb-induced size effects in ZnO nanocrystallites", Journal of Raman Spectroscopy, 32 (2001) 979-981.

[20] A.K. Pradhan, K. Zhang, G.B. Loutts, U.N. Roy, Y. Cui and A.J. Burger, "Structural and spectroscopic characteristics of ZnO and ZnO:Er^{3+} nanostructures", Journal of Physics: Condensed Matter, 16 (2004) 7123–7133.

[21] J. Yang, X. Li, J. Lang, L. Yang, M. Wei, M. Gao, X. Liu, H. Zhai, R. Wang, Y. Liu and J. Cao, "Synthesis and optical properties of Eu-doped ZnO nanosheets by hydrothermal method", Materials Science in Semiconductor Processing, 14 (2011) 247–252.

[22] C. Y.Park, C. Y. You, K. R. Jeon and S. C. Shin, "Charge-carrier mediated ferromagnetism in Mo-doped In_2O_3 films", Applied Physics Letters, 100 (2012) 222409.

[23] D. W. Sheel and J. M. Gaskell, "Deposition of fluorine doped indium oxide by atmospheric pressure chemical vapour deposition Thin", Solid Films, 520 (2011) 1242–1245.

[24] M. Józefowicz and W. Piekarczyk, "Preparation of In_2O_3 single crystals by chemical vapour transport method", Materials Research Bulletin, 22 (1987) 775-780.

[25] M. Dieterle, G. Weinberg and G. Mestl, "Raman spectroscopy of molybdenum oxides Part I. Structural characterization of oxygen defects in $MoO_{3-x}$by DR UV/VIS, Raman spectroscopy and X-ray diffraction", Physical Chemistry Chemical Physics, 4 (2002) 812-821.

[26] M. Dieterle and G. Mestl, "Raman spectroscopy of molybdenum oxides Part II. Resonance Raman spectroscopic characterization of the molybdenum oxides Mo_4O_{11} and MoO_2", Physical Chemistry Chemical Physics, 4 (2002) 822-826.

[27] I. R. Beattie and T. R. Gilson, "Oxide phonon spectra", Journal of the Chemical Society A, (1969) 2322-2327.

[28] G. Mestl, N. F. D. Verbruggen, E. Bosch and H. Knozinger, "Mechanically Activated MoO_3. 5. Redox Behavior", Langmuir, 12 (1996) 2961–2968.

[29] G. M. Ramans, J. V. Gabrusenoks, A. R. Lusis and A. A. Patmalnieks, "Structure of amorphous thin films of WO_3 and MoO_3", Journal of Non-Crystalline Solids, 90 (1987) 637-640.

[30] M. A. Py and K. Maschke, "Intra- and interlayer contributions to the lattice vibrations in MoO_3", Physica B, 105 (1981) 370-374.

[31] H. Al-Kandari, A. M. Mohamed, F. Al-Kharafi and A. Katrib, "XPS-UPS, ISS characterization studies and the effect of Pt and K addition on the catalytic properties of $MoO_{2-x}(OH)_y$ deposited on TiO_2", Journal of Electron Spectroscopy and Related Phenomena, 184 (2011) 472-78.

[32] J. Baltrusaitis, B. M. Sanchez, V. Fernandez, R. Veenstra, N. Dukstiene, A. Roberts and N. Fairley, "Generalized molybdenum oxide surface chemical state XPS determination via informed amorphous sample model", Applied Surface Science, 326 (2015) 151-161.

[33] B. M. Sánchez, T. Brousse, C. R. Castro, V. Nicolosi and P. S. Grant, "An investigation of nanostructured thin film α-MoO_3 based supercapacitor electrodes in an aqueous electrolyte", Electrochimica Acta, 91 (2013) 253-260.

[34] D. O. Scanlon, G. W. Watson, D. J. Payne, G. R. Atkinson, R. G. Egdell and D. S. L. Law, "Theoretical and Experimental Study of the Electronic Structures of MoO_3 and MoO_2", Journal of Physical Chemistry C, 114 (2010) 4636-4645.

3. Optical Properties

3.1. Optical properties of the crystalline ZnO thin film

3.1.1 Band gap energy calculation

The optical transmittance and reflectance spectra of the ZnO:Eu sprayed thin films grown on glass substrates are shown in Figure III.1. The spectra show a relatively high transmittance coefficient of all thin films, in visible and IR region. This can be attributed to less scattering effects, structural homogeneity and an improvement of the crystalline state. Besides, the presence of the interference phenomenon indicates a smooth and homogeneous surface of all observed films The film thickness was measured using the following formula [1– 3]:

$$d = \frac{1}{2} \frac{\lambda_{max}^p \lambda_{max}^{p+1}}{n(\lambda_{max}^p)\lambda_{max}^{p+1} - n(\lambda_{max}^{p+1})\lambda_{max}^p} \qquad (III.1)$$

where $n(\lambda_{max}^p)$ is the refractive index associated with the wavelength giving maximum reflection of order p (p integer) in the area of low absorption value and it can be calculated by using the following relationship [1–5]:

$$n(\lambda_{max}) = \sqrt{n_0 n_s \frac{1 + \sqrt{R_{max}}}{1 - \sqrt{R_{max}}}} \qquad (III.2)$$

n_0 and n_s are respectively the optical indices of air (n_0=1) and substrate; the thickness of the obtained ZnO thin films are in 0.5– 0.52 mm domain.

The fundamental absorption edge of the films corresponds to the transitions of electrons from the valence band to the conduction band. This can be used to calculate the effect of the doping concentration on the optical band gap of the films. The absorption coefficient can be expressed by [6, 7]:

$$\alpha = \frac{1}{d} Ln \frac{(1-R)^2}{T} \qquad (III.3)$$

In the case of a direct transition, the absorption coefficient and optical band gap are related to hv by the following relation [8-11]:

$$\alpha h \nu = B(h\nu - E_g)^p \qquad (III.4)$$

where A is a constant, hv is the photon energy, Eg is the optical band gap and p is a number which has the value of 1/2 or 2 for direct and indirect transitions, respectively. From the $(\alpha.h\nu)^2$ plots versus the photon energy hv which, we can obtain the optical band gap E_g by a linear fit (Figure III.2). The calculated values of E_g of undoped and Eu-doped ZnO are summarized in table III.1.

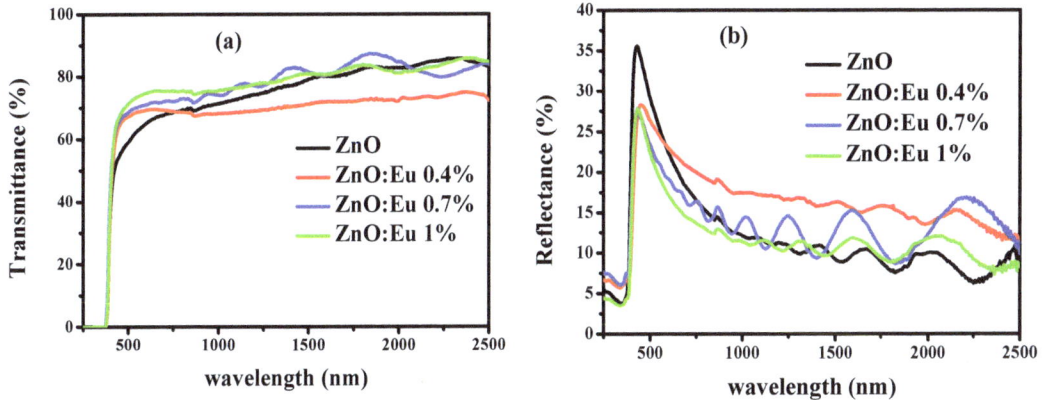

Figure III.1 Transmittance and reflectance spectra of ZnO:Eu thin films

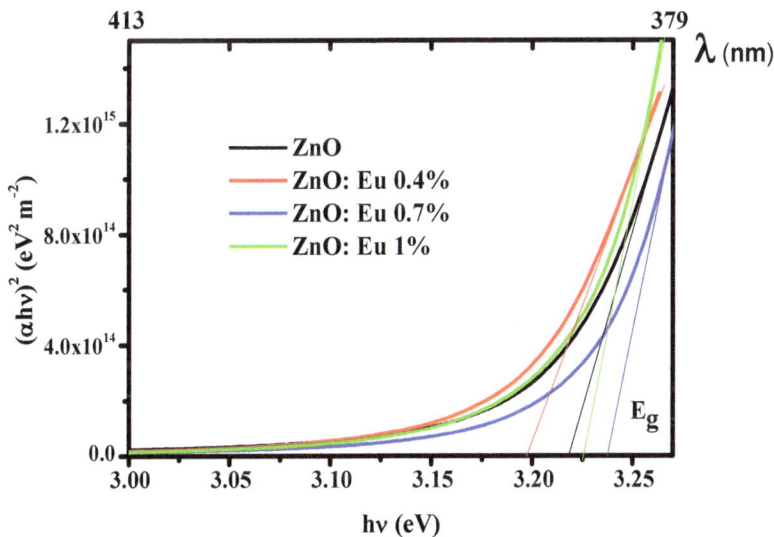

Figure III.2 Plot of $(\alpha h\nu)^2$ versus the photon energy of ZnO:Eu thin films.

In addition, optical absorption by defects appears at energy lower than optical gap. The interactions with phonons increases and therefore the formation of band tailing in the band gap. Mott *et al.* [12] noted that unlike crystalline structures where the fundamental edge is mainly determined by conduction and valence levels, ion-doped binary semiconductor compounds present a particular optical absorption edge profile. In these oxide materials, the absorption coefficient profile increases exponentially with the photon energy near the energy gap. This variation results in "blurring" of the valence–conduction bands and narrows slightly the band gap by appearance of the so-called Urbach tailing. Regarding these interactions, α is written as another exponential function of photon energy which follows the empirical Urbach law [13–15]:

$$\alpha = \alpha_0 \exp\left(\frac{h\nu}{E_U}\right) \qquad \text{(III.5)}$$

where α_0 is a constant and E_U is the Urbach energy which characterizes the slope of the exponential limit, Figure III.3. The value of E_U is obtained from the inverse of the slope of Lnα vs. hν [2] using the following relation [14-16]:

$$E_U = \left[\frac{d\,Ln\alpha}{d\,h\nu}\right]^{-1} \qquad \text{(III.6)}$$

E_U values are listed in Table III.1. It is found that the Urbach energy value decreases with the Eu content. This suggests the decreases of defects and may be related to disappear of interstitial zinc atom by Europium doping. This doping causes indeed a possible reorganization of ZnO structure of such films

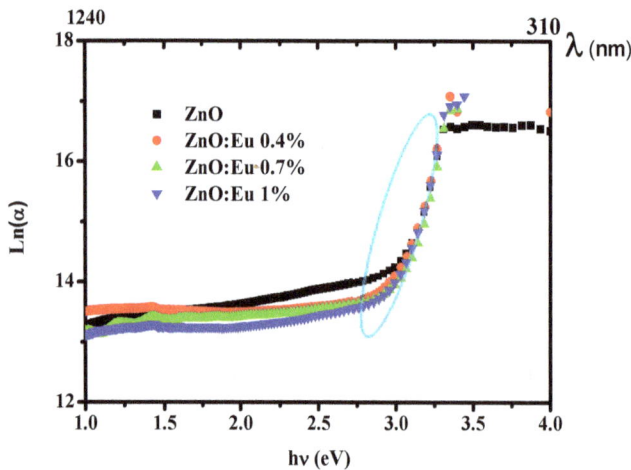

Figure III.3
Plot of Ln(α) versus hν for different content of europium doped ZnO sprayed thin films (y= [Eu³⁺]/[Zn²⁺] (0≤y≤1 at %).

Table III.1 The band gap and Urbach energy value of ZnO:Eu thin films

	Eg (eV)	Eu (meV)
ZnO	3.210	113
ZnO:Eu 0.4%	3.190	113
ZnO:Eu 0.7%	3.240	101
ZnO:Eu 1 %	3.230	104

3.1.2. Opto-thermal investigation

Opto-thermal properties of ZnO:Eu thin films were investigated through the Amlouk–Boubaker opto-thermal expansivity Ψ_{AB} [17– 22] values. This parameter has been defined by Eq. (III.7):

$$\Psi_{AB} = \frac{D}{\hat{a}} \qquad (III.7)$$

where D is the thermal diffusivity, as deduced from the related literature [21–22], and $\hat{a}$ is the effective absorptivity. The effective absorptivity $\hat{a}$ is defined as the mean normalized absorbance weighted by I $(\lambda)_{AM1.5}$, the solar standard irradiance, were λ is the normalized wavelength:

$$\hat{a} = \frac{\int_{\lambda_{min}}^{\lambda_{max}} I(\lambda)_{AM1.5} \times a(\lambda)d\lambda}{\int_{\lambda_{min}}^{\lambda_{max}} I(\lambda)_{AM1.5} \, d\lambda} \qquad (III.8)$$

where λ_{min} and λ_{max} are the limit of the visible spectrum and where: I $(\lambda)_{AM1.5}$ is the Reference Solar Spectral Irradiance (corresponding to solar irradiation AM1.5). Table III.2 summarizes the calculated values of Ψ_{AB} of ZnO:Eu spray thin films. From the calculated values of the opto-thermal expansivity, it can be seen that the Eu doping slightly decreases the opto-thermal expansivity. The maximum value of Ψ_{AB} was obtained for 0.7% Eu doping level. In comparison with the previous work it was found that Ψ_{AB} of ZnO: Eu are higher than Ψ_{AB} of those obtained earlier for ZnO: In (12.9 $10^{-12}m^3s^{-1}$). ZnO: Yb (8.2 $10^{-12}\,m^3s^{-1}$) and ZnO: Mo (15.2 $10^{-12}\,m^3s^{-1}$) [9]. This property is interesting for possible optoelectronic applications of ZnO:Eu thin films by reducing the absorption in solar cells.

Table III.2 Opto-thermal expansivity Ψ_{AB} values with Eu doping ZnO content.

Sample	ZnO	ZnO:Eu 0.4%	ZnO:Eu 0.7%	ZnO:Eu 1%
$\Psi_{AB}(10^{-12}m^3s^{-1})$	20	16	18	17

3.1.3. Refractive index and extinction coefficient

The refractive index (n) and the extinction coefficient (k) are the most important optical parameters of materials and their applications. Thus, it is important to determine these two optical parameters of the deposited thin films. The optical characteristics of dispersion n (λ) and k(λ), for values of the wavelength λ between 300 and 1800 nm, have been calculated using the optical experimental measurements and the method of Belgacem et al. [2] and Bathe et al. [23]. The present research is based on the solution of the following system of nonlinear equations:

$$R_{exp}(n, k, \lambda) - R_{theo}(n, k, \lambda) = 0 \qquad (III.9)$$
$$T_{exp}(n, k, \lambda) - T_{theo}(n, k, \lambda) = 0 \qquad (III.10)$$

where R_{theo} and R_{exp} represent the theoretical and experimental reflectance, and T_{theo} and T_{exp} represent rather the theoretical and experimental transmittance, respectively. The plots of n(λ) and k(λ) are presented in Figure III.4.

For all films, the values of the extinction coefficient (Figure III.4-a) in the ultraviolet region are larger than those in the visible domain. The refractive index (Figure III.4-b) has hyperbolic trend and it decreases with the wavelength along the Cauchy distribution [9,16,24]. This was observed for all thin films of ZnO:Eu.

$$n = A + \frac{B}{\lambda^2} \qquad \text{(III. 11)}$$

where A and B are the Cauchy's parameters and λ is the wavelength of the light used, implying that the films have normal dispersion for the entire range of wavelength studied. The values of A and B are given in Table III.3.

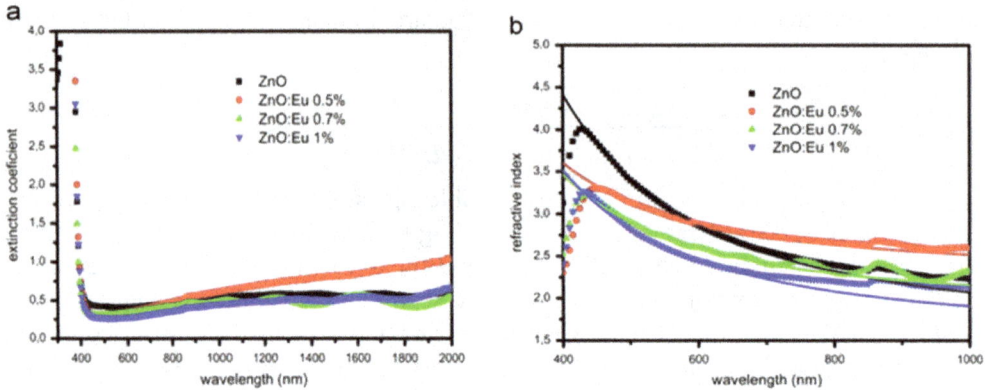

Figure III.4 Extinction coefficient (a) and variation of refractive index (b) with wavelength for different content of europium in the spray solution (y= [Eu^{3+}]/[Zn^{2+}] (0≤y≤1 at %)

Likewise, the hyperbolic decrease of the refractive index in the low absorption spectral range, was used to obtain some other optical constants based on Wemple–DiDomenico single-oscillator model [24–26]:

$$n^2 = 1 + \frac{E_0 E_d}{E_0^2 - E^2} \qquad \text{(III. 12)}$$

where E=hυ is the incident photon energy, E_0 is the energy of the effective dispersion oscillator, and E_d is the dispersion energy. Wemple–DiDomenico parameters are given in Table III.3. From these calculated values of both Cauchy and Wemple and Di-Dominico parameters, it was found that the particularity of 1% Europium doping level in terms of Cauchy parameters A, B, oscillator dispersion (E_0, E_d) energy are related to Wemple and Di-Dominico model.

Table III.3 Values of Cauchy (A, B) and Wemple and Di-Dominico parameters for different content of europium in the spray solution (y= [Eu^{3+}]/[Zn^{2+}] (0≤y≤1 at %).

	A	B	E_0(eV)	E_d(eV)
ZnO	1.623	0.446	2.816	9.408
ZnO :Eu 0.4%	2.311	0.206	3.745	18.949
ZnO :Eu 0.7%	1.900	0.250	3.544	13.453
ZnO :Eu 1%	1.597	0.307	3.288	9.899

3.1.4 Determination of complex dielectric functions and oscillator energy

The complex dielectric constant $\varepsilon(\omega)$ where $\varepsilon(\omega)=\varepsilon_1(\omega)-i\varepsilon_2(\omega)$ characterizes the optical properties of the solid material. Figure III.5 shows the variation of ε_1 and ε_2 vs wavelength.

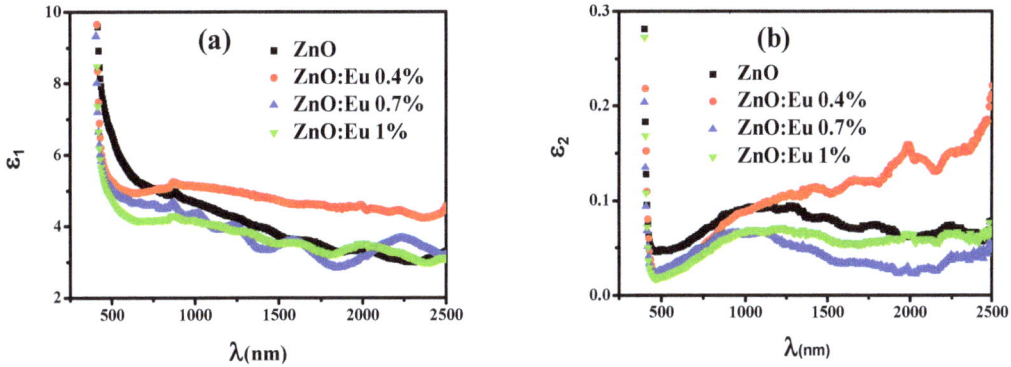

Figure III.5 Variation of the real part (a) and imaginary part (b) of complex dielectric constant with wavelength for different Eu content.

The real and imaginary parts of dielectric constant for pure and doped ZnO with Eu content are also determined by the following relations [27]:

$$\varepsilon(\lambda) = (n(\lambda) - ik(\lambda))^2 = \varepsilon_1(\lambda) - i\varepsilon_2(\lambda) \quad \text{(III. 13)}$$

$$\varepsilon_1(\lambda) = n(\lambda)^2 - k(\lambda)^2 \quad \text{(III. 14)}$$

$$\varepsilon_2(\lambda) = 2n(\lambda)k(\lambda) \quad \text{(III. 15)}$$

These results can be used to determine the optical constants: ε_∞, ω_p and τ, which represent the dielectric constant at high frequencies, the pulse plasma and relaxation time, respectively, by using the following relations [28–30]:

$$\varepsilon_1(\lambda) \approx \varepsilon_\infty - \frac{\varepsilon_\infty \omega_p^2}{4\pi^2 c^2} \lambda^2 \quad \text{(III. 16)}$$

$$\varepsilon_2(\lambda) = 2nk \approx \frac{\varepsilon_\infty \omega_p^2}{8\pi^3 c^3 \tau} \lambda^3 \quad \text{(III. 17)}$$

$$\omega_p^2 = \frac{4\pi N e^2}{\varepsilon_\infty m_e^*} \quad \text{(III. 18)}$$

where c is the light velocity and m_e^* is the effective mass. The calculated values of these constants are presented in Table III.4. From the results listed in Table III.4, after doping ZnO with europium, the lowest value of ε_∞ is obtained for 1 at% content of Eu, which the pulse plasma ω_p and relation time τ are 6.256×10^{14} rd s^{-1} and 0.319×10^{-14} s, res[ectively.

Table III.4 Values of the dielectric constant at high frequencies(ε_∞), the pulse plasma (ω_p) relaxtion time (τ) deduced from optical measurements for different content of europium in the spray solution (y= [Eu^{3+}]/[Zn^{2+}] (0≤y≤1 at %).

	ε_∞	ω_p (10^{12}rd s^{-1})	τ (10^{-14}s)
ZnO	5.10	555	4.511
ZnO :Eu 0.4%	6.63	312	0.038
ZnO :Eu 0.7%	6.24	861	2.228
ZnO :Eu 1%	4.99	625	0.319

3.2. Optical properties of crystalline In$_2$O$_3$: Eu thin films

3.2.1. Reflectance and transmission spectra

The optical transmission spectrum of Eu-doped In$_2$O$_3$ thin films grown on glass substrates are shown in Figure III. 6.

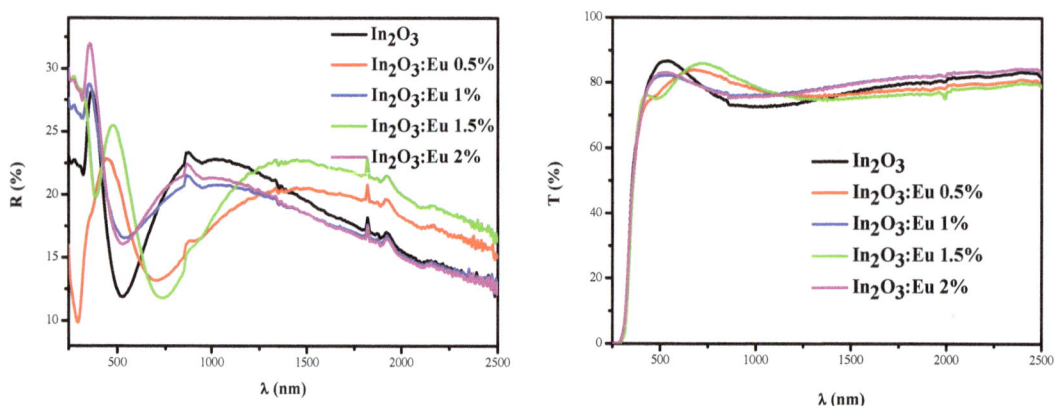

Figure III.6 Reflection and transmission spectra of In$_2$O$_3$:Eu sprayed thin films

The average value of transmittance coefficient in the transparency range is found to be 70–85%, while the reflectance coefficient remains inside a narrow interval (10–30%). In the visible region of solar spectrum, both reflectance and transmission spectra show the presence of the interference phenomena which indicates a smooth and homogeneous surface of the all Eu-doped In$_2$O$_3$ films.

3.2.2. Effect of Eu content on the absorption and optical band gap

The absorption coefficient α of the prepared thin films is determined from transmittance (T) measurement using the following expression [11]:

$$\alpha = \frac{1}{d} Ln \frac{1}{T} \qquad (III.\,19)$$

where d is the film thickness. The optical absorption edge was analyzed by equation (III-4) from which we can deduce the optical band gap E_g. In Figure III.7, $(\alpha h\nu)^2$ curves as a function of incident photon energy (hν) is plotted. It has been observed that these plots were linear over a wide range of photon energies indicating that E_g has a direct transition.

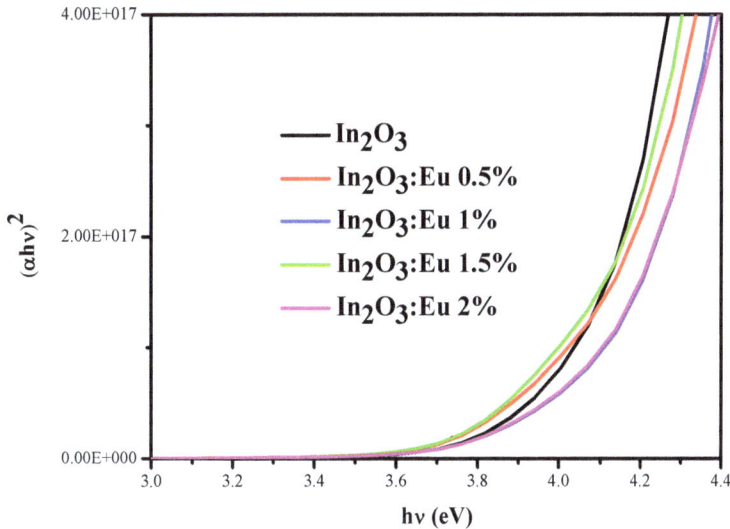

Figure III.7 $(\alpha h\nu)^2$ vs hν for In$_2$O$_3$:Eu prepared thin films by spray pyrolysis for different content of europium (y= [Eu^{3+}]/[In^{3+}] (0≤y≤2 at %)

All values of E_g are summarized in table III-5, which shows a dependency of band gap energy with the europium content. This phenomenon may be related to the europium incorporation into In$_2$O$_3$ matrix. It was found that a minimum value of band gap (E_g = 4.09 eV) was obtained for 1.5% doping level. On the other hand, the presence of defects in such thin films could be understood by the spreading of state densities in the band gap performed by the empirical Urbach model. For Eu-doped indium oxide thin films, the plot of Ln(α) versus the incident photon energy for various Eu content are reported in Figure III.8.

Table III.5. Band gap (E_g) and Urbach (E_u) energies of Eu-doped In$_2$O$_3$ thin films for different content of europium (y= [Eu^{3+}]/[In^{3+}] (0≤y≤2 at %)

	E_g(eV)	E_u(meV)
In$_2$O$_3$	4.10	278
In$_2$O$_3$: Eu 0.5%	4.17	285
In$_2$O$_3$: Eu 1%	4.15	313
In$_2$O$_3$: Eu 1.5%	4.09	289
In$_2$O$_3$: Eu 2%	4.10	309

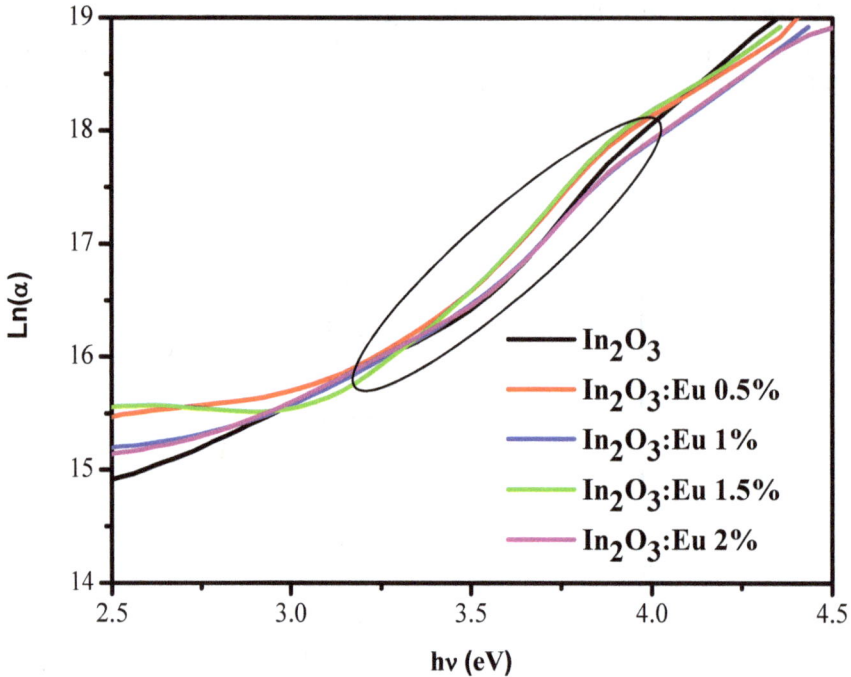

Figure III.8 Plot of Ln(α) versus hν

By using the Urbach Law equation (III-5), we obtained the Urbach energy E_u which was calculated for different europium content. All values of E_u are summarized in table III-5. This suggests that defects and impurities from interstitial indium atom disappear depending on the doping concentration level. Thus, the doping causes a reorganization of the structure as shown in section II, that the best crystallinity is obtained for an europium content (y= [Eu^{3+}]/[In^{3+}]) of 1.5 at%. This result is consistent with other studies reported by Mott and Davis [12]. It is noted that, unlike the crystalline structures, in the fundamental superiority is mainly determined by conduction and valence levels, ion-doped binary semiconductor compounds present a particular optical absorption edge profile. In these materials, the absorption coefficient profile increases exponentially with the photon energy near the energy gap.

3.2.3. Refractive index and extinction coefficient

The plots of $n(\lambda)$ and $k(\lambda)$ are presented in Figure III.9. It is found that the refractive index has hyperbolic decrease as a function of wavelength of incident photon. This variation is modeled by Cauchy law (III.11), where A and B values for In$_2$O$_3$: Eu-sprayed thin films are given in Table III.6.

In addition to the Cauchy extrapolation described above, which was done to provide reasonable values for refractive index in the spectral range of low absorption, we have also carried out some other optical constants based on Wemple–Di-Domenico single-oscillator model. All values of E_0 and E_d are listed in table III.6.

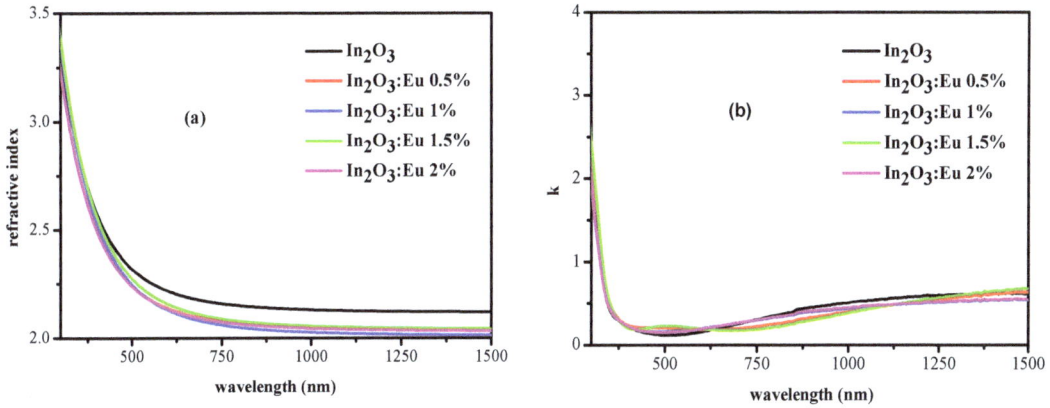

Figure III.9 Refractive index (n) and extinction coefficient (k) of Eu doped In_2O_3 thin films.

Table III.6 Optical constants of Cauchy law and E_0 and E_d energies.

	A	B (nm^2)	E_0 (eV)	E_d (eV)
In_2O_3	2.101	30237.62	4.797	15.544
In_2O_3 :Eu 0.5%	2.017	31786.78	4.720	13.910
In_2O_3:Eu 1%	1.990	38332.15	4.451	12.412
In_2O_3 :Eu 1.5%	2.020	37649.29	4.509	13.159
In_2O_3 : Eu 2%	2.014	31302.34	4.749	13.593

In fact, in low absorption domain, we can approximate n^2 to ε_1. In this range, ε_1 is given by the following equation [31, 32]:

$$\varepsilon_1 = 1 + \omega_p^2 \sum \frac{f_n}{(\omega_n^2 - \omega^2)} \qquad (III.20)$$

where f_n is the oscillator strength corresponding to the transition dipole at frequency ω_n. In the framework of the single oscillator model, we consider a single oscillator that has an angular frequency ω_0 (energy E_0) and therefore neglect the effect of the other frequencies. In the spectral study, a single oscillator can be considered, which has the most dominant contribution and whose strength is denoted as f_0. Also the relationship can be reduced by introducing energy parameters E_0 and E_d to the equation (III.12). The variation of $(n^2-1)^{-1}$ vs. $(h\nu)^2$ allows us to obtain the E_0 and E_d constants. Their values are listed in Table III.6. It is found that values of both Cauchy and Wemple Di-Dominoco parameters are strongly dependent on europium doping level. Furthermore, the dispersive energy (E_d) has a minimum value of ~12.5 eV for 1% doping level. This situation is a good indicator that by using the Eu doping, the dispersion energy inside In_2O_3 thin films is reduced.

3.2.4. Complex dielectric functions

The fundamental spectrum of the electron excitation in the films is described by using a frequency dependence of the complex electronic dielectric constant equations (III.13-14-15). For all the samples, ε_1 has a linear variation function of the square of the wavelength (Figure III.10) in the infrared range, while ε_2 is practically linear with λ^3.

This behavior was used to evaluate ε_∞, ω_p and τ, high frequency dielectric constant, plasma frequency and relaxation time, respectively, through relations III-16, III-17 and III-18. All values of ε_∞, ω_p and τ are given in table III.7. From this study, it is shown that the high plasma pulsation (ω_p) has a maximum value at 1.5 at% Eu doping which indicates a high interaction with optical radiation.

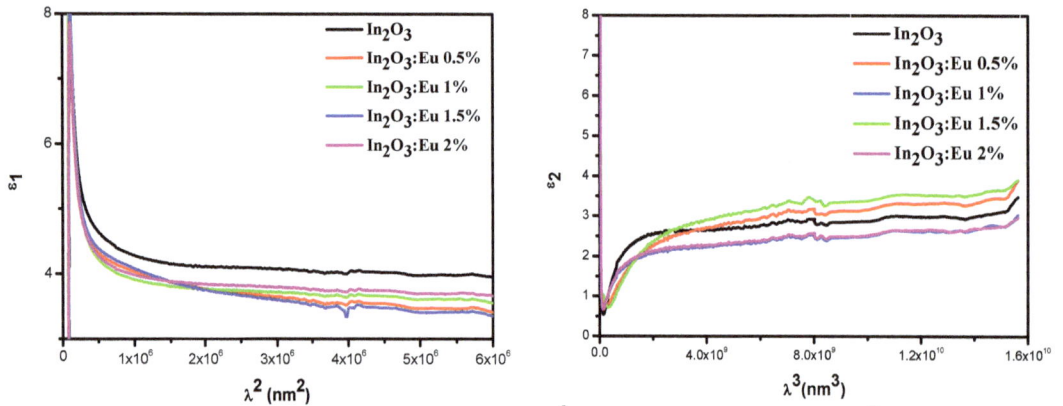

Figure III.10 Variation of the real part with λ^2 and imaginary part with λ^3 of complex dielectric constant of Eu-doped In_2O_3 thin films grown for different content of europium.

Table III.7 Optical constants of Eu-doped In_2O_3 thin films.

	ε_∞	ω_p $(10^{14}$ rad s$^{-1})$	$\tau(10^{-16}$ s$)$
In_2O_3	4.201	1.907	5.786
In_2O_3 : Eu 0.5%	3.974	3.078	7.543
In_2O_3 : Eu 1%	3.876	2.220	5.980
In_2O_3 : Eu 1.5%	4.141	3.957	1.275
In_2O_3 : Eu 2%	3.938	2.056	5.447

3.3. Optical investigation on MoO₃:Eu thin films

3.3.1. Reflection and transmission spectra of MoO₃:Eu

The optical transmission spectra of undoped and Eu-doped MoO_3 thin films grown on glass substrates are shown in Figure III.11. The average values of transmittance of thin films in the visible range were in the 40 - 85% range, while the reflectance remains in the 10%–30% range. These results are consistent with those reported by Gesheva *et al.* [70].

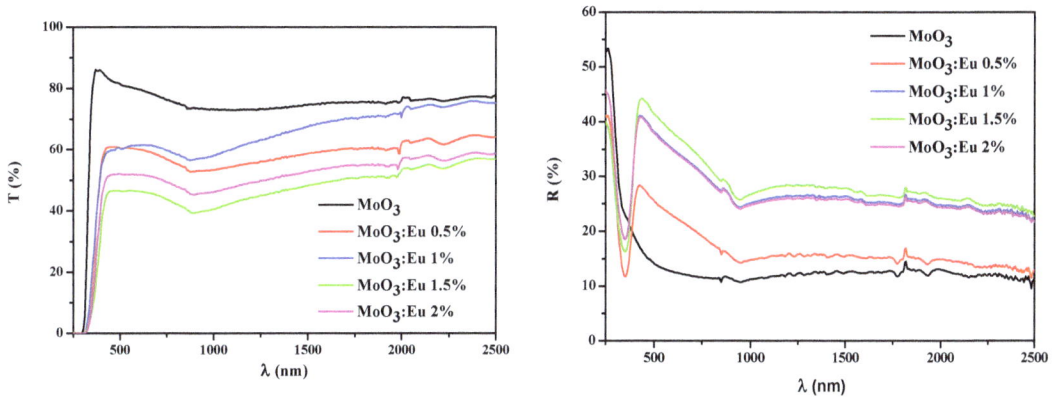

Figure III.11. Transmission and reflection spectra of MoO_3:Eu sprayed thin films

3.3.2. Absorption coefficient of MoO_3:Eu

The absorption coefficient α of these films was obtained from transmittance and reflectance data using the relationship presented in (III.3). The optical absorption edge was analyzed by equation (III.4). In Figure III.12, $(\alpha h \upsilon)^2$ values are plotted versus a photon incident energy $h\upsilon$. It is noted that the plots of $(\alpha h \upsilon)^2$ versus $h\upsilon$ are linear over a wide range of photon energies indicating that there is a direct transition. The extrapolation of these plots (straight lines) on the energy axis give the energy band gap (E_g) values which are presented in Table III.8. It is found that E_g values after doping are less than that the one before incorporation of the europium atoms. Also, a minimum value of band gap for film having Eu content equal to 1.5% was observed. This is consistent with the XRD analysis described previously in chapter II. Indeed, the best crystallinity and a minimum number of defaults such as dislocation, strain and air inclusion is reached for the film doped with 1.5% Eu doping.

3.3.3. Band gap and Urbach energy of MoO_3:Eu

The defects on the prepared thin films suggested above are interpreted by the spreading of state densities in the band gap that can be qualified by the empirical model of Urbach using relations (III.5) and (III.6).

The band gap (E_g) and Urbach (E_u) energies of undoped and Eu-doped MoO_3 thin films values are presented in Table III.8.

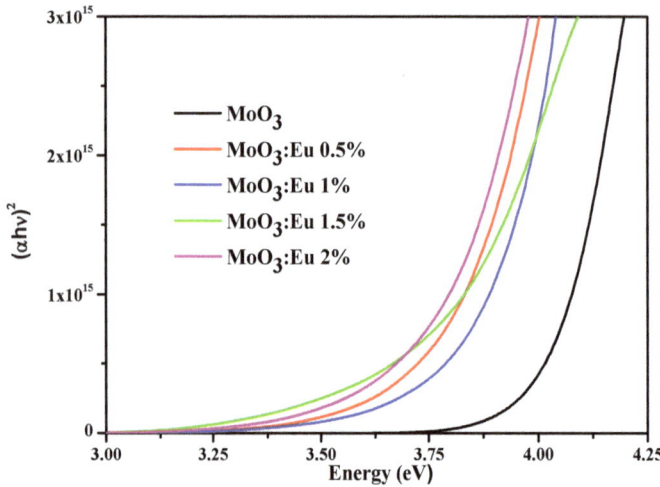

Figure III.12
$(\alpha h v)^2$ vs hv for MoO_3:Eu thin films

Table III.8 Optical band gap E_g and Urbach energy for different content of europium in the spray solution (y= $[Eu^{3+}]/[Mo^{6+}]$ ($0 \leq y \leq 2$ at %).

	E_g(eV)	E_u (meV)
MoO$_3$	4.05	163
MoO$_3$: Eu 0.5%	3.79	350
MoO$_3$: Eu 1%	3.87	305
MoO$_3$: Eu 1.5%	3.76	498
MoO$_3$: Eu 2%	3.85	369

The variation of Urbach energy (E_u) value with the doping concentration suggests that the presence of defects and impurities depends on the doping level. Doping causes a certain reorganization of the structure of the Eu doped films. These results are consistent with those reported previously by Mott *et al.* [79] who found out that unlike crystalline structures where the fundamental edge is mainly determined by conduction and valence levels, ion-doped binary semiconductor compounds present a particular optical absorption edge profile. In these oxide materials, the absorption coefficient profile increases exponentially with the photon energy near the energy gap (Figure III.13).

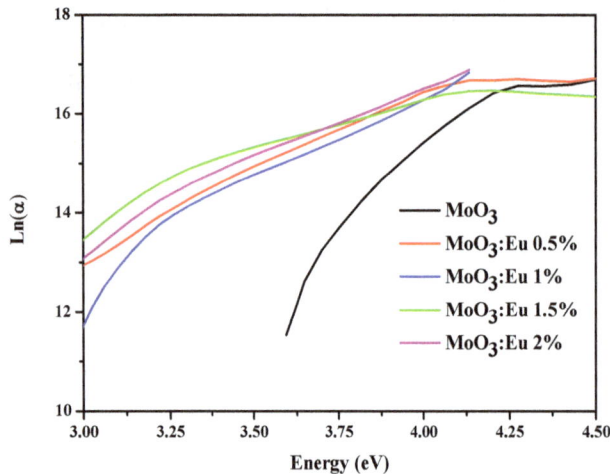

Figure III.13 *Ln(α)* vs *hv* spectra of MoO$_3$:Eu sprayed thin films.

3.4. Discussion of optical results

From this optical study, we can conclude the following relevant findings:

• The band gap energy value of Eu doped MoO_3 thin films decreases with the increase of Eu content. However, for both ZnO and In_2O_3 thin films, Eu doping element does not affect Eg values of such materials.

• The same remark has been found for Urbach energy. Indeed, Eu values for undoped or doped ZnO is in the order of 113 meV while for undoped or doped In_2O_3 thin films is in the order of 300 meV. On the contrary, Eu value increases from 160 to 370 meV for MoO_3 thin films. This may be be of interest for possible applications of the three binary oxides as sensors.

• The optical properties of such oxides will be discussed in terms of their photoluminescence measurements as well as their surface morphologies.

3.5. Conclusion

The effect of europium doping element on the optical properties of ZnO, In_2O_3 and MoO_3 thin films prepared by spray was presented in this section. This study reveals that Eu-doping enhances the optical properties. The best crystallinity was obtained for the following conditions: for ZnO:Eu 1 at%, for In_2O_3:Eu 1.5% and for MoO_3:Eu 1.5at%.

Optical properties show a high transmittance around 80% in the visible and near infrared regions for ZnO and In_2O_3. Calculated band gap is in the range of 3.21, 4.1 and [3.76-4.05] eV respectively for ZnO, In_2O_3 and MoO_3, respectively. All these experimental results confirmed the efficiency of europium doped zinc oxide, indium oxide and molybdenium oxide as a transparent conductive oxide.

Eu-doped ZnO, In_2O_3 and MoO_3 thin films could be very promising materials with a great potential in the field of optoelectronics.

3.6. Reference

[1] B.D. Cullity, "Elements of X-ray Diffraction", Addison-Wesley Publishing Company, 1978.

[2] S. Belgacem, R. Bennaceur, J. Saurel, J. Bougnot, "Propriétés optiques des couches minces de SnO_2 et $CuInS_2$ airless spray", Revue de Physique Appliquée, 25 (1990) 1245–1258.

[3] D.A. Minkov, "Calculation of the optical constants of a thin layer upon a transparent substrate from the reflection spectrum", Journal of Physics D: Applied Physics, 22 (1989) 199.

[4] R. Swanepoel, "Determination of the thickness and optical constants of amorphous silicon", Journal of Physics E: Scientific Instruments, 16 (1983) 12–14.

[5] S. Belgacem, J.M. Saurel, J. Bougnot, "The optical properties of sprayed CdS Thin films", Thin Solid Films, 92 (1982) 199–209.

[6] A. Amlouk, K. Boubaker, M. Bouhafs, M. Amlouk, "Optimization of transparent conducting oxide ZnO compound thickness in terms of four alloys thermo-physical performance aggregates", Journal of Alloys and Compounds, 509 (2011) 3661–3666.

[7] Y. Narendar, G.L. Messing, "Mechanisms of phase separation in gel-based synthesis of multicomponent metal oxides", Catalysis Today, 35 (1997) 247–255.

[8] X.Y.Li, H.J.Li, Z.J.Wang, H.Xia, Z.Y.Xiong, J.X.Wang, B.C.Yang, "Effect of substrate temperature on the structural and optical properties of ZnO and Al-doped ZnO thin films prepared by dc magnetron sputtering", Optics Communications, 282 (2009) 247–252.

[9] A. Boukhachem, S. Fridjine, A. Amlouk, K. Boubaker, M. Bouhafs, M. Amlouk, "Comparative effects of indium/ytterbium doping on, mechanical and gas-sensitivity-related morphological, properties of sprayed ZnO compounds", Journal of Alloys and Compounds, 501 (2010) 339–344.

[10] L.P. Peng, L. Fang, X.F. Yang, Y.J. Li, Q.L. Huang, F. Wu, "Effect of annealing temperature on the structure and optical properties of In-doped ZnO thin films", Journal of Alloys and Compounds, 484 (2009) 575-579.

[11] J.I. Pankove, "Optical Processes in Semiconductors", Prentice-Hall Inc, Englewood Cliffs, 1971.

[12] N.F. Mott, E.A. Davis, "Electronic Processes in Non-Crystalline Materials", Clarendon Press, Oxford, UK, 1979.

[13] F. Urbach, "The Long-Wavelength Edge of Photographic Sensitivity and of the Electronic Absorption of Solids", Physical Review, 92 (1953) 1324.

[14] W. Martienssen, "Über die excitonenbanden der alkalihalogenidkristalle", Journal of Physics and Chemistry of Solid, 2 (1957) 257-267.

[15] A. Boukhachem, R. Boughalmi, M. Karyaoui, A. Mhamdi, R. Chtourou, K. Boubaker, M. Amlouk, "Study of substrate temperature effects on structural, optical, mechanical and opto-thermal properties of NiO sprayed semiconductor thin films", Materials Science and Engineering: B, 188 (2014) 72–77.

[16] K. Boubaker, A. Chaouachi, M. Amlouk, H. Bouzouita, "Enhancement of pyrolysis spray disposal performance using thermal time-response to precursor uniform deposition", The European Physical Journal - Applied Physics, 37 (2007) 105–111.

[17] C. Bouzidi, N. Sdiri, A. Boukhachem, H. Elhouichet, M. Férid, "Impedance analysis of $BaMo_{1-x}W_xO_4$ ceramics, Superlattices and Microstructures", 82 (2015) 559–573.

[18] A. Belhadj, O. Onyango, N. Rozibaeva, "Boubaker Polynomials Expansion Scheme-Related Heat Transfer Investigation Inside Keyhole Model", Journal of Thermophysics and Heat Transfer, 23 (2009) 639–640.

[19] S. A. H. A. E. Tabatabae, T. Zhao, O. Awojoyogbe, F. Moses, "Cut-off cooling velocity profiling inside a keyhole model using the Boubaker polynomials expansion scheme", Heat Mass Transfer, 45 (2009) 1247–1251.

[20] A. Belhadj, J. Bessrour, M. Bouhafs, L. Barrallier, "Experimental and theoretical cooling velocity profile inside laser welded metals using keyhole approximation and Boubaker polynomials expansion", Journal of Thermal Analysis and Calorimetry, 97 (2009) 911–915.

[21] S. Fridjine, M. Amlouk, "A new parameter: an abacus for optimizing PVT hybrid solar device functional materials using the Boubaker Polynomials Expansion Scheme", Modern Physics Letters B, 23 (2009) 2179–2191.

[22] M. Ben Amor, A. Boukhachem, K. Boubaker, M. Amlouk, "Structural, optical and electrical studies on Mg-doped NiO thin films for sensitivity applications", Materials Science in Semiconductor Processing, 27 (2014) 994-1006.

[23] S. R. Bathe, P. S. Patil, "Electrochromic characteristics of fibrous reticulated WO_3 thin films prepared by pulsed spray pyrolysis technique", Solar Energy Materials and Solar Cells, 91 (2007) 1097-1101.

[24] B. Ouni, A. Boukhachem, S. Dabbous, A. Amlouk, K. Boubaker, M. Amlouk, "Some transparent semi-conductor metal oxides: Comparative investigations in terms of Wemple–

DiDomenico parameters, mechanical performance and Amlouk–Boubaker opto-thermal expansivity", Materials Science in Semiconductor Processing,13 (2010) 281–287.

[25] M. Di-Domenico Jr., M. Eibschütz, H.J. Guggenheim, I. Camlibel, "Dielectric behavior of ferroelectric $BaMF_4$ above room temperature", Solid State Communication, 16 (1969) 1119-1122.

[26] S.H. Wemple, "Interband optical transition strengths in SiO_2, GeO_2, SnO_2 and TiO_2", Solid State Communication, 12 (1973) 701-704.

[27] A.Bouzidi, N.Benramdane, H.Tabet-Derraz, C.Mathieu, B.Khelifa, R. Desfeux, "Effect of substrate temperature on the structural and optical properties of MoO_3 thin films prepared by spray pyrolysis technique", Materials Science and Engineering: B, 97 (2003) 5–9.

[28] A.K. Walton, T.S. Moss, "Determination of Refractive Index and Correction to Effective Electron Mass in PbTe and PbSe", Proceedings of the Physical Society, 81 (1963) 509-513.

[29] M. Sesha Reddy, K.T. Ramakrishna Reddy, B.S. Naidu, P.J. Reddy, "Optical constants of polycrystalline $CuGaTe_2$ films", Optical Materials, 4 (1995) 787-790.

[30] J.I. Pankove, "Optical Processes in Semiconductors", Prentice-Hall Inc, Englewood Cliffs, 1971, p. 92.

[31] M. Aven and J. S. PRENER, Eds. North-Holland, Amsterdam; Interscience (Wiley), New York, "Intrinsic exciton absorption, in Physics and chemistry of II–VI compounds" (eds), 1967 862.

[32] M. Altarelli and N. O. Lipari, "Exciton dispersion in semiconductors with degenerate bands", Physical Review B, 15 (1977) 4898-4906.

[33] K.A. Gesheva, T. Ivanova and A. Szekeres, "Optical properties of chemical vapor deposited thin films of molybdenum and tungsten based metal oxides", Solar Energy Materials and Solar Cells, 76 (2003) 563-576.

[34] N. F. Mott, E.A. Davis, "Electronic Processes in Non-Crystalline Materials", Clarendon Press, Oxford, UK, 1979.

4. Photoluminescence characterization techniques and results

4.1 Principles of Photoluminescence

Photoluminescence (PL) analysis is a simple and nondestructive characterization tool. It gives an insight on the electron transfer mechanism, as well as on the identification of impurity and defect states depending on their optical activity, due to its unique sensitivity to discrete electronic states which are near surfaces and interfaces. A photoluminescence spectrum is a measure of spontaneous emission intensity at a fixed excited wavelength as a function of the detection energy from a material under optical excitation. The absorbed or emitted light is related to the difference between the initial electronic state and a high energy state. The intensity of the emitted light is related to the contribution of the radiative recombination. The emission spectrum is generally used to identify surface, interface, and impurity levels. The PL signal intensity provides information on the quality of surfaces and interfaces, the transition energies, which can be used to determine electronic energy levels. Also, it gives relative rates measures of radiative and non-radiative recombination. Photons are absorbed and electronic excitations are created when a laser light hits a material.

Eventually, these excitations relax and the electrons return to the fundamental state. When radiative relaxation occurs, the emitted light is called photoluminescence.

4.2 Photoluminescence spectrometer

The excitation source, sample cell, photoluminescence spectrometer detector are the main parts of a photoluminescence spectrometer. Usually, a deuterium or xenon lamp is used to excite the molecules of the sample. An excitation light passes through a monochromator, which allows passing the light of only a selected wavelength with a value that depends on the optical band gap of the analyzed compound. The photoluminescence is detected by a photomultiplier tube. The excitation and the photoluminescence spectra are found by scanning of the excitation monochromator and the fluorescence monochromator, respectively.

PL measurements of europium doped ZnO, In_2O_3 and MoO_3 sprayed thin films were carried out by using Perkin Elmer LS55 equipment. For the Eu doped oxide thin films, the photoluminescence properties were studied under 325 nm excitation wavelength at room temperature.

4.2.1 Photoluminescence of ZnO:Eu

Figure IV.1 shows PL spectra of europium doped of ZnO excited using a wavelength of 350 nm. We observed a broad emission peak in the range of 400-600 nm. This green emission is associated with recombination of delocalized electrons at singly occupied oxygen vacancies with deep trapped holes. The red curve corresponds to the sum of all these emissions. We remark the emission sum is in good agreement with experimental curve (black curve). It seems that the wideband is a result of the emission peaks which occurred at 444 nm, 491 nm, 502 nm and 523 nm.

Furthermore, the effect of percentages of doping on the PL properties of ZnO thin films was studied. We observed that new peaks appear after doping, as shown in Figure IV.2. Usually, the photoluminescence of ZnO is deep-level emission, being related to structure defects and impurities [1]. For a wide gap oxide nanomaterial, oxygen vacancies induce the formation of other energy levels in the band gap. Zhang *et al.* [2] demonstrated by theoretic calculation that electron hole radioactive recombination leads to the green luminescence in nano ZnO. Wu *et al.* [3] reported that the photoluminescence of as-grown, oxidized, and deoxidized ZnO nanostructures were investigated and the results provided an indirect proof supporting the oxygen vacancies phenomenon. Thus, the density of free exciton in ZnO nanofilms affects the intensity of ultraviolet emission [4]. After doping, the increase of PL intensities with the increase of Eu content in ZnO results in the trapping of photo-generated electrons by Eu^{3+} ions, and therefore an enhancement of charge separation in the ZnO thin films is observed.

The second peak corresponds to the visible band emission which is mainly related to the intrinsic defects of ZnO lattice such as oxygen vacancies, antisite oxygen, interstitial oxygen, interstitial zinc, zinc vacancies and antisite zinc [5]. Moreover, it can be generally attributed to electron transfer from the conductor band to defect levels (or impurity levels) in the forbidden band or from defect levels to the valence band or between different defect levels (or impurity levels).

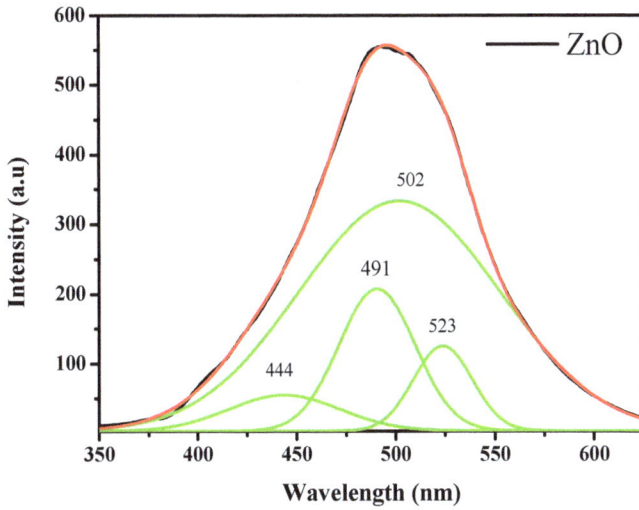

Figure IV.1
Photoluminescence of
ZnO undoped thin film

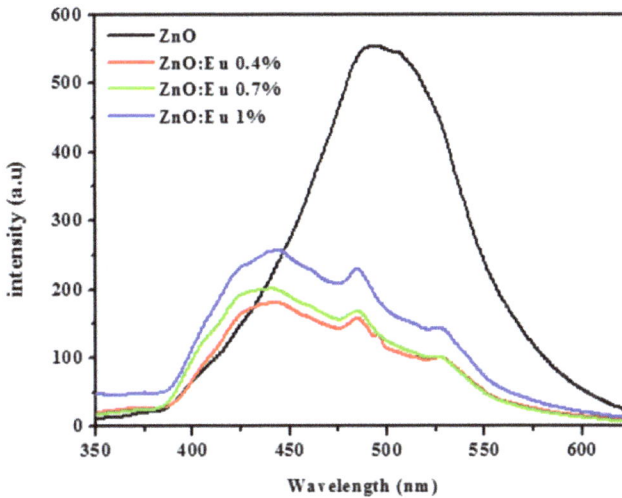

Figure IV.2
Photoluminescence of
Eu doped ZnO thin films for
different Eu content

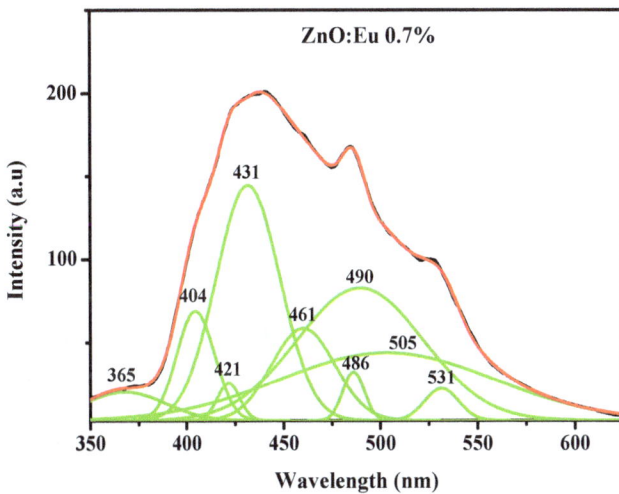

Figure IV.3
Photoluminescence of
ZnO:Eu 0.7%

4.2.2 Photoluminescence study of In$_2$O$_3$:Eu

Generally, the photoluminescence of In$_2$O$_3$ compound is due to the presence of traps in indium oxide thin films. It is well known that In$_2$O$_3$ thin films have several types of defects levels such as indium interstitial, oxygen and indium vacancies [6].

The photoluminescence spectra of as-grown, In$_2$O$_3$ thin films were measured under the excitation at 350 nm. Figure IV.4 shows PL spectrum obtained from as-grown In$_2$O$_3$ thin films. The peak centered at 394 nm is originated from the free exciton emission around 3.14 eV from the wide-band-gap of In$_2$O$_3$. The stable blue light was also observed at 423, 436 and 487 nm. This large blue-near green emission may be related to lattice defects or oxygen deficiencies in the indium oxide thin films as claimed by Kaleemulla *et al.* [7]. In fact, they reported that photoluminescence process can be produced as follow: electrons located in donor level can be attracted by a hole positioned in acceptor level. So a trapped exciton is formed leading to photoluminescence emission [8]. A strong and broad PL emission spectrum recorded from In$_2$O$_3$ nanofibres is reported with its maximum intensity centered at 470 nm [9-10]. Wu *et al.* [3] reported that emission peaks observed in the visible range may be attributed to the recombination of carrier concentration that occurs between valence band and oxygen vacancies acting as donor levels.

In the literature, In$_2$O$_3$ nanowires have PL peaks centered at 425, 429, 442, 460 nm [11], 416 and 435 nm [12] and the nanocubes have a strong PL peak centered at 450 nm [13]. The photoluminescence peaks in the visible emission would be attributed to the oxygen vacancies.

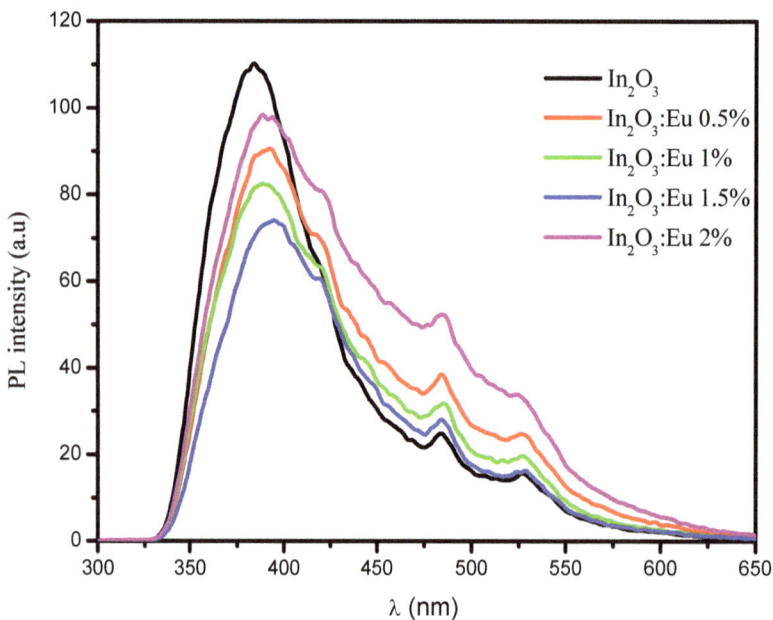

Figure IV.4
Photoluminescence of Eu doped In$_2$O$_3$

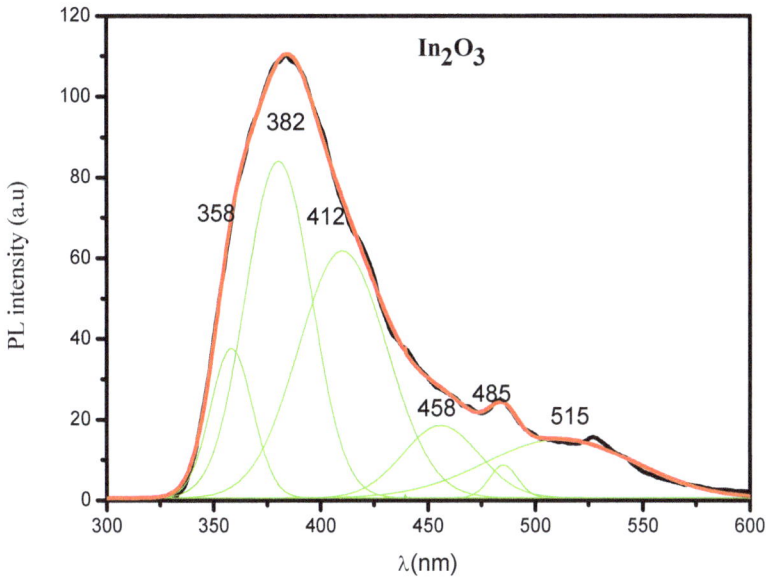

Figure IV.5
Photoluminescence of undoped In_2O_3 thin film

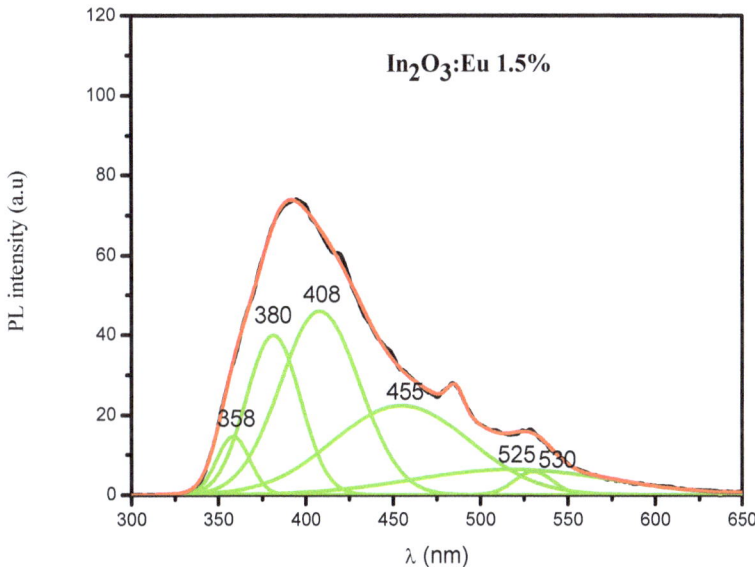

Figure IV.6
Photoluminescence of 1.5% Eu doped In_2O_3

4.2.3 Photoluminescence study of MoO_3

MoO$_3$ thin films exhibit PL emission peaks positioned at wavelengths 393, 420 nm, 451 nm, 486 nm, 510 nm, 532nm and 587nm, labeled as P1, P2, P3, P4, P5, P6 and P7, respectively. These luminescence peaks of molybdenum oxide may be attributed to the radiative decay of self-trapped excitons, linked to certain lattice defects such as oxygen vacancies, to other complex defect centers such as the clusters of oxygen vacancies, and to the low valence Mo^{6+} ion associated with the transfer of charge from oxygen vacancies to Mo ions [14, 15].

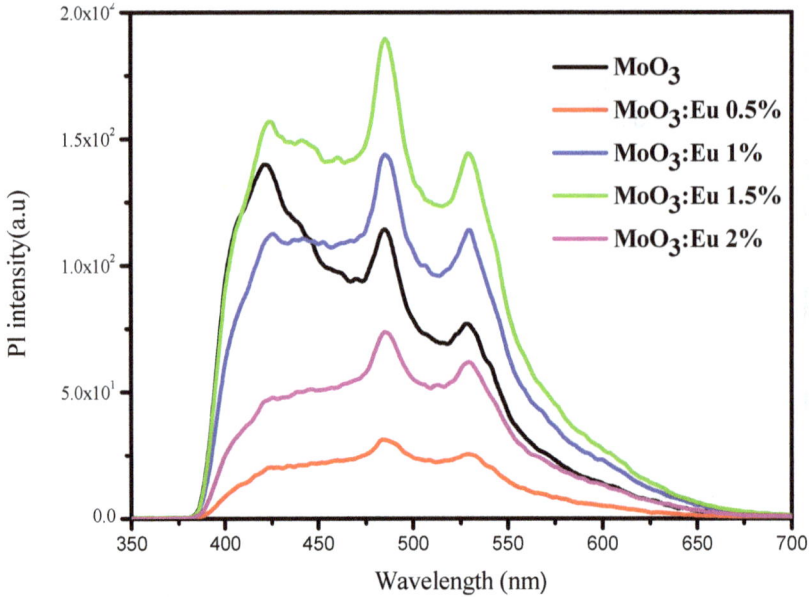

Figure IV.7. Photoluminescence of MoO$_3$:Eu

Gaussian convoluted form PL spectra of MoO$_3$ (Figure IV.8) with peaks labeled as P1, P2, P3, P4, P5, P6 and P7, assigned to the free exciton recombination are in accordance with earlier reported results. The transitions located at 451, 486, 510, 532 and 587 nm may be attributed, in the crystal field model, to Mo^{6+} d–d band transition of a heavily distorted polyhedron (Mo–O) in an octahedral crystal system [16, 17].

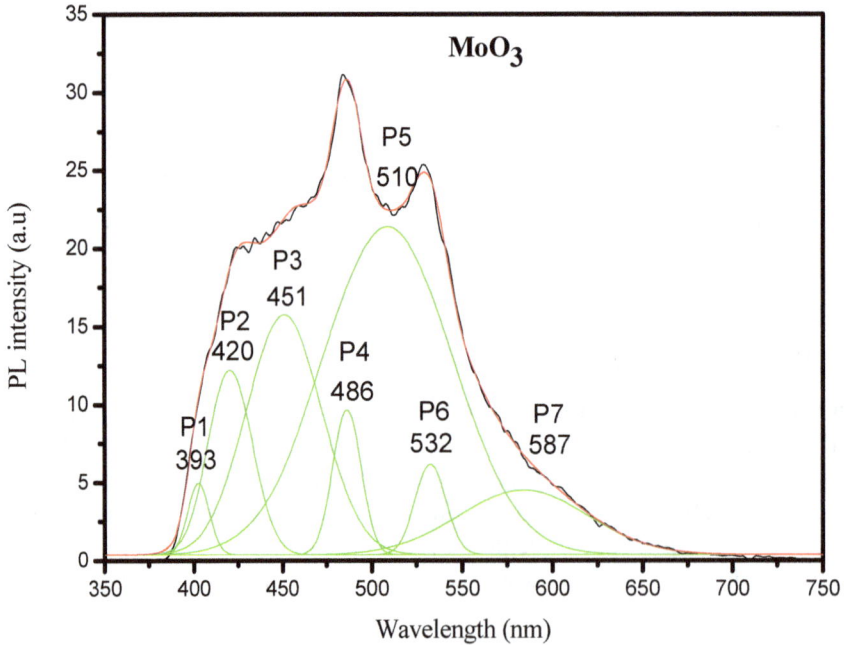

Figure IV.8. Photoluminescence of MoO$_3$

4.3. Scanning Electron Microscope (SEM) observations

4.3.1. Scanning Electron Microscope (SEM)

The operation of scanning electron microscope is based on the electron emission produced by the cathode and the detection of signals resulting from the interaction of these electrons with the sample. The gun produces a beam electron through a tungsten filament heated by a current. This beam is accelerated by the high voltage created between the filament and the anode. It is then focused on the sample by a series of electromagnetic lens. Controlled via computer, the SEM operator can adjust the beam to control magnification as well as to determine the surface area to be scanned. Further, a high vacuum should be applied to make the most accurate measurement.

We used a SEM Philips apparatus (Figure IV.9) to study the surface morphology of zinc oxide, indium oxide and molybdenum oxide doped with different contents of rare earth element (Eu). Then, we can obtain the effect of doping on the surface topography of such oxides. The scanning electron microscopy is composed by the following component (Figure IV.9):

- Electron gun: used to produce electrons due to heating a tungsten filament by thermoelectric field, the lifetime of tungsten filament is of some hundred hours.
- Condenser lenses: to focus the beam to a point spot.
- Deflection coils: that imposes the beam to do a scanning movement of the sample surface.
- Detectors: divided into secondary electron detectors and back scattered electrons detectors.
- High voltage generator: used to supply electron gun.
- Sweep generator: used to supply deflection coils.

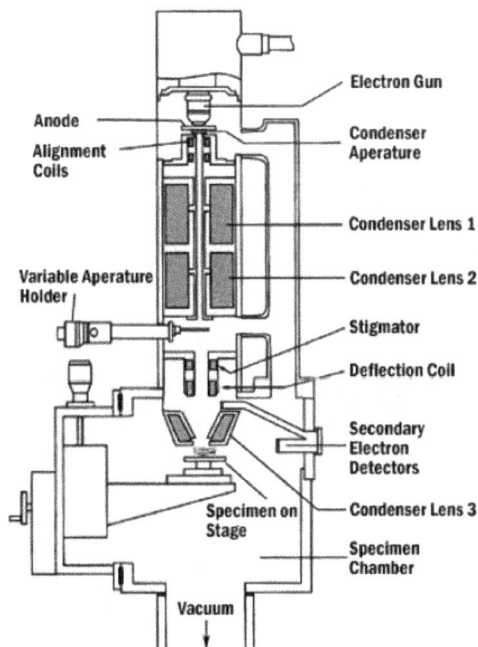

Figure IV.9 Experimental set-up of scanning electron microscopy (SEM)

4.3.1.1. SEM images of ZnO:Eu

Figure IV.10 shows that all the film surfaces are compact, without any defect zones or cracks and with a uniform grain distribution. Also, the grain growth followed c-axis which is perpendicular to the substrate plane, which is found as a polar axis [18]. This result is in agreement with the XRD results presented in chapter II. By using Eu doping, the morphology and the geometry of the nanofilms was not affected by the Eu concentration, it remains polygonal for all films, and only the grains size varies. For ZnO: Eu 1% film, there is an expansion of the grains size, likewise, it exhibits a rough surface compared to other doped nanofilms. Whereas for ZnO: nanofilms, the grains size decreases. The average diameter size measured from SEM data is found ranging from 50 to 110 nm while, the calculated crystallites size from XRD investigations is lower than such values. This might be due to the agglomeration of the smaller grains for clustered formation occurring during endothermic reaction between precursors to form these oxide films.

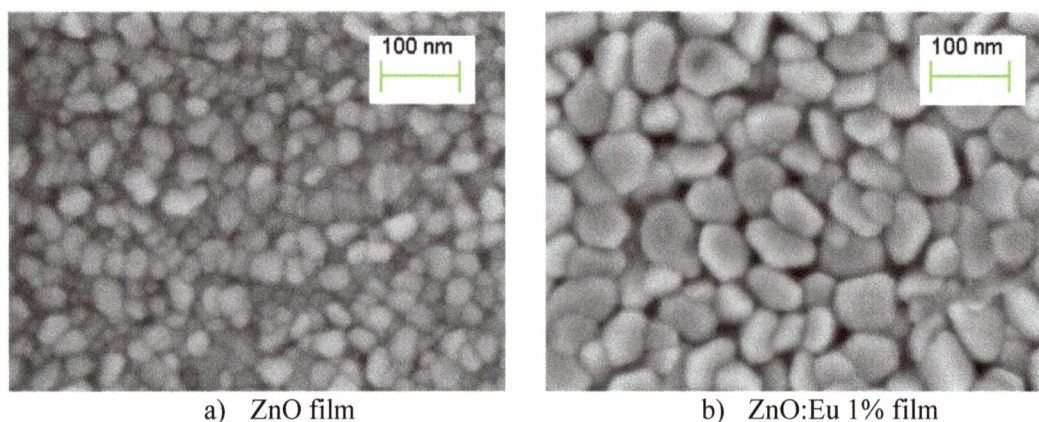

a) ZnO film   b) ZnO:Eu 1% film

Figure IV.10 SEM observations of Eu-doped ZnO thin films

4.3.1.2. SEM images of In$_2$O$_3$:Eu

SEM observations allow us to obtain microscopic information on the surface structure of both undoped and Eu-doped In$_2$O$_3$. Indeed, these micrographs (Figure IV.11) reveal that all the film surfaces are rough. These perturbed surfaces are probably due to very small droplets resulting from the spray pyrolysis technique that vaporize above the glass substrates and condense as microcrystallites with various dimensions (50–200 nm). Doping with Eu (1%) of In$_2$O$_3$ has a little effect to improve the surface morphology of In$_2$O$_3$. On the contrary Eu content (2 at%) increases the size of the crystallites on the film surface.

4.3.1.3. SEM images of MoO$_3$:Eu

The surface morphology of undoped and Eu doped MoO$_3$ spayed thin films were provided by scanning electron microscopy. Figure IV.12 shows SEM micrographs of undoped MoO$_3$ and Eu–doped MoO$_3$ thin films. It can be seen that SEM observation concerning undoped MoO$_3$ film (Figure IV.12-(a)) depicts a clearly disturbed surface with randomly oriented islets-like (with various sizes) showing a rough surface morphology.

a) In_2O_3 b) In_2O_3 : Eu 1%

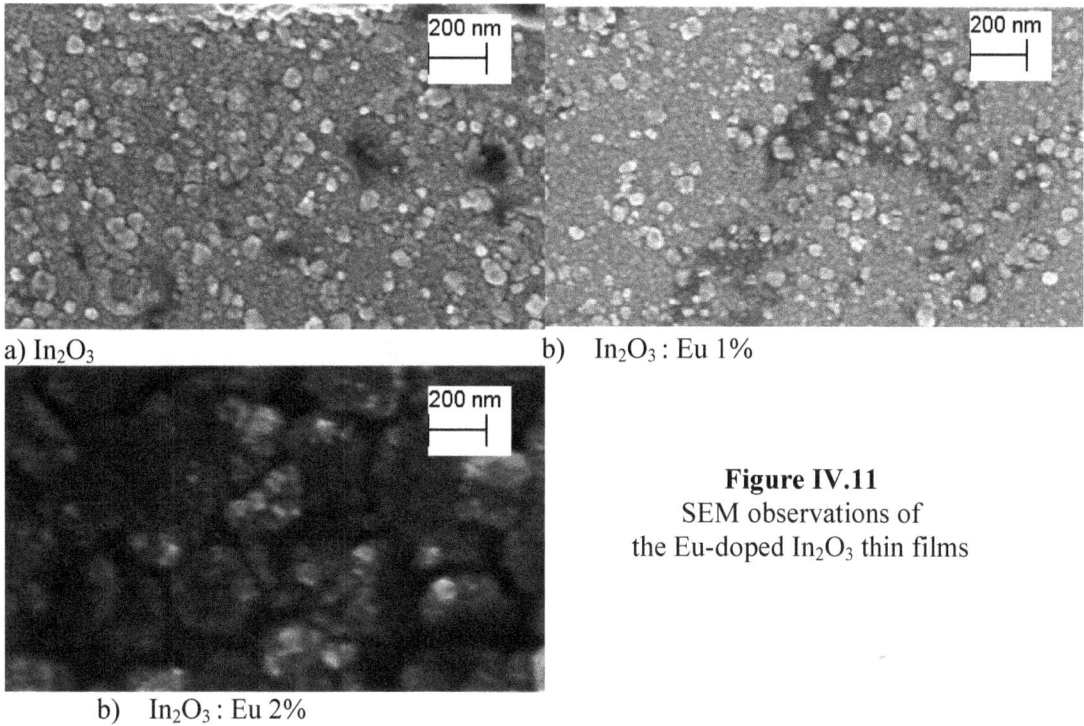

Figure IV.11
SEM observations of
the Eu-doped In_2O_3 thin films

b) In_2O_3 : Eu 2%

On the contrary, with 0.5 Eu content seems to have an effective role on the film crystallinity (Figure IV.12-(b)) with the appearance of a coral-like structure as plates having different shapes like. This result agrees with those reported by Imawan *et al.* [19]. Also, this finding is more apparent on SEM surface observation related to film prepared using 2% of Eu (Figures IV.12-(d) and (e)). Indeed, from 1.5 at% of europium content, (Figure IV.12-(c)) seems to lead to an improvement of the coalesce of such plates to form parallelepiped and relatively regular ones. These novel plates having average dimensions (varying from 1x2x0.2 to 2x4x0.3 μm^3) are lying onto the surface perpendicularly to the glass substrate plane, Figures IV.12-(c), (d) and (e). This phenomenon increases the specific surface of especially film having 2% of Eu as doping level. The latest doped film could play therefore an important role in possible sensitivity applications (gas sensors, photocatalysis, etc).

4.4. Photocatalysis tests

The photocatalytic activity of semiconductors is based on the production of electron-hole pair (e, h). This depends on two processes: the first one is the competition between the photo-generation of electron-hole pair and the recombination or capture of photo-generated electrons and the second one is the competition between the recombination of electrons and the transfer of charge to the surface. The recombination time of the photo-generated carriers and the increase of their transfer rates are both favorable to the improvement of the photocatalytic efficiency.

a) MoO₃

b) MoO₃: Eu 0.5%

c) MoO₃: Eu 1%

d) MoO₃: Eu 1.5%

e) MoO₃: Eu 2%

Figure IV.12 SEM images of MoO_3: Eu grown for different europium content in the spray solution
(y= $[Eu^{3+}]/[Mo^{6+}]$ (0≤y≤2 at %)

4.4.1. Possible Photocatalytic Mechanisms

Under sunlight irradiation, there is a generation of electron-hole pairs: the photo generated electrons are excited from the valence band to the conduction one or from impurity levels to the conduction band or between different defect levels that create holes in the valence band. The methylene blue (MB) adsorption occurs through the coulomb interaction since the organic dye has cationic configuration and OH⁻ ions present on the surface of the catalyst. The generated holes react with OH⁻ ions and the trapped electrons react with the O_2 absorbed molecule to create the OH· radicals and the superoxide anion radicals (O_2^-) respectively in the aqueous solution. The formed (O_2^-) and $(OH·)$ radicals attack C-S⁺=C functional group of MB dye which is attached through coulomb interaction on the surface of catalysts [34] and degrade this organic dye into H_2O and CO_2 and other no toxic materials compounds and products such as: NO_3^-, SO_4^{2-} and H⁺. In the following, we detailed the mechanism which occur during the degradation of MB dye due to Eu doped oxides thin films.

➢ **For ZnO:Eu**

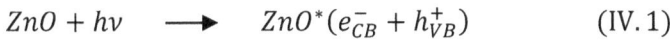
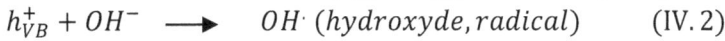

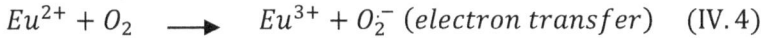
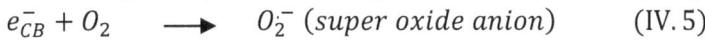
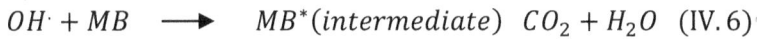
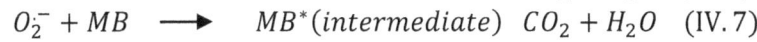

$$ZnO + h\nu \longrightarrow ZnO^*(e_{CB}^- + h_{VB}^+) \qquad (IV.1)$$
$$h_{VB}^+ + OH^- \longrightarrow OH· \ (hydroxyde, radical) \qquad (IV.2)$$
$$Eu^{3+} + e_{CB}^- \longrightarrow Eu^{2+} (electron\ trapping) \qquad (IV.3)$$
$$Eu^{2+} + O_2 \longrightarrow Eu^{3+} + O_2^- \ (electron\ transfer) \qquad (IV.4)$$
$$e_{CB}^- + O_2 \longrightarrow O_2^- \ (super\ oxide\ anion) \qquad (IV.5)$$
$$OH· + MB \longrightarrow MB^*(intermediate)\ CO_2 + H_2O \quad (IV.6)$$
$$O_2^- + MB \longrightarrow MB^*(intermediate)\ CO_2 + H_2O \quad (IV.7)$$

where e_{CB}^- and h_{VB}^+ are the electron in the conduction band and the electron vacancies in the valence band, respectively.

➢ **For In₂O₃:Eu**

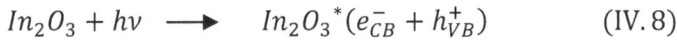
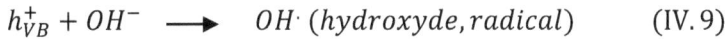

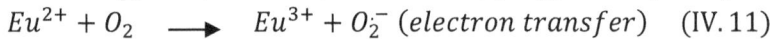
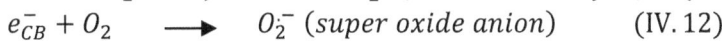
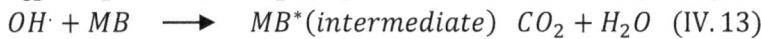
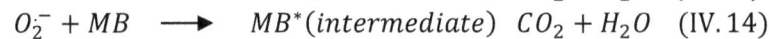

$$In_2O_3 + h\nu \longrightarrow In_2O_3^*(e_{CB}^- + h_{VB}^+) \qquad (IV.8)$$
$$h_{VB}^+ + OH^- \longrightarrow OH· \ (hydroxyde, radical) \qquad (IV.9)$$
$$Eu^{3+} + e_{CB}^- \longrightarrow Eu^{2+} (electron\ trapping) \qquad (IV.10)$$
$$Eu^{2+} + O_2 \longrightarrow Eu^{3+} + O_2^- \ (electron\ transfer) \qquad (IV.11)$$
$$e_{CB}^- + O_2 \longrightarrow O_2^- \ (super\ oxide\ anion) \qquad (IV.12)$$
$$OH· + MB \longrightarrow MB^*(intermediate)\ CO_2 + H_2O \quad (IV.13)$$
$$O_2^- + MB \longrightarrow MB^*(intermediate)\ CO_2 + H_2O \quad (IV.14)$$

➢ **For MoO₃ :Eu**

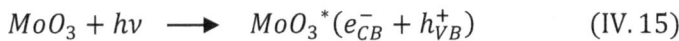
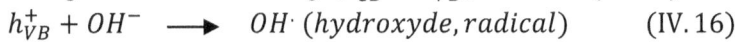

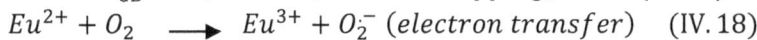
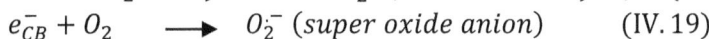

$$MoO_3 + h\nu \longrightarrow MoO_3^*(e_{CB}^- + h_{VB}^+) \qquad (IV.15)$$
$$h_{VB}^+ + OH^- \longrightarrow OH· \ (hydroxyde, radical) \qquad (IV.16)$$
$$Eu^{3+} + e_{CB}^- \longrightarrow Eu^{2+} (electron\ trapping) \qquad (IV.17)$$
$$Eu^{2+} + O_2 \longrightarrow Eu^{3+} + O_2^- \ (electron\ transfer) \qquad (IV.18)$$
$$e_{CB}^- + O_2 \longrightarrow O_2^- \ (super\ oxide\ anion) \qquad (IV.19)$$
$$OH· + MB \longrightarrow MB^*(intermediate)\ CO_2 + H_2O \quad (IV.20)$$
$$O_2^- + MB \longrightarrow MB^*(intermediate)\ CO_2 + H_2O \quad (IV.21)$$

4.4.2. Photocatalytic activity test using Eu doped In₂O₃ thin films

Photocatalytic activities were evaluated by the degradation of an organic dye, methylene blue (MB), in an aqueous solution under ultraviolet light irradiation for 1 h. The UV light was obtained using a lamp (Northen Electronic, Wigan, Lancs). A volume of 30 ml of aqueous solution of MB (3 mg l^{-1}) was placed in a vessel, and a In₂O₃ : Eu film photocatalyst with the area of 4x4 cm^2 was placed into the solution. Prior to irradiation, the solution was magnetically stirred in the dark for 30 min to establish an adsorption-desorption equilibrium.

MB decomposition tests with and without In₂O₃ films for different races of doping were carried out using UV–Vis absorption spectra, as shown in Figure IV. 13. It is well known that the most observed absorption peak of MB solution is at 660 nm, which is attributed to its monomer [30]. Figure IV. 13 indicates that samples with cubic phase presented higher intrinsic activities for the photodegradation of MB as compared to samples in rutile one.

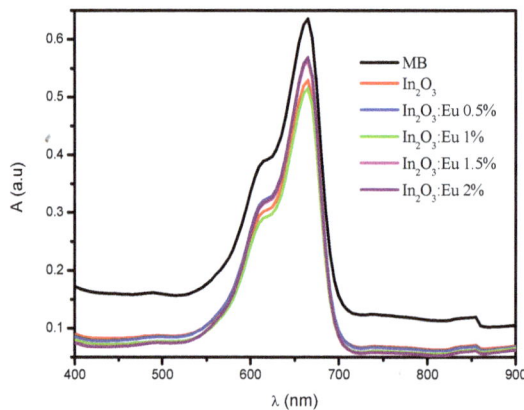

Figure IV.13 Absorption spectra of MB solution under UV light

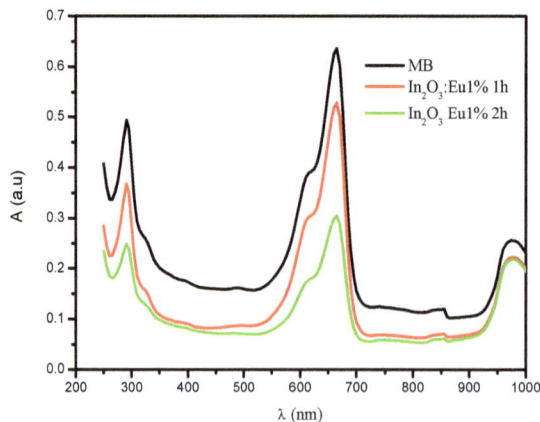

Figure IV.14 Time-dependent absorption spectra of MB solution under UV light for In₂O₃:Eu 1%

4.4.3 Photocatalytic activity test of MoO₃

UV light was obtained using 2 lamps (Philips Germicidal Ultraviolet-C) in parallel with 16 W total powers. A volume of 25 ml of aqueous solution of MB (3 mg L^{-1}) was placed in a vessel, and a MoO₃: Eu film photocatalyst with area of 1x3 cm² was placed in the solution. Prior to irradiation, the solution was magnetically stirred in the dark for 30 min to establish an adsorption-desorption equilibrium. MB decomposition testing with and without MoO₃ films for different race of doping were carried out using UV-vis absorption spectra, as shown in Figure IV.15. The two absorption peaks found at 609 nm and 660 nm correspond to methylene blue [20]. The peak intensity located at 660 nm decreases with the presence of Eu content, which is a strong evidence of the decomposition of methylene blue which is attributed to that of its monomer [21]. After 1 h, about 50% of MB amount was degraded by MoO₃:Eu 1.5%, film while the value for MoO₃ film one was approximately 35%.

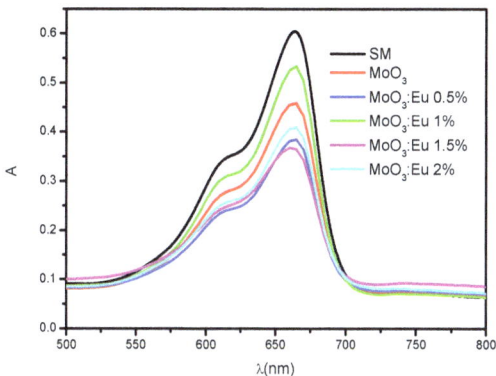

Figure IV.15 Absorption spectra of MB solution under UV light

4.5. Discussion

Reviewing the literature on ZnO, In₂O₃ and MoO₃ oxides, it was found that the oxygen vacancies play an important role in the photocatalytic process. The mechanism of the photocatalysis of the three binary oxides doped with Eu element is presented in Figure IV.16. On the other hand, Figure IV.17 shows the schematic diagram of the localization of oxygen vacancy level in the band gap energy of ZnO, In₂O₃ and MoO₃ materials. It is noted that these levels are lying near to conduction band. Parallel to the oxygen vacancy, Urbach energy E_u is assumed as the principal factor to reveal the powerful character of oxide against irradiation. In MoO₃, the structure defect can be seen by means of SEM figures. It is pointed out the presence of large specific thin film surfaces of both undoped and Eu doped MoO₃ sprayed films.

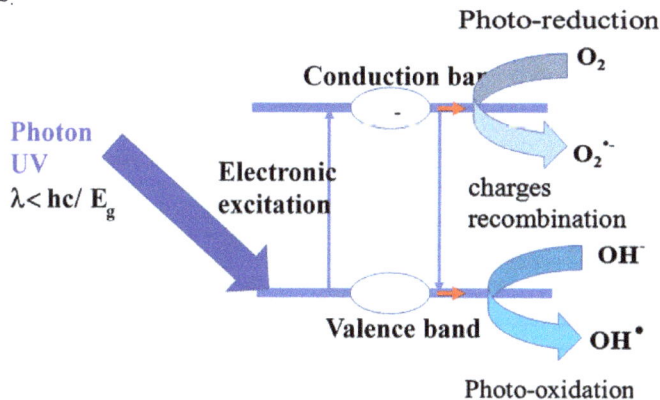

Figure IV.16 Mechanism of photocatalytic studies

$E_c = 3.2$ eV

2.9

$E_v = 0$ eV

2.75	2.42
2.49	
	1.94
2.43	
2.34	

2.51

2.16
1.82

ZnO

References: our work   [13]   [14]

$E_c = 4$ eV

2.9   2.77
            2.69
2.67

2.52   2.56

In$_2$O$_3$

$E_v = 0$ eV
References: our work   [14]

$E_c = 3.7$ eV

2.91   2.77
            2.69
2.71

2.52   2.56
2.40

MoO$_3$

$E_v = 0$ eV
References :   our work   [14]

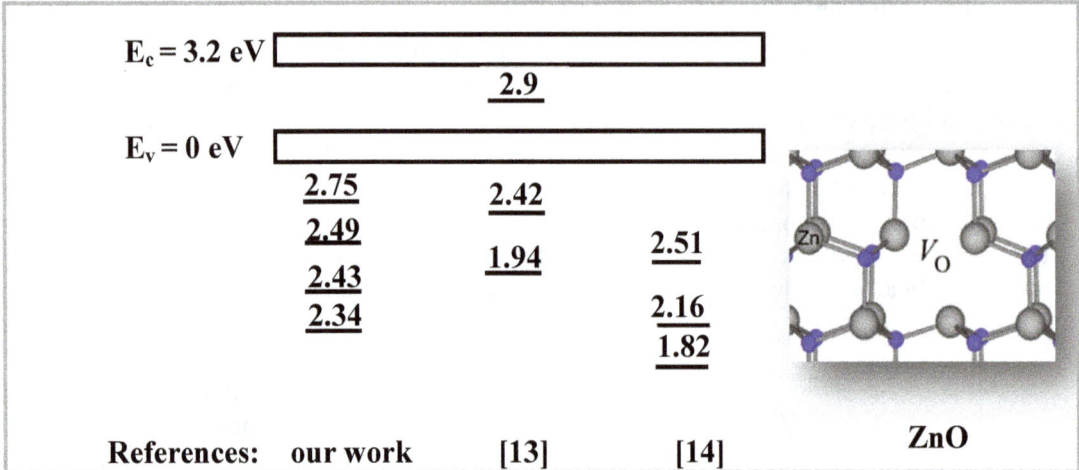

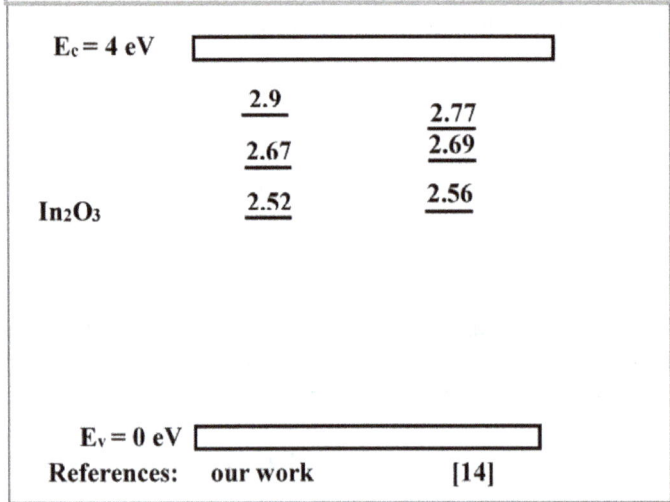

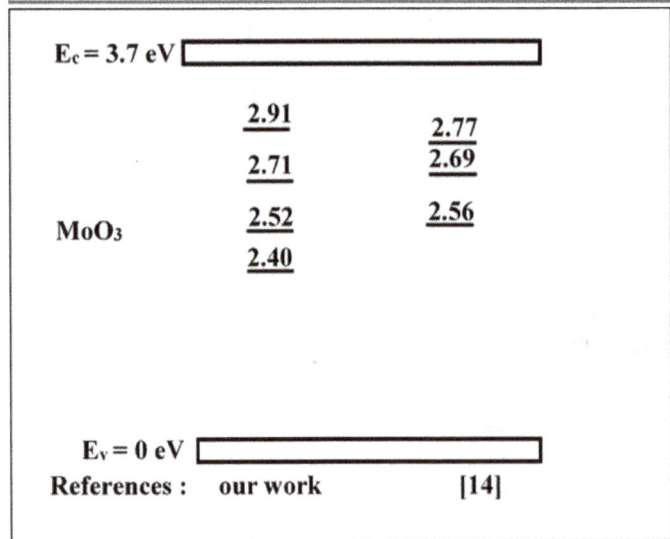

Figure IV.17 Schematic diagram of the localization of oxygen vacancy level in the band gap energy of ZnO, In$_2$O$_3$ and MoO$_3$

Table IV.1 Gap and Urbach energies for optimum Eu doping levels for ZnO, In$_2$O$_3$ and MoO$_3$

	ZnO: Eu 0.7%	In$_2$O$_3$:Eu 1.5%	MoO$_3$:Eu 1.5%
E_g (eV)	3.24	4.09	3.76
E_u (meV)	101	289	498

On the other hand, PL emission mechanism involving oxygen vacancies and interstitial Europium doping in oxides has been described above. This mechanism is illustrated in the band gap narrowing schematic diagram of related states in the blue- and red-light regions shown in Figure 5. The new electronic transition mechanism represents main energy states between the conduction band (E_c) and the valence band (E_v) caused by different amounts of electrons filled up the oxygen vacancy-induced donor levels.

We can conclude from the convolution of the photoluminescence figures that the presence of several peaks corresponds to the trap levels in the gap. They can be attributed to oxygen vacancies or to metallic donor levels.

By comparing the band structure of the three oxides we note that we have more defected states in the gap for MoO$_3$ than those for In$_2$O$_3$ and ZnO oxides. Moreover, from Table IV.1 it is clear that the highest value of Urbach energy is obtained for MoO$_3$ which agrees well with the results regarding the photoluminescence study.

4.6. Conclusion

The study on photocatalytic properties of the Eu doped oxides thin films against MB dye, revealed that two factors have a strong influence on the photocatalytic efficiency: the first is the specific surface of the film and second is related to oxygen deficiency inside the oxide film. The latest is generally reveled by PL measurements.

It was found that Eu doped MoO$_3$ thin films exhibit the appropriate conditions to give noticeably degradation phenomenon against this dye. These results are interesting since a low cost-effective method has been used to prepare Eu doped thin films, which can paves the way for possible use of such films in various applications such as photocatalysis and gas sensor.

4.7. References

[1] R. Dingle, "Luminescent transitions associated with divalent copper impurities and the green emission from semiconducting Zinc Oxide", Physical Review Letters, 23 (1969) 579-581.

[2] S. B. Zhang, S. H. Wei and A. Zunge, "Intrinsic n-type versus p-type doping asymmetry and the defect physics of ZnO", Physical Review B, 63 (2001) 075205-7.

[3] P. Wu, Q. Li, X. Zou, W. Cheng, D. Zhang, C. Zhao, L. Chi and T. Xiao, "Correlation between photoluminescence and oxygen vacancies in In$_2$O$_3$, SnO$_2$ and ZnO metal oxide nanostructures", Conference Series, 188 (2009) 012054-8.

[4] M. M. Khan, S. A. Ansari, J. Lee, and M. H. Cho, "Highly visible light active Ag@ TiO2 nanocomposites synthesized using an electrochemically active biofilm: a novel biogenic approach", Nanoscale, 5 (2013) 4427-4435.

[5] N. Han, X. Wu, L. Chai, H. Liu, Y. Chen, "Counterintuitive sensing mechanism of ZnO nanoparticle based gas sensors", Sensors and Actuators B: Chemical, 150 (2010) 230-238.

[6] V.G.Myagkov, I.A.Tambasov, O.A.Bayukov, V.S.Zhigalov, L.E.Bykova, Yu.L.Mikhlin, M.N.Volochaev and G.N.Bondarenko, "Solid state synthesis and characterization of ferromagnetic nanocomposite Fe–In_2O_3 thin films", Journal of Alloys and Compound, 612 (2014) 189–194.

[7] S. Kaleemullaa, A. Sivasankar Reddy, S. Uthanna, P. Sreedhara Reddy, "Physical properties of In_2O_3 thin films prepared at various oxygen partial pressures", Journal of Alloys and Compounds, 479 (2009) 589–593.

[8] B. Yahmadi, N. Kamoun, C. Guasch, R. Bennaceur, "Synthesis and characterization of nanocrystallized In_2S_3 thin films via CBD technique", Materials Chemistry and Physics, 127 (2011) 239–247.

[9] C. Liang, G. Meng, Y.Lei, F. Phillipp, and L. Zhang, "Catalytic Growth of Semiconducting In_2O_3 Nanofibers", Advanced Materials, 13 (2001) 1330-1333.

[10] M. Jothibas, C. Manoharan, S. Ramalingam, S. Dhanapandian and M. Bououdina, "Spectroscopic analysis, structural, microstructural, optical and electrical properties of Zn-doped In_2O_3 thin films", Spectrochimica Acta Part A: Molecular and Biomolecular Spectroscopy, 122 (2014) 171–178.

[11] M. J. Zheng, L. D. Zhang, G. H. Li, X. Y. Zhang, and X. F. Wang, "Ordered indium-oxide nanowire arrays and their photoluminescence properties", Applied Physics Letters, 79 (2001) 839-841.

[12] H. Cao, X. Qiu, Y. Liang and and Q. Zhu, "Room-temperature ultraviolet-emitting In_2O_3 nanowires", Applied Physics Letters, 83 (2003) 761-763.

[13] Q. Tang, W. Zhou, W. Zhang, S. Ou, K. Jiang, W. Yu and Y. Qian, "Size-Controllable Growth of Single Crystal $In(OH)_3$ and In_2O_3 Nanocubes", Crystal Growth Design, 5 (2005) 147.

[14] M. Itoh, K. Hayakawa and S. Oishi , "Optical properties and electronic structures of layered MoO_3 single crystals", Journal of Physics: Condensed Matter, 13 (2001) 6853-6864.

[15] Y. Zhoa, J. Liu, Y. Zhou, Z. Zhang, Y. Xu, H. Naramoto and S. Yamamoto, "Preparation of MoO_3 nanostructures and their optical properties", Journal of Physics: Condensed Matter, 15 (2003) L547-L552.

[16] M. Dieterle and G. Mestl, "Raman spectroscopy of molybdenum oxides Part II. Resonance Raman spectroscopic characterization of the molybdenum oxides Mo_4O_{11} and MoO_2", Physical Chemistry Chemical Physics, 4 (2002) 822-826.

[17] M. Labanowaka, "Paramagnetic defects in MoO_3—revisited", Physical Chemistry Chemical Physics, 1 (1999) 5385-5392.

[18] S. Baruah, S.S. Sinha, B. Ghosh, S.K. Pal, A. Raychaudhuri, J. Dutta, Photoreactivity of ZnO nanoparticles in visible light: Effect of surface states on electron transfer reaction, Journal of Applied Physics, 105 (2009) 074308.

[19] C. Imawan, F. Solzbacher, H. Steffes, E. Obermeier, "Gas-sensing characteristics of modified-MoO_3 thin films using Ti-overlayers for NH_3 gas sensors", Sensors and Actuators B: Chemical, 64 (2000) 193-197.

[20] A. Janotti and C.G. Van de Walle, "Oxygen vacancies in ZnO", Applied Physics Letters, 87 (2005) 122102-3.

[21] A. Janotti and C.G. Van de Walle, "New insights into the role of native point defects in ZnO", Journal of Crystal Growth, 287 (2006) 58-65.

5. Conclusions

This chapter presents a comprehensive study on the physical properties of Europium doped ZnO, In_2O_3 and MoO_3 thin films prepared by a cost-effective spray pyrolysis technique on glass substrates at 450°C. The physical properties were mainly discussed and presented in terms of doping content, as follows:

- The X- ray diffraction analysis revealed that the ZnO thin films are predominantly orientated along (002) orientation which is a characteristic of the würtzite structure. In_2O_3 thin films are rather predominantly orientated along two orientations (004) and (222) which are characteristic of the cubic structure. However, XRD patterns of undoped and Eu-doped MoO_3 sprayed thin films show the presence of (020), (040) and (060) peaks which characteristic of α-MoO_3 orthorhombic structure. Also, the lattice parameters, interplanar spacing, grain size and microstrain values were examined for each oxide.

- Raman investigations showed the presence of E_2(Low) mode for the zinc oxide, which involves Zn sub-lattice motion and E_2(high) mode, which is attributed to the vibration of oxygen atoms and characterizes the würtzite structure of Eu doped ZnO thin films with high crystallinity and with a preferential orientation of crystallites along c-axis. Regarding In_2O_3 sprayed thin film, Raman spectrum shows indeed the expected vibrational modes located at 106, 131, 305 and 366 cm^{-1} corresponding to (E_g) modes which are the signatures of cubic In_2O_3 structure. For Eu doped MoO_3, peaks located at 162, 200, 221, 247, 290, 340, 379, 474, 668, 819 and 994 cm^{-1} correspond to orthorhombic α-MoO_3 phase. Besides, XPS analysis confirmed the existence of Eu doped element in the host matrix, which reveals that they were successfully doped and incorporated into α-MoO_3 phase.

- From the transmittance and reflection measurements, it is found that all oxide thin films exhibit a high transparency within the visible range with an average transmittance lying between 75% and 90%, and the reflectance is less than 30%, hence the absorbance decreases with the doping for these films. In addition, the optical band gap energy as well as the Urbach energy were calculated from the optical absorption data, which are deduced from the optical transmission and reflection measurements. Indeed, the band gap energy value of Eu doped MoO_3 thin films decreases with the increase of Eu content. However, for both ZnO and In_2O_3 thin films, Eu doping element does not affect Eg values of such materials. Also, E_u values for undoped or Eu doped ZnO is in the order of 113 meV, while for undoped or doped In_2O_3 thin films is of the order of 300 meV. On the contrary, E_u value increases from 160 to 498 meV for MoO_3 thin films, which proved the existence of more localized states in the band gap energy.

- Photoluminescence investigations indicate that Eu doping leads to the decrease of emission intensity in the near band-edge. Second, the SEM micrographs indicated that the films have homogeneous grains distribution for ZnO and In_2O_3 thin films. However, undoped and Eu–doped MoO_3 thin films suggest a clearly disturbed surface with randomly oriented islets-like (with various sizes) showing a rough surface morphology.

- In summary, Eu doped MoO_3 thin films showed high photocatalytic activity for the degradation of Methylene Blue dye under UV irradiation. The enhanced photocatalytic activity of all oxides films by Eu doping is mainly attributed to the excellent electron trapping properties for photo-generated electrons and enhancement of charge separation.

These results are interesting since a low-cost effective spray pyrolysis method has been used to prepare europium doped ZnO, In_2O_3 and MoO_3 thin films and pave the way for possible use of such films in various applications such as in photocatalysis for water filtration and purification, as gas and bio-sensors.

Growth and Analyses of Sprayed Indium Oxide Thin Films for Optoelectronic Devices

Nasreddine Beji, Mehdi Souli, Najoua Kamoun-Turki

Université Tunis El Manar, Faculté des Sciences de Tunis, Département de Physique, LR99ES13
Laboratoire de Physique de la Matière Condensée (LPMC), 2092 Tunis Tunisie, Tunisia

Abstract: The optoelectronic industry, which is based on semiconductors, showed an increased interest in the last decade. Especially transparent conductive oxide (TCO) compound are widely used to manufacture many optoelectronic devices such as thin films transistor, light emitting diodes, flat panel displays, touch panels and essentially in photovoltaic applications. The growth of TCO compound with promising properties using a low cost effective chemical method is a challenge. This chapter covers the effect of doping on physical properties of In_2O_3 films. It is structured into five chapters. The first part of the chapter is focused on physical properties of In_2O_3 thin layers, and various characterization techniques including the Rietveld method, and describes the spray pyrolysis technique as well as the experimental conditions used to prepare doped In_2O_3 thin films i.e. In_2O_3:Fe, In_2O_3:Zn, In_2O_3:Mo and In_2O_3:Eu. Next, the effects of doping atoms (Zn, Fe, Mo and Eu) on the physical properties of In_2O_3 thin films are discussed. To improve these properties, the authors provided a comprehensive discussion on the physical properties correlated with the heat treatment under nitrogen atmosphere on doped In_2O_3 thin films using various temperatures and annealing times.

Table of Contents

1. Introduction

The field of optoelectronics is an important branch in industry that combines both electronics and optics. Optoelectronic devices find various applications in telecommunications, military services, medical field, and automatic control systems. The optoelectronic industry, which is based on semiconductors, received a special attention and interest in the last decade. Especially, transparent conductive oxide (TCO) compounds (e.g. ZnO, SnO_2, In_2O_3, etc...) are widely used to manufacture many optoelectronic devices such as thin films transistor, light emitting diodes, flat panel displays, touch panels and essentially in photovoltaic applications. In fact, as energy demands around the world increase, the need for a renewable energy source that will not harm the environment become more interesting. Some studies indicate that the global energy demand will almost triple by 2050. The photovoltaics (PV) industry is one of several ways to meet the need. Among these TCO semiconductor, doped In_2O_3 compound have attracted particular attention because of its promising properties to be used as optical window or transparent electrode in solar cells. In fact, doped indium oxide thin films are characterized by their high transmission, good chemical stability high conductivity and in some case high mobility.

This chapter is dedicated to the TCO In_2O_3 thin film obtained by spray pyrolysis and doped with transitional metals such as iron, zinc and molybdenum, and rare earth element such as europium. In the first part, the physical properties of In_2O_3 thin films are presented according to the literature and their applications area. Also, various characterization techniques and the Rietveld method are presented, as well as the spray pyrolysis technique and the preparation of doped indium oxide thin layers. In the second section, the effect of iron doping and annealing under nitrogen atmosphere on the structural, optical and electrical properties of In_2O_3 thin film are investigated. The third section presents the influence of zinc doping on the physical properties of sprayed In_2O_3 thin layers, and the heat treatment under nitrogen atmosphere. The fourth section presents a study on the structural, optical and electrical properties of molybdenum doped indium oxide thin films. The fifth section presents a study on the structural, optical and electrical properties of europium doped indium oxide thin films and the and heat treatment on the physical properties of In_2O_3 thin layers.

1.1. Transparent conductive oxide (TCO) materials

Transparent conductive oxide materials have attracted much attention due to their potential and promising properties. Among them, we find high transmission and low resistivity. Owing to such properties, TCO compounds and especially In_2O_3 thin films have been widely used in optoelectronics devices [1] such as photovoltaic devices [2], thin films transistor [3], light emitting diodes [4] and in gas sensing applications [5]. TCO materials include several kinds of semiconductors such as: SnO_2 [6], ZnO [7] and In_2O_3 [8,9]. It is well known as displayed in Figure I.1 that ZnO thin films have been intensively investigated in the last decades. Further, toxicity is a primordial factor for the safety of a researcher. Since ZnO thin films is a toxic material, then as alternative thin films In_2O_3 seems to exhibit interesting properties. Indeed, indium oxide thin films have certain properties that performed it a useful compound for instance: stability, non-toxicity, high transparency in the visible and near infrared region, large optical band gap, and high refractive index (in order of 2). All of these properties lead to conclude that In_2O_3 is a potential candidate for high performance especially in optoelectronics devices.

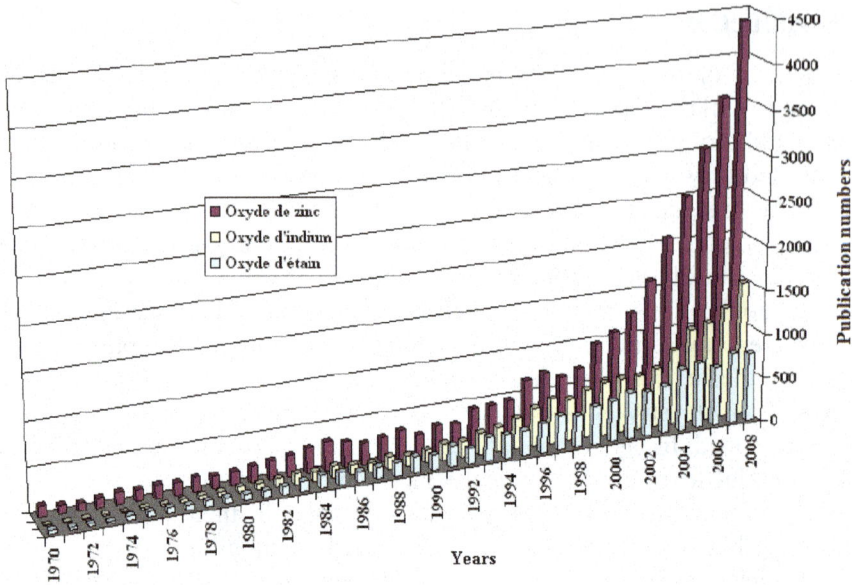

Figure I.1: Publication numbers of ZnO, In_2O_3 and SnO_2 per year since 1970 to 2008 [10].

1.2. Applications area of In_2O_3 thin films

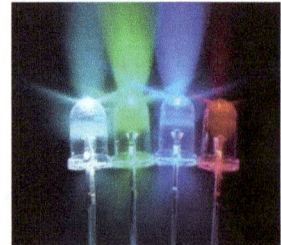

Due to its interesting properties as cited in previous paragraph, indium oxides can be used in several research areas such: solar cells [2], thin-film transistor (TFT) [3], light emitting diodes [4] and gas sensors [5] and antireflective coatings [11]. In the next paragraph, we have summarized the different indium oxide applications.

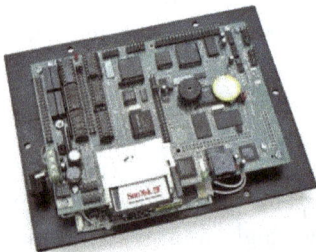

Thin films transistor   **Solar cells**   **Light emitting diodes**

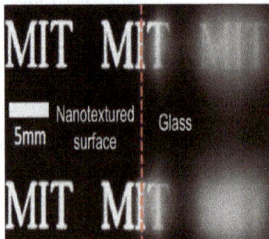

Anti-reflection coatings   **Gas sensors**

Figure I.2: Various applications of indium oxide thin films.

Thin films transistor (TFT): A thin-film transistor (TFT) is a special kind of field-effect transistor made by depositing thin films of an active semiconductor layer as well as the dielectric layer and metallic contacts over a supporting (but non-conducting) substrate. A common substrate is glass, because the primary application of TFTs is in liquid-crystal displays. This differs from the conventional transistor, where the semiconductor material typically is the substrate, such as a silicon wafer.

Solar cells In_2O_3 thin films are used as transparent electrodes in solar cells. For this, a high optical transmission and also good electrical conductivity should be obtained. The high optical transmission is needed to enable an effective transport of photons to the active layer however the decrease of resistivity leads to a loss of charge transport photo generated. These two properties are related to the free carrier concentration n_v. Indeed, the transmission is inversely proportional whereas the conductivity is proportional to the carrier concentration.

Light emitting diodes (LED): It is a semiconductor device with a p-n junction able to emit light for a direct polarization. For a suitable voltage applied on LED compound placed in direct contact, electron from n-type semiconductor moves to p-type semiconductor and conversely holes from p-type semiconductor moves to n-type semiconductor. Thus a recombination between electrons and holes occurs which leads to a photon emitting. The output from an LED can be ranged from red to blue-purple. Another type of LED called an infrared-emitting diode (IRED), emit infrared (IR).

Anti-reflection coasting: Anti-reflection coating is a type of optical equipment used at the top surface of optical elements to decrease and even suppress the reflection such as lenses, camera, telescope and solar cells. It consists of dielectric transparent thin layers with a special thickness chosen in order to destructive interference and then causes the reflected wave from semiconductor surface. TCO compounds reflect infrared radiation due to the free carriers. Anti-reflection coating is based on the creation of destructive interference between the waves reflected at both interfaces air-reflected thin films and reflected thin films-compounds. For this, both reflected waves should be out of phase with the same amplitude.

Gas sensors: The electrical properties of TCO can be changed due to the presence of certain gases. The adsorbed molecule can capture a free electron since the gas can be absorbed at the films surface or into the grain boundaries. Therefore, a reduction of electrical conductivity could be obtained. The ratio of the resistivity before and after the presence of gas leads to determine the gas sensor sensitivity. Among the different performance required of gas sensors (cost, ease of use, reproducibility, etc) one can generally emphasize the need to obtain the best compromise between sensitivity and stability over time. Current research focuses its efforts on obtaining the best compromise.

1.3. Physical properties of In_2O_3 thin films

Due to its chemical, optical, and electronic distinctive properties, indium oxide has attracted more and more attention in optoelectronic devices and photovoltaic cells used as transparent conductive oxide. To the development and characterization of thin films of In_2O_3 it is essential to recall the basic physical properties of indium oxide thin layers.

1.3.1. Structural properties:

The indium oxide in bulk form solid is yellow and its melting point is around $1913\,°\,C$ [12]. The density of In_2O_3 is 7.12 g/cm^3. It crystallized generally in cubic structure. The space group is Ia3 and the lattice parameter is a = 1.012 nm [13]. The preferred orientation changes with the deposition parameters. In fact, the most founded preferred orientations are (222) and (400) plans [9,14]. In the body centered cubic structure of indium oxide, each indium atom is surrounded by four oxygen atoms.

The presence of vacancies induced a slight shift of the surrounded cations. There are two types of arrangement of these vacancies as shown in Figure I.3: a) Site In1: indium atom is surrounded by an oxygen octahedron distorted with a trigonal form, and b) Site In2: octahedron formed by the oxygen atoms is more distorted and smaller than the site symmetry In1. The In1 / In2 ratio is equal to = 1:3

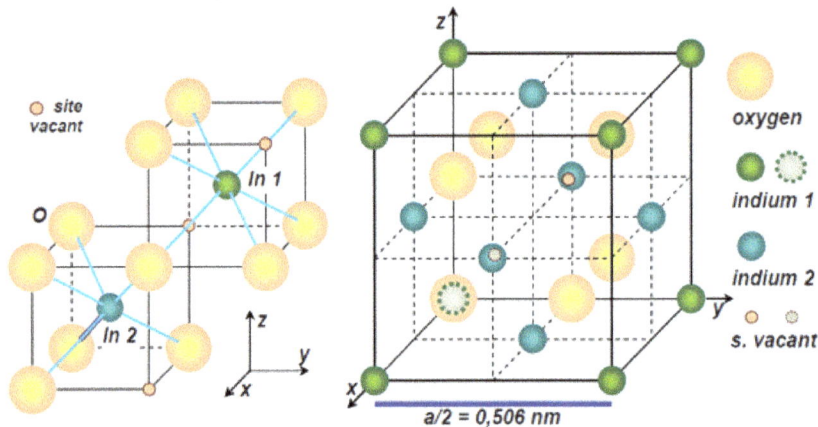

Figure I.3: Body centered cubic structure of In_2O_3 compound [15]

1.3.2. Optical properties:

Indium oxide thin films are characterized by a high transmission is the transparency zone. The optical transmittance (T) in the visible region and I-Red is about 70 - 90% for In_2O_3 films deposited by various processing techniques. Moreover, In_2O_3 thin films exhibit large band gap energy (E_g). The optical gap (E_g) and transmission values of undoped In_2O_3 thin films synthesized by different processing techniques are summarized in the following table.

As shown in Table I.1, In_2O_3 thin films grown using varying methods, exhibit a high transmission and large band gap energy (greater than 3 eV). However, there is a difference in the obtained values between Raj *et al.* [14] and Prathap *et al.* [16] despite of they have used the same elaboration technique. This can be due to fact that growth parameters used to elaborate In_2O_3 films are different for each work.

1.3.3. Electrical properties:

In Table I.2, some electrical properties of In_2O_3 thin films for different deposition techniques are summarized:

Table I.1: Energy band gap (E_g) and average transmission of undoped In_2O_3 thin films deposited using different techniques.

Elaboration technique	Sol gel	Spin coating	Spray	Spray	Activated reactive evaporation
E_g (eV)	3.75	3.94	[3.53 - 3.68]	[3.10 - 3.58]	[3.41 - 3.45]
Transmission (%)	90	-	84	80>	80
References	[1]	[2]	[14]	[16]	[17]

Table I.2: Electrical parameters transmission of undoped In_2O_3 thin films deposited using different techniques.

Elaboration technique	Spray	Spray	Sol gel	Activated reactive evaporation
Mobility μ ($cm^2V^{-1}s^{-1}$)	43	37	9	/
Resistivity ρ ($\Omega.cm$)	$1.28*10^{-3}$	$0.69*10^{-3}$	2.52	$9.7*10^{-3}$
Volume free carrier concentration N (cm^{-3})	$1.5*10^{20}$	$1.75*10^{20}$	$4.25*10^{17}$	$2.2*10^{20}$
References	[18]	[14]	[1]	[17]

Indium oxide thin films, exhibit a low resistivity in the order of 10^{-3} $\Omega.cm$. This result may be due to high volume free carrier concentration N. The obtained results in terms of resistivity, free carrier and mobility permits us to conclude that the elaboration technique affect extremely the electrical properties of In_2O_3 layers.

1.4. Effect of doping on physical properties of In_2O_3 thin films

Various doping atoms can be inserted into semiconductor lattice structure to enhance its physical properties and performed them to be appropriate for certain applications. Different reports have deals with the effect of doping ratio on physical properties of In_2O_3 thin films. The doping atoms served as a control of structural, optical and electrical properties of indium oxide thin films. It is well known that pure In_2O_3 thin films don't have low resistivity. Doping is one way to decrease the electrical resistivity and so to enhance the optoelectronic properties of elaborated In_2O_3 material. The doping atoms as mentioned in the literature include tin (Sn), molybdenum (Mo), antimony (Sb), chromium (Cr), tungsten (W), zinc (Zn), iron (Fe) and zirconium (Zr). As shown from **Table I.3**, that the resistivity decreases significantly with doping process reaching for some works 10^{-4} $\Omega.cm$. On the other hand, the band gap energy E_g is always greater than 3 eV. Moreover, the nature of the dopant has an impact on the obtained values as shown in the Table I.3. For example, Mo doped In_2O_3 films exhibit a grain size in the range of [42-102] nm. However, Cr doped indium oxide films, have a low grain size in the order of 12 nm. One notes that, all doped films have a high volume carrier concentration n_v. On the other hand, the Hall mobility μ is extremely affected by varying the doping element. Indeed, for example, w-doped In_2O_3 layers show Hall mobility in the order of 89 $cm^2V^{-1}s^{-1}$. However, for Ga-doped indium oxide films, μ is of around 1.4 $cm^2V^{-1}s^{-1}$.

Table I.3: Electrical parameters transmission of undoped In_2O_3 thin films deposited using different techniques.

Doping atoms	Sn	Mo	W	Sb	Ga	Zr	Cr
Elaboration technique	Electron beam evaporation	Spray	Reactive plasma	Spray	Sol gel	Spray	Sol gel
Grain size (nm)	6 -13	42 - 102	#	8.3 – 14.1	#	15-25	12
Band gap (eV)	3.61 - 3.89	3.69 – 3.82	3.60- 3.83	3.90	#	3-3.5	3.76
$\rho * (10^{-3})$ (Ω.cm)	#	81	#	100	#	0.64	3.5 $x10^3$
$n_v * (10^{20})$ (cm^{-3})	#	1.9	1.6	#	#	2.5	1.9
μ ($cm^2V^{-1}s^{-1}$)	#	34	89	#	1.4	73.54	1.43
References	[19]	[20]	[21]	[22]	[23]	[24]	[25]

1.5. Effect of annealing on physical properties of In_2O_3 thin films

The quality of TCO thin films depends on structural and especially optoelectronic properties. These properties can be controlled apart from doping effect by means of appropriate heat treatment. Indeed, annealing process greatly influenced the electrical parameters and optical transmission. The annealing variables such as temperature and duration have an effect on doping atoms, stress induced by doping and as a result, the carrier and optical band gap can be affected. Indeed, the annealing in air atmosphere lead to coalescence between grains which lead to an increase of grain size as expected from Flores-Mendoza *et al.* [26]. However, the same authors have mentioned that a reduction in optical band gap values may ascribed to a larger grain size. On the other hand, Cho *et al.* [27] have reported the effect of rapid thermal annealing temperature (RTA) on physical properties of In_2O_3 thin films. The annealing process was applied for 1 min at temperatures ranging between 400 and 600°C under vacuum. They revealed that annealed films at higher annealing temperature display a higher value of mobility followed by an increase of free carrier concentration. Moreover, Senthilkumar *et al.* [28], have founded an enhancement of the structural, optical and electrical properties of ITO films annealed in air. In fact, they have founded a decrease of electrical resistivity ρ. Authors have explained that such decrease in "ρ", especially for 400°C, can be ascribed to the enhancement of crystalline structure.

Xu *et al.* [29] have been reported also the effect of annealing in air atmosphere for 10 min on physical properties of In_2O_3 thin films. The authors show that an increase trend of carrier density until a certain annealing temperature and above this temperature, they noted a drastic decrease of carrier concentration. Such behavior was explained as follow: the origin of carrier density increase can be attributed to the activation of donors. However, the observed drastic decrease could be due to the reduction of oxygen vacancies. Indeed, since the annealing process was performed in air atmosphere, then the oxygen vacancies decline with increasing annealing temperature. Therefore, both competition processes have been occurred during heat treatment. The authors inferred that for increasing annealing temperature an inactivation of donor atoms followed by a reduction of oxygen vacancies have been take place.

1.6. Characterization techniques

The physical properties of indium oxide thin films have been investigated using the above characterized techniques:

- **X-ray diffraction (XRD):** using an automated Bruker D8 advance X-Ray diffractometer with CuK_α radiations for 2θ values over 10-80°. The wavelength, accelerating voltage and current were respectively, 1.5418 Å, 40 kV and 20 mA
- **Raman spectrometer:** HORIBA Jobin Yvonet_LABRAM HR
- **Atomic force microscopy (AFM):** Standard Veeco Dimension 3100, tapping mode
- **Energy dispersive X-rays spectroscopy (EDX)** coupled to SEM
- **Spectrophotometer** with a Perkin–Elmer Lambda 950 spectrophotometer operating at room temperature
- **Hall Effect measurements** used in the Van Der Pauw configuration

X-ray diffraction and Raman spectrometer have been employed for structural analysis. However, for morphological treatment, atomic force microscopy (AFM), and energy dispersive X-rays spectroscopy (EDX) have been used. The optical properties were also analyzed by spectrophotometer and electrical properties were studied using Hall Effect measurements in the Van Der Pauw configuration.

X-ray diffraction "XRD": X-rays were discovered for the first time by Wilhelm Röntgen. It is a well-known and nondestructive technique which is based on Bragg law where:

$$\lambda.n = 2d_{hkl} \sin\theta_{hkl}$$

where λ: wavelength of the X-ray beam; d_{hkl}: distance between two reticular planes of the same (hkl) family; θ_{hkl}: X-ray incidence angle of (hkl) reticular planes (hkl); n: diffracted angle.

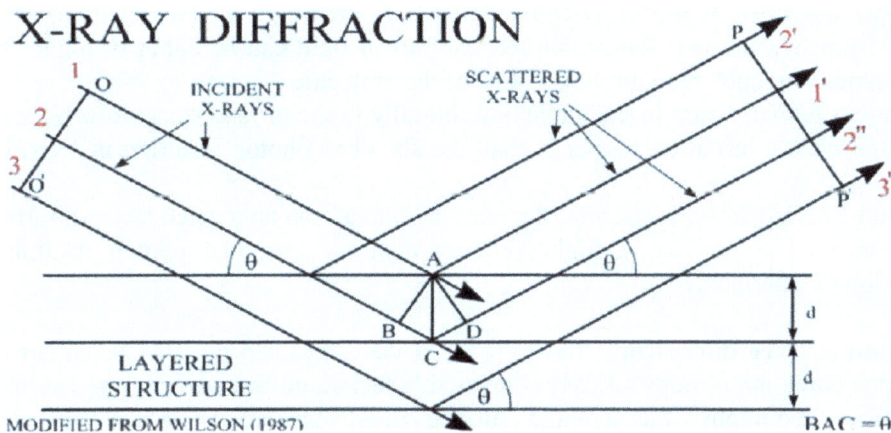

Figure I.4: Diffracted angle according to Bragg law

It is used to get information of crystalline structure of diffracted materials. One can deduce the different phase present in the sample from the different parameters such as: the diffracted peaks position, the preferred orientation plan and intereticular distance which are found in JCPDF Card.

The production of X-rays is achieved following the above points:

✓ Heating a tungsten filament by thermoelectric field

✓ As a result, an emission of electron from the cathode to anode due to heating has been detected

✓ Then, an acceleration of the electron into the anode direction due to the appearance of electric field has been take place. The origin of such electric field comes from the high potential difference between the anode and the cathode

✓ The collision between the accelerated electrons with the target excites the atoms of the target and the relaxation of those atoms product X-ray patterns.

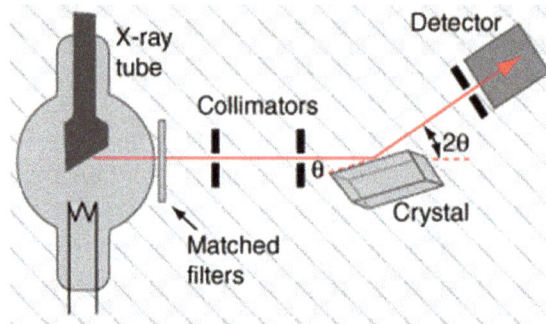

Figure I.5: X-Ray Diffraction device

Raman Spectroscopy: Raman spectroscopy is a nondestructive spectroscopic technique used to observe vibrational and rotational modes in a system. If the sample is irradiated with a monochromatic light, different phenomena can be produced such as absorption, reflection or scattering. Two kinds of scattering can be occurred: Rayleigh scattering and Raman scattering.

Rayleigh scattering: resulted from scattering process without a change in the wavelength described by Lord Rayleigh. A change in the frequency (wavelength) of the light is called Raman scattering. Raman shifted photons of light can be either of higher or lower energy, depending upon the vibrational state of the molecule.

Stokes Raman scattering: the material, initially in the ground state, absorbs energy and the emitted photon has a lower energy than the absorbed photon resulting in a creation of a phonon.

Anti-Stokes Raman scattering: the material, already in an excited state, absorbs energy and the emitted photon has a higher energy than the absorbed photon resulting in an annihilation of a phonon.

Atomic force microscopy "AFM": AFM were invented by IBM Scientists in 1986. The atomic force microscopy (AFM) is a suitable technique from which one can determine the surface topography, the average surface roughness (RMS), grain size and their distribution. AFM is based on interactions that can occur between the atoms of the tip and those of the sample. These interactions may result from Van der Waals forces, electrostatic forces, magnetic forces or ionic repulsion forces. The system is constituted by a flexible cantilever with a metallic tip (tip radius of about 10 nm) and a piezoelectric tube which is placed below the sample to displaced it in the three directions. The detection of the cantilever deflection is provided by an optical system whereas the constant deflection of the cantilever is provided by a control system. Further, AFM displays a very high-resolution with a resolution on the order of fractions of a nanometer better than that of SEM.

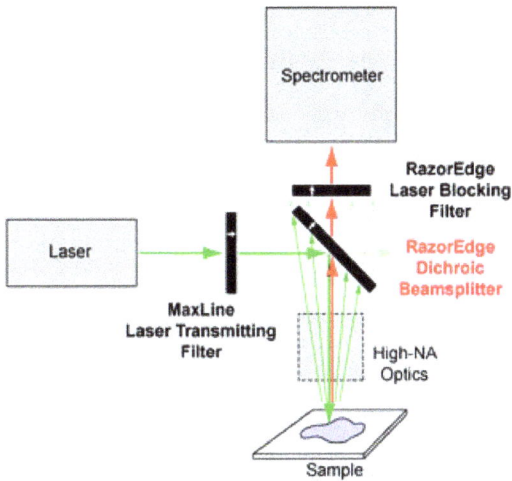

Figure I.6: Raman spectroscopy

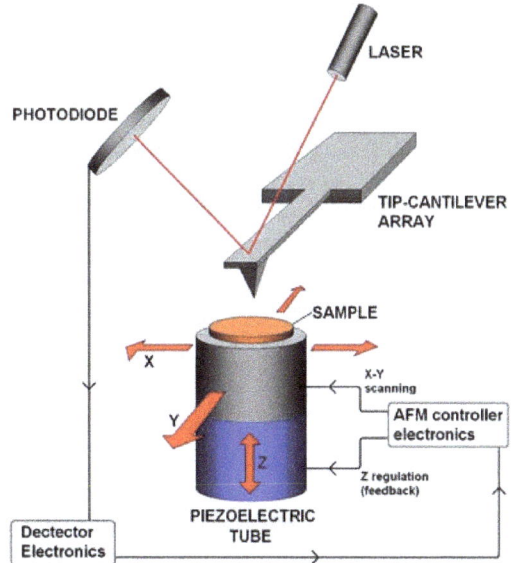

Figure I.7: Experimental set-up of atomic force microscopy

There are 3 modes of AFM depending on the tip motion nature:
• Contact mode: the tip is dragged along the surface of the sample
• Tapping mode: the cantilever oscillate up and down at or near its resonance frequency
• Non-contact mode: the tip of the cantilever does not contact the sample surface. The cantilever is instead oscillated at either its resonant frequency

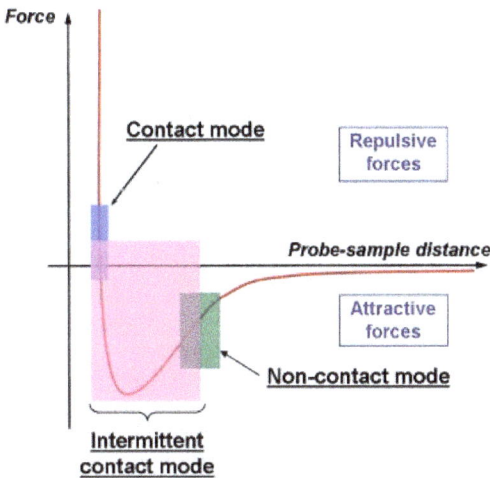

Figure I.8:
Atomic force microscopy modes

Fluorescence spectrometer: The fluorimeter is used to measure the fluorescence. The excitation source, sample cell, fluorescence detector are the main parts of Fluorimeter. Usually a deuterium or xenon lamp is used for excitation of molecules of the samples. An excitation light passes through a monochromator which allows passing of the light of only a selected wavelength. The fluorescence is detected by a photomultiplier tube. The excitation spectrum and the fluorescence spectrum are found by scanning of the excitation monochromator and the fluorescence monochromator respectively.

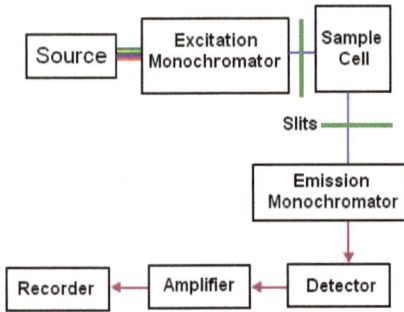

Figure I.9: Experimental set up of fluorimeter

Scanning electron microscopy (SEM): The functioning of microscope is based on the electron emission produced by the cathode and the detection of signals resulting from the interaction of these electrons with the sample the gun produces a beam electron through a tungsten filament heated by a current. This beam is accelerated by the high voltage created between the filament and the anode. It is then focused on the sample by a series of electromagnetic lens. Controlled via computer, the SEM operator can adjust the beam to control magnification as well as determine the surface area to be scanned. Further, a high vacuum should be applied to make the most accurate measurement. The scanning electron microscopy is composed by the following component:

- electron gun: used to produce electrons due to heating a tungsten filament by thermoelectric field, the lifetime of tungsten filament is of some hundred hours
- condenser lens: intended to focus the beam to a point spot
- deflection coils: and impose the beam to do a scanning movement of the sample surface
- detectors: divided into secondary electron detectors and back scattered electrons detectors
- high voltage generator: used to supply electron gun
- sweep generator supplies deflection coils

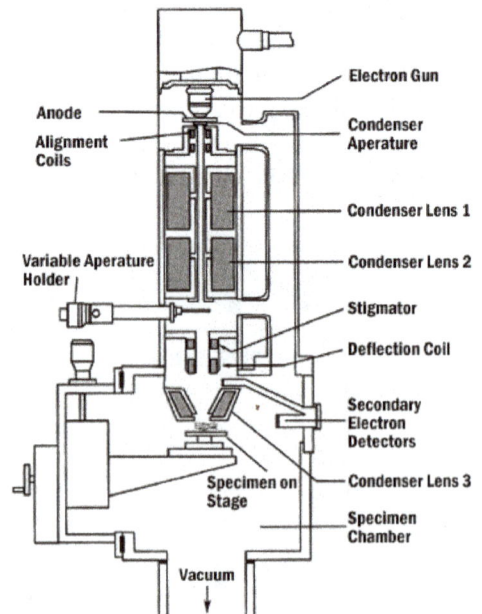

Figure I.10: Experimental set-up of scanning electron microscopy (SEM)

Energy dispersive X-rays spectroscopy (EDX): The Energy Dispersive Microscopy "EDS" is a technique coupled to SEM system. One can determine from this technique the chemical composition of the layers and their respective content in atomic percentage or weight percentage. The first step consists to excite the layers. The release of the electron can be achieved by the emission of X photons. Since energy states of an atom are specific to each chemical element, therefore the energy emitted by X photon is also specific to each element in turn.

Spectrophotometer: A spectrophotometer is commonly used for the measurement of transmittance and reflectance of several types of samples such us solution and solids. However they can also be designed to measure the diffusivity on any of the listed light ranges that usually cover around 200 nm - 2500 nm using different controls and calibrations. Within these ranges of light, calibrations are needed on the machine using standards that vary in type depending on the wavelength of the photometric determination. Spectrophotometry uses photometers that can measure a light beam's intensity as a function of its color (wavelength) known as spectrophotometer. The transmission and reflection measurements were made using a Perkin Elmer Lambda 950 spectrophotometer. This device includes a monochromator suitable for several types of measures in a total spectral range [250-2500] nm. This spectrophotometer is computerized, after processing the data obtained of the spectra of transmittance and reflectance directly to a computer. The absorption coefficient is deduced from these spectra.

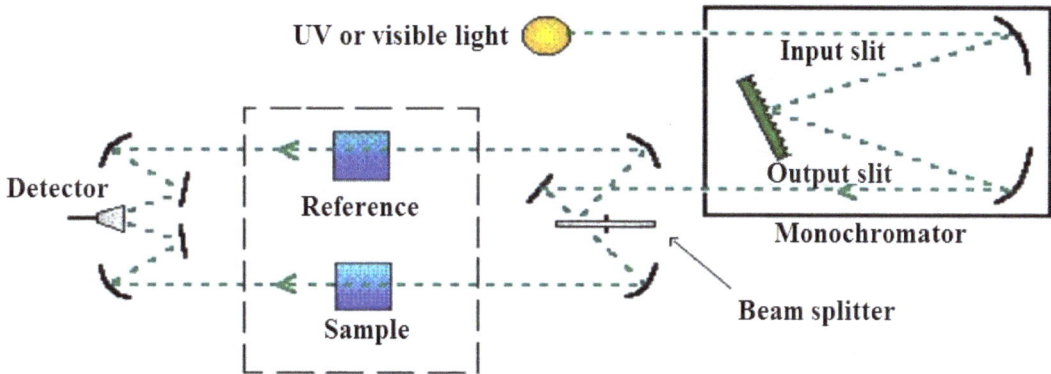

Figure I.11: Optical path in spectrophotometer

Hall Effect measurements: The Hall effect was discovered in 1874, by E. H. Hall and can be used to measure the mobile carrier density and the sign of the charge carriers. The importance of the Hall effect is supported by the need to determine accurately carrier density, electrical resistivity, and the mobility of carriers in semiconductors. The Hall effect provides a relatively simple method for doing this. Because of its simplicity, low cost, and fast turnaround time, it is an indispensable characterization technique in the semiconductor industry and in research laboratories. Furthermore, two Nobel prizes (1985, 1998) are based upon the Hall effect. An electric current is passed through the semiconductor block via contacts at either end while a magnetic field $\vec{B}$ is applied perpendicular to both the surface and the electric current as shown in Figure I.12. The Hall voltage V_H is recorded perpendicular to the direction of current flow $\vec{J}$. The force acting on the moving charge carriers in a magnetic field is the Lorentz force $\vec{F}$.

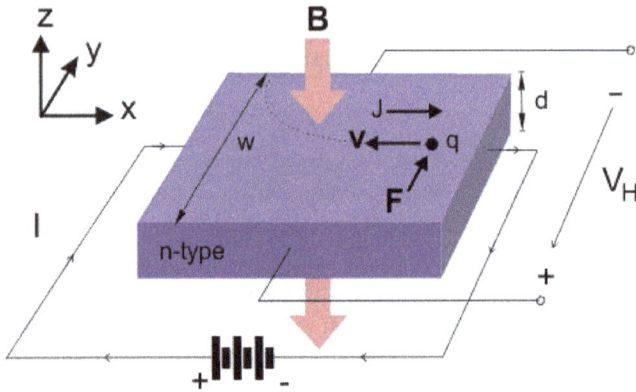

Figure I.12:
Schematic diagram of the Hall effect for an n-type semiconductor

The force $\vec{F}$ acting on a particle of electric charge q with instantaneous velocity $\vec{v}$, due to an external electric field $\vec{E}$ and magnetic field $\vec{B}$, is given by:

$$\vec{F} = q\,(\vec{E} + \vec{v} \wedge \vec{B}) \quad (1)$$

$\vec{F}$, $\vec{v}$ and $\vec{B}$ form a right-handed Cartesian co-ordinate system. Since we have chosen $\vec{v}$ to be perpendicular to $\vec{B}$. As a result, $\vec{F}$ is also perpendicular to both $\vec{v}$ and $\vec{B}$:

$$\begin{cases} \vec{E} = E_x\vec{i} + E_y\vec{j} \\ \\ \vec{B} = -B_z\vec{k} \end{cases} \qquad \vec{v} = -v_x\vec{i}$$

According to to equation (1), the Lorentz force $\vec{F}$ can be written as follow:

$$\vec{F} = F_x\vec{i} + F_y\vec{j} + F_z\vec{k} = qE_x\vec{i} + q(E_y - v_x B_z)\vec{j} + 0\vec{k}$$

$$\text{So} \begin{cases} F_x = qE_x \\ F_y = q(E_y - v_x B_z) \end{cases}$$

The movement of charge carriers sets up an electric field E_y in the opposite direction to the Lorentz force F and this field continues to grow as the charge carrier's move to one side of the semiconductor until the Lorentz force become equal to zero. In this situation we have:

$$\begin{aligned} F_y &= q(E_y - v_x B_z) = 0 \\ E_y - v_x B_z &= 0 \qquad (2) \\ E_y &= v_x B_z \qquad (3) \end{aligned}$$

E_y can be also deduced from the Hall voltage V_H and the width of the sample w:

$$E_y = -V_H w \qquad (4)$$

The current density J_x is be expressed as a function of the velocity v ($v = v_x$) and the carrier's density n by:

$$J_x = qnv_x \implies v_x = \frac{J_x}{qn} \quad (5)$$

According to **(3)** and **(5)**, the electric field E_y can be written:

$$E_y = \frac{J_x B_z}{qn}$$

So, the Hall voltage V_H is determined:

$$V_H = -\frac{E_y}{w} = -\frac{J_x B_z}{qnw}$$

$$V_H = -\frac{E_y}{w} = \frac{J R_H B_z}{w} \quad (J = J_x)$$

$$\mathbf{R_H} = -\frac{1}{qn}$$

where R_H is the Hall constant which is negative if the majority carriers are electrons and positive if the majority carriers are holes.

In general:

$$\begin{cases} R_H = -\dfrac{1}{qn} \ (\text{electrons}) \\ R_H = \dfrac{1}{qn} \quad (\text{holes}) \end{cases}$$

Thus, the sign of R_H give us the conductivity type (n or p) of semiconductor. The mobility μ is also measured. However, the resistivity ρ of the sample is calculated using the Van der Pauw technique.

Rietveld Refinement [30-32]: All compounds can be identified by using a database of diffraction patterns. The purity of a sample can also be determined from its diffraction pattern, as well as the composition of any impurities present. A diffraction pattern can also be used to determine and refine the lattice parameters of a crystal structure. The Rietveld method is used to refine the crystal structure model of a material. It can be used for quantitative phase, lattice parameter and crystallite size calculations, determine atom positions and microstrain. The Rietveld method for structural refinement of powder diffraction data has been developed over the last four decades and has proved indispensable in solving crystal structures. The process involves minimizing the difference between a crystallographic model and experimental data, via a least squares refinement; such intricate modeling of the height, width and position of Bragg reflections in an X-ray diffraction pattern can yield a lot of information about the crystal structure of a material.

a) Powder diffraction: Powder diffraction is a scientific technique using X-ray, neutron, or electron diffraction on powder or microcrystalline samples for structural characterization of materials. The intensity of a spectrum in a powder diffractometer is calculated using:

$$I_i{}^{calc} = S_F \sum_{j=1}^{Nphases} \frac{f_j}{v_j^2} \sum_{k=1}^{Npeaks} L_k \left| F_{K,j} \right|^2 \Omega_j (2\theta_i - 2\theta_{k,j}) P_{k,j} A_j + bkg_i$$

where: the **bkg$_i$** is the background term. In the case of Rietveld method, the background is generally a polynomial function in 2θ expressed as follow:

$$bkg(2\theta_i) = \sum_{n=0}^{N_b} a_n (2\theta_i)^n$$

N_b and a_n are respectively the polynomial degree and the polynomial coefficients.

$S_F \sum_{j=1}^{Nphases} \frac{f_j}{V_j^2}$ is the scale factor. For each phase j, the scale factor S_j is expressed by:

$$S_j = S_F \frac{f_j}{V_j^2}$$

S_F, f_j and V_j^2 are respectively the beam intensity, phase volume fraction and phase cell volume.

L_k is the Lorentz-polarization factor for each peak "k" which depends on the instrument. It is extremely affected by the Bragg angle θ. L_k is a combination between the Lorenz factor and the polarization factor. The Lorentz factor reflects geometric correction of the diffraction. However, the polarization factor can be explained as follow: the incident beam is unplorized. So it may the sum of two plane - polarized components

$$L_K = \frac{1 + \cos^2 2\theta}{\sin^2 \theta \cos \theta}$$

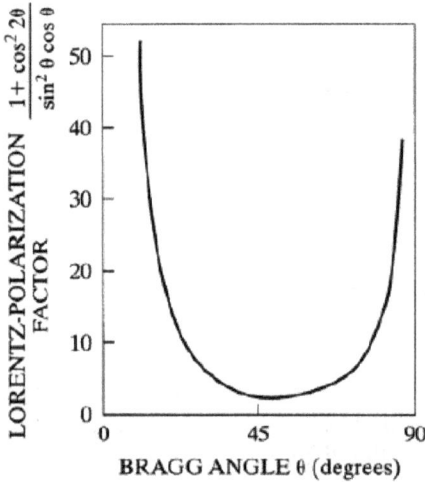

Figure I.13:
Lorentz-polarization factor

$F_{K,j}$ is the structure factor which expressed by:

$$|F_{K,j}|^2 = m_k \left| \sum_{n=1}^{N} f_n e^{-B_n \frac{\sin^2 \theta}{\lambda^2}} (e^{2\pi i(hx_n + ky_n + hz_n)}) \right|^2$$

where h, k and l are the Miller indices; m_k is the multiplicity corresponding to the k reflection; B_n is the temperature factor; N = the number of atoms; x_n, y_n and z_n are the coordinates of the nth atom; f_n is the atomic scattering factor; A_j is the absorption factor; $P_{k,j}$

is the preferred orientations which can be expressed using the March-Dollase formula:

$$\mathbf{P_{K,j}} = \frac{1}{m_k} \sum_{k=1}^{m_k} (P_{MD}^2 \cos^2\alpha_n + \frac{\sin^2\alpha_n}{P_{MD}})^{-\frac{3}{2}}$$

where: $\mathbf{P_{MD}}$ is the March-Dollase parameter; α_n is the angle between the preferred orientation vector and the crystallographic plane (hkl)

$\Omega_j(2\theta_i - 2\theta_{k,j})$ is the profile shape function which will be study with detail in next section.

Profile shape function: The most used profile shape function for X-ray diffraction is the Pseudo – Voigt function h(2θ). It is a convolution of Gaussian and Lorentzian components.

$$h(2\theta) = \int_{-\infty}^{+\infty} G(2\theta')L(2\theta - 2\theta')d(2\theta')$$

Among the approximations used to calculate the Void function, h(2θ) can be calculated using the following expression:

$$\Omega(2\theta) = \eta L(2\theta, H_L) + (1 - \eta)G(2\theta, H_G)$$

where G and L are respectively the Gaussian and Lorentzian components of the full-width at half-maximum (FWHM) respectively H_G and H_L. The parameter η is defined as the pic diffraction shape:

$$\eta = \frac{H}{\beta}$$

where β is the integral width and H is the Fourier transform of the Pseudo – Voigt function h(2θ). η exhibits two limit values, η = 0 (Gaussian limit) or η = 1 (Lorentzian limit). η is related to the integral width β according to the following equation:

$$\beta = \eta\beta_L + (1 - \eta)\beta_G$$

where β_L and β_G are the integral width of respectively the Lorentzian and Gaussian components.

The average crystallites <D> size and microstrain $<\sigma^2>^{1/2}$ can be determined by these equations:

$$<D> = \frac{\lambda}{\beta \cos\theta}$$

$$<\sigma^2>^{\frac{1}{2}} = \frac{1}{2\sqrt{2\pi}} \frac{\beta_G}{\tan\theta}$$

where λ is the X-ray wavelength. The refinement of XRD spectra by Rietveld method uses the Caglioti equation to determine the full-width at half-maximum (FWHM) H for a Bragg reflection at 2θ:

$$H^2 = U.\tan^2\theta + V.\tan\theta + W \quad \textbf{Caglioti Formula}$$

where U, V and W are the Caglioti coefficients.

a. **Rietvled refinement (fitting procedure):** The fitting procedure is carried out by adopting the Marquardt least squares method for the minimization of the difference between the experimental and calculated diffraction patterns. The Rietlved method is used to minimize the following expressions using the non linear least square method:

$$\mathbf{WSS} = \sum_{i=1}^{N} [\mathbf{w_i}(\mathbf{I_i^{exp}} - \mathbf{I_i^{calc}})]^2$$

where N is the number of experimental points and $\mathbf{w_i} = \dfrac{1}{\sqrt{I_i^{exp}}}$

The quality of the fit is controlled by the GOF (Goodness of Fit) which expressed by:

$$\mathbf{GOF} = \frac{\mathbf{R_{wp}}}{\mathbf{R_{exp}}}$$

where R_{wp} and R_{exp} are respectively the weighted residual error and the expected error.

$$\mathbf{R_{wp}} = \sqrt{\frac{\sum_{i=1}^{N}[w_i(I_i^{exp} - I_i^{calc})]^2}{\sum_{i=1}^{N}[w_i I_i^{exp}]^2}}$$

$$\mathbf{R_{exp}} = \sqrt{\frac{(N-P)}{\sum_{i=1}^{N}[w_i I_i^{exp}]^2}} \quad ; \text{P is the number of parameters}$$

To obtain a good refinement, the Goodness of Fit should be lower than 2. A better refinement is obtained when GOF is practically close to 1.

b. **MAUD software:** The XRD patterns of films were refined using MAUD program which is based on the Rietveld method. The procedure consists in modeling the diffraction patterns by analytical functions. In fact, the goal of this fitting is to reduce the difference between experimental and theoretical diffraction patterns. So, some parameters are refined during the refinement process in order to obtain a Goodness of Fit (GOF) near to 1 as we can as possible. After fitting of XRD patterns some micro-structural and structural parameters can be deduced such us:

- Different phases detected in the sample with their percentages. The MAUD software gives us the opportunity to confirm the presence of some undesirable phases detected by the XRD data.
- Lattice structure
- Lattice parameters
- Microstrain $<\sigma^2>^{1/2}$
- Average grain size $<d>$

In the Figure I.14, we have summarized refinement results before and after fitting by Maud Sotware. From Figure I.14 (a) to Figure I.14 (b), the experimental (dots) and the calculated (full line) spectrum become practically superimposed. This result can be easily detected from the spectrum given below of each figure. For a good fitting this spectrum is looks to a bold line as evident in the Figure I.14 (b).

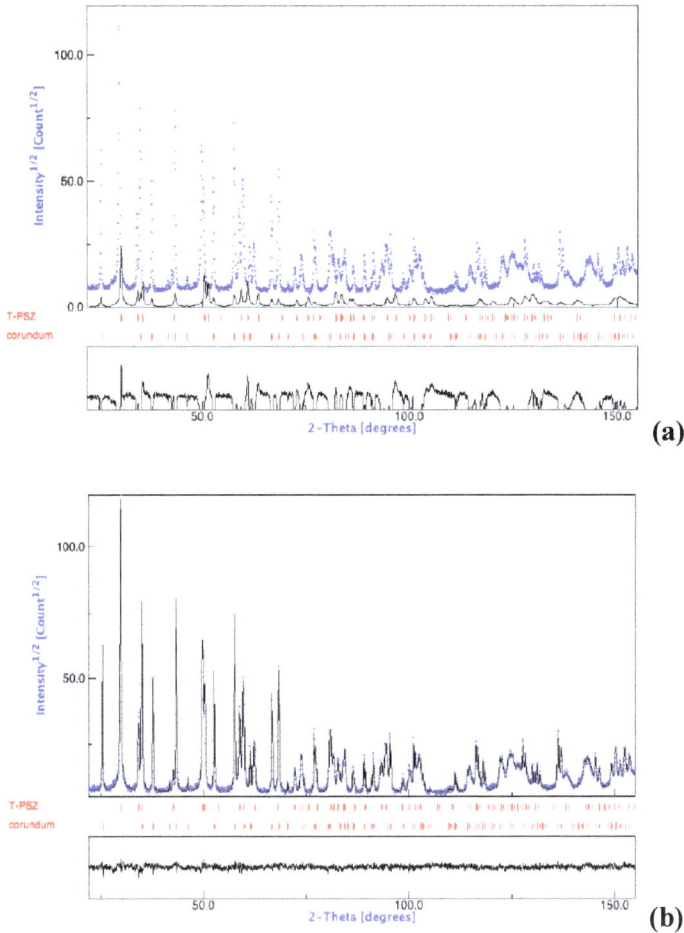

Figure I.14: Rietveld refinement of XRD spectrum before **(a)** and after fitting **(b)** The difference between the experimental (dots) and the calculated (full line) spectrum is given below

In previous work, we have synthesized one among the most important TCO material, which is In_2O_3 film. In fact, some of the growth parameters like indium concentration $[In^{3+}]$ and substrate temperature T_s were optimized. A best crystallinity is obtained for $[In^{3+}] = 0.04$ mol/l and $T_s = 500°C$. Undoped films show an electrical resistivity in the order of 650.20 x 10^{-3} Ω.cm [33] which is not appropriate for a good TCO material. So, we have doped In_2O_3 with some metal transition and rare earth elements to enhance physical properties and especially electrical ones. On the other hand, as known, indium tin oxide compound (ITO) is the most used transparent conductive oxide among doped In_2O_3 films. In order to find other doped In_2O_3 compound which can replace ITO material in optoelectronic applications, we have thought to optimize In_2O_3 by other doped element. In fact, we have used iron, zinc, molybdenum and europium to dope indium oxide compound. Therefore, in this research we aim to grow promising sprayed TCO In_2O_3 thin films with high transparency and low electrical resistivity ($\sim 10^{-3}$ Ω.cm). In addition to doping effect, the annealing conditions such as temperature and duration have an impact on the physical properties of In_2O_3 thin films. So, annealing process can be in turn a good way to reduce more the electrical resistivity and in some cases to increase the film's transparency. It is thus important to acquire more attention to heat treatment effect on doped indium oxide thin films. An annealing process under nitrogen atmosphere (N_2) was then applied on each optimized doped material.

1.7. Deposition techniques used to grow In_2O_3 thin films

A thin film is a solid layer having thickness ranging from nanometers to several micrometers. Electronic semiconductor devices and optical coatings are the most important applications benefiting from thin-film growth. Thin films are elaborated using many deposited techniques such as: Spray pyrolysis, sol-gel, chemical vapor deposition (CVD), atomic layer deposition (ALD), chemical bath deposition (CBD), pulsed laser ablation, DC magnetron sputtering, molecular beam epitaxy (MBE), vacuum thermal evaporation. These methods can be classified in two main categories: Chemical or physical. In the figure below, we have classified some deposited techniques.

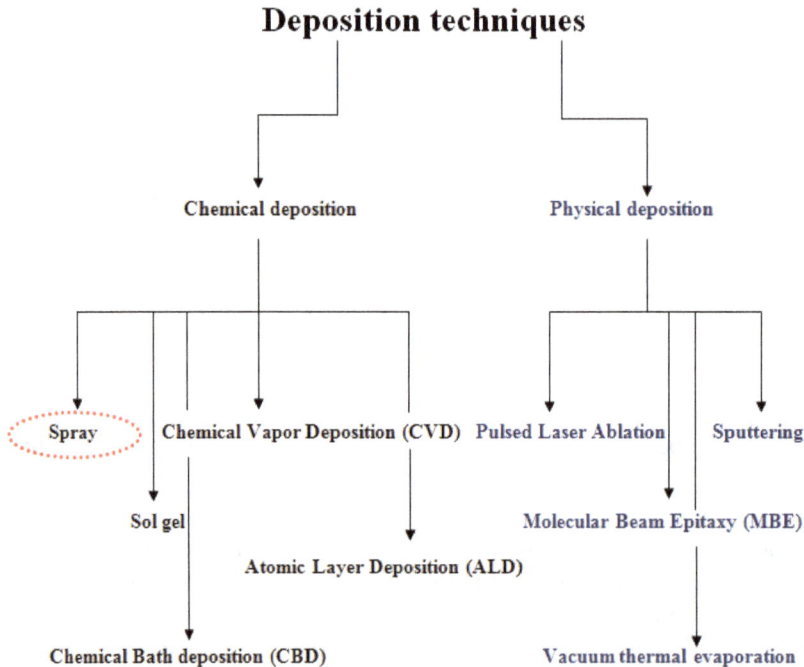

Figure I.15: Classification of techniques deposition

1.7.1. Spray pyrolysis technique

In our work, indium oxide thin films are prepared by the pulverization technique in liquid phase (spray). The experimental setup involves a heating system for the substrate and a nozzle fixed on a two-dimensional moving table allowing to pulverize the whole isothermal zone containing the heated substrates [34]. The experimental set up is explained with more detail in Figure I.16.

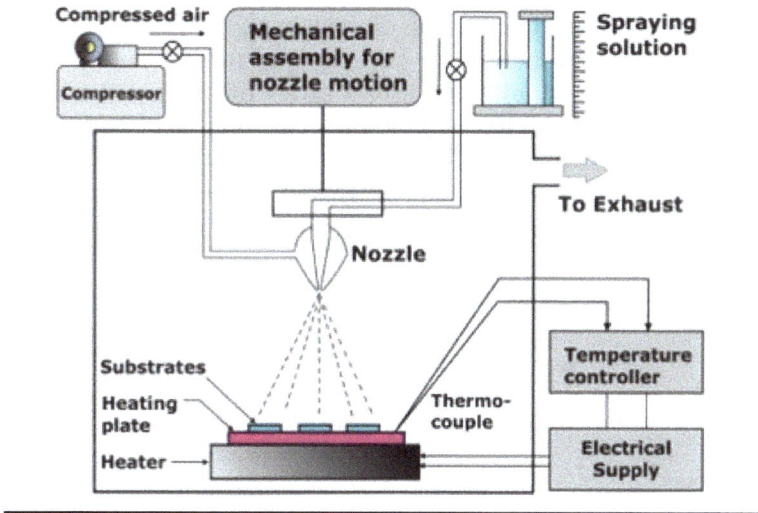

Figure I.16: Schematic of spray pyrolysis showing its various parts

The different step for the growth thin films using spray pyrolysis technique is summarized in Figure I.17 [35]: atomization of the precursor solution, aerosol transport of the droplet, decomposition of the precursor to initiate film growth.

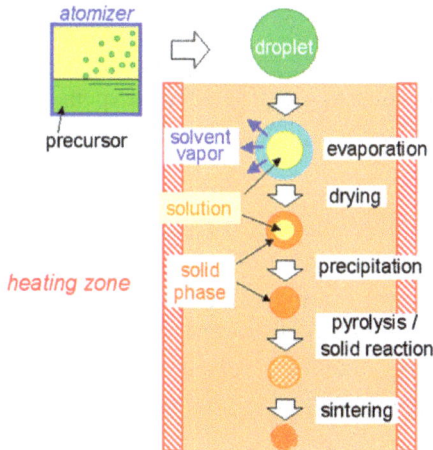

Figure I.17:
Schematic of material particle preparation by spray pyrolysis

The spray pyrolysis equipment consist of:

Heated plate: The heating plate is a cuboid block with gray slot to ensure a quite perfect contact with the substrates. The heating process of the plate is achieved using cylindrical resistors. Seven resistors (Vulcanic mark) are placed in parallel and attached to temperature regulator. This thermocouple which is placed in the hot plate, is connected also to the temperature regulator to give us the temperature of the plate. As we know, there is a loss of energy by Joule effect which leads to obtain a substrate temperature T_s smaller then the regulator temperature T_r. the variation of the regulator temperature T_r with the substrate temperature T_s is shown in the Figure I.18.

The pump: The suction pump controls the solution flow rate. To determine the solution flow rate (FR), we vary the position of the pump graduation "G_p". For each position G_p, we measure the quantity of the volume (V_{sp}) sprayed for 3 minutes. We can deduce the solution flow rate FR. FR = V_{sp} / 3 min. So, a series of measurements is obtained. The variation of the position "G_p" with the solution flow rate "FR" is plotted in Figure I.19. This curve is almost linear with an error coefficient value of about 1%.

Nozzle: The solution is sprayed onto the heated substrate by a nozzle which is fixed into an atomizer. The atomizer allows pulverizing the whole isothermal zone containing the heated substrates.

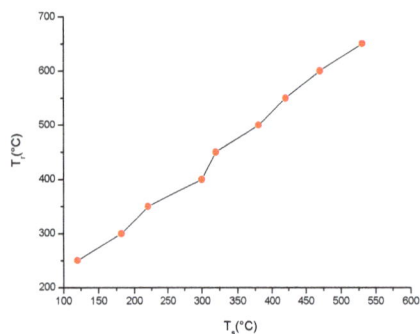

Figure I.18:
Calibration curve of the regulator temperature
T_r versus the substrate temperature T_s

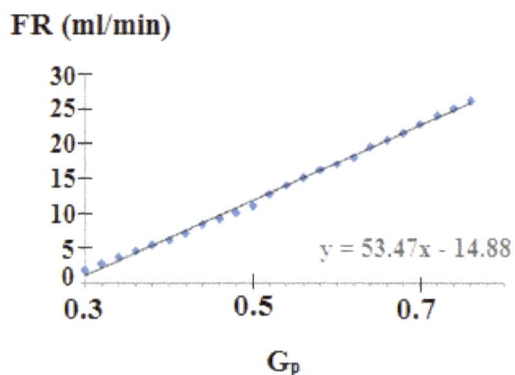

Figure I.19:
Calibration curve.

Advantage of spray pyrolysis technique: Chemical spray pyrolysis technique exhibit several advantages for the growth of thin films such as simple, safe, inexpensive, and it can be adapted easily for production of large-area films.

1.7.2. Preparation of indium oxide thin films

Indium oxide thin films are prepared on glass substrates by the pulverization of an aqueous solution. In_2O_3 is doped with iron (**Fe**), zinc (**Zn**), molybdenum (**Mo**) and europium (**Eu**). The starting precursors and the other deposition parameters are listed with details in **Table I.4** and **Table I.5**. The formation of In_2O_3 material is the result of the endothermic chemical reaction. Only the doping precursor is changed. However, all other deposition parameters were kept constant.

Table I.4: Experimental details for the growth of undoped indium oxide thin films

Solution Composition to get undoped In_2O_3	• $InCl_3$ powder/$[In^{3+}]$ = 0.04 mol.l^{-1} • 250 cm^3 of bi-distilled water (H_2O)
Substrate temperature T_s (°C)	• **500 °C**
Solution flow rate	• **2.5 ml.min^{-1}**
Distance nozzle - sample	• **28 cm**
Carrier gas	• **Compressed air**
Substrate	• **Microscope slide**

Table I.5: Doping precursors and atomic concentration y for each doped In_2O_3 layers

Doped layers	Source of the dopant	Atomic concentration y (at.%)
In_2O_3:Fe	**FeCl$_2$ powder**	$y = \frac{[Fe^{2+}]}{[In^{3+}]} = 0, 2, 4$ **and 6**
In_2O_3:Zn	**Zinc acetate dehydrate (Zn (CH$_3$COO)$_2$.2H$_2$O) powder**	$y = \frac{[Zn^{2+}]}{[In^{3+}]} = 0, 1, 2, 3, 4, 5$ **and 6**
In_2O_3:Mo	**Molybdenum (V) chloride (MoCl$_5$) powder**	$y = \frac{[Mo^{6+}]}{[In^{3+}]} = 0, 1, 3, 5$ **and 7**
In_2O_3:Eu	**Europium (III) oxide Eu$_2$O$_3$ powder**	$y = \frac{[Eu^{3+}]}{[In^{3+}]} = 0; 0.1; 0.3$ **and 0.5**

1.8. Conclusion

Due to its interesting properties, indium oxide can be used in several applications. It has been shown that In_2O_3 thin films are highly dependent on processing conditions. Indium oxide compound crystallizes generally into body centered cubic structure. Moreover, some structural, optical and electrical properties of In_2O_3 thin films were displayed in this chapter. A description of characterization techniques applied on sprayed In_2O_3 thin films were also revealed. The Rietveld method was also explained. Maud software which is based on Rietveld method allow us to determine some structural parameters including grain size, lattice parameters, microstrain and phases amounts. Also, the description of the elaboration technique "Chemical Spray Pyrolysis" of indium oxide thin films and its deposition process are investigated.

1.9. References:

[1] L.N. Lau, N.B. Ibrahim and H. Baqiah; Applied Surface Science 345 (2015) 355–359

[2] M.A.Majeed Khan, W.Khan, M.Ahamed, M.Alhoshan; Materials Letters 79 (2012) 119–121

[3] Z. Yuan, X. Zhu, X. Wang, X. Cai, B. Zhang, D. Qiu and H. Wu; Thin Solid Films 519 (2011) 3254–3258

[4] Y.H. Shin, S.B. Kang, S. Lee, J.J. Kim and H.K. Kim; Organic Electronics 14 (2013) 926

[5] N.G. Pramod and S.N. Pandey; Ceramics International 41 (2015)527–532

[6] S. Sujatha Lekshmy, Georgi P. Daniel and K. Joy; Applied Surface Science Volume 274, June 2013, Pages 95–100

[7] Y. Wang, H. Huang, X. Meng, F. Yang, J. Nan, Q. Song, Q. Huang, Y. Zhao and X. Zhang; Journal of Alloys and Compounds 636 (2015) 102–105

[8] M. Jothibas, C. Manoharan, S. Ramalingam, S. Dhanapandian and M. Bououdina; Spectrochimica Acta Part A: Molecular and Biomolecular Spectroscopy 122 (2014) 171–178

[9] S. Parthiban, V. Gokulakrishnan, K. Ramamurthi, E. Elangovan, R. Martins, E. Fortunato and R. Gancsan; Solar Energy Materials & Solar Cells 93 (2009) 92–97

[10] Thesis: Jerome Garnier. Elaboration de couches minces d'oxydes transparents et conducteurs par spray cvd assiste par radiation infrarouge pour applications photovoltaiques. Engineering Sciences. Arts et Metiers ParisTech, 2009. French

[11] G. Oh, K. S. Lee and E. K. Kim; Current Applied Physics 15 (2015) 794-798

[12] https://en.wikipedia.org/wiki/Indium(III)_oxide

[13] G. Korotcenkov, V. Brinzari, A. Cerneavschi, A. Cornet,J. Morante, A. Cabot and J. Arbiol, Sensors and Actuators B 84 (2002) 37–42

[14] A. Moses Ezhil Raj, K.C. Lalithambika, V.S. Vidhya, G. Rajagopal, A. Thayumanavan,

M. Jayachandran and C. Sanjeeviraja, Physica B 403 (2008) 544–554

[15] Thesis: Contribution à l'élaboration de couches minces d'Oxydes Transparents Conducteurs (TCO)

[16] P. Prathap, Y.P.V. Subbaiah, M. Devika and K.T. Ramakrishna Reddy; Materials Chemistry and Physics 100 (2006) 375–379

[17] S. Kaleemulla, A. Sivasankar Reddy, S. Uthanna and P. Sreedhara Reddy; Optoelectronics and Advanced Materials – Rapid Communications Vol. 2, No. 12, December 2008, p. 782 – 787

[18] J. Joseph Prince, S. Ramamurthy, B. Subramanian, C. Sanjeeviraja and M. Jayachandranc, Journal of Crystal Growth 240 (2002) 142–151

[19] V. Senthilkumar, P. Vickraman, M. Jayachandran and C. Sanjeeviraj; Vacuum 84 (2010) 864–869

[20] P. Prathap, N. Revathi, K.T. Ramakrishna Reddy and R.W. Miles; Thin Solid Films 518 (2009) 1271–1274

[21] F. Meng, J. Shi, Z. Li, Y. Cui, Z. Lu and Z. Feng; Solar Energy Materials & Solar Cells 122 (2014)70–74

[22] N.G. Pramod, S.N. Pandey; Ceramics International, 40 (2014) 3461-3468

[23] S. Jeong, J-Y. Lee, M-H. Ham, K. Song, J. Moon, Y-H. Seo, B-H. Ryu and Y. Choi; Superlattices and Microstructures, 59 (2013) 21-28

[24] C.Manoharan, M.Jothibas, S.Johnson Jeyakumar and S.Dhanapandian; Spectrochimica Acta Part A: Molecular and Biomolecular Spectroscopy, 145 (2015) 47-53

[25] H. Baqiah, N.B. Ibrahim, M.H. Abdi, S.A. Halim; Journal of Alloys and Compounds, 575 (2013) 198-206

[26] M.A. Flores-Mendoza, R. Castanedo-Perez, G. Torres-Delgado, J. Márquez Marín, O. Zelaya-Angel; Thin Solid Films 517 (2008) 681–685

[27] S. Cho; Microelectronic Engineering 89 (2012) 84-88

[28] V. Senthilkumar and P. Vickraman; Current Applied Physics 10 (2010) 880–885

[29] Z. Xu, P. Chen, Z. Wu, F. Xu, G. Yang, B. Liu, C. Tan, L. Zhang, R. Zhang and Y. Zheng; Materials Science in Semiconductor Processing 26 (2014) 588–592

[30] Rietveld H. M., J. Appl. Cryst. **2** (1969) 65.

[31] Wiles D.B. and Young R.A., J. Appl. Cryst. **14** (1981) 149.

[32] Lutterotti L. MAUD CSD Newletter (IUCR) **24** (2000).

[33] N. Beji, M. Souli, M. Ajili, S. Azzaza, S. Alleg, N. Kamoun Turki; Superlattices and Microstructures 81 (2015) 114–128

[34] M. Ajili, M. Castagné and N. Kamoun Turki; Superlattices and Microstructures 53 (2013) 213–222

[35] Lado Filipovic, Member, IAENG, Siegfried Selberherr, Giorgio C. Mutinati, Elise Brunet, Stephan Steinhauer, Anton Kock, Jordi Teva, Jochen Kraft, J̈org Siegert, and Franz Schrank; Proceedings of the World Congress on Engineering 2013 Vol II, WCE 2013, July 3 - 5, 2013, London, U.K

2. Study on the physical properties of iron doped In$_2$O$_3$ thin films

Iron doped indium oxide thin films have been used mainly for gas sensor applications [1,2] and to study ferromagnetic properties [3–5]. The investigation of iron doped indium oxide thin films in photovoltaic applications was not widely explored in the literature. In our knowledge only few reports have studied microstructure and optical properties of In_2O_3:Fe thin films like Ibrahim *et al.* [6] which have elaborated iron doped indium oxide thin layers by sol gel technique. So the main contribution of this chapter is to investigate the influence of Fe-doping on the structural, optical and electrical properties of In_2O_3 thin films elaborated by spray.

2.1. Structural properties

2.1.1. X-Ray Diffraction

XRD patterns of the doped In_2O_3 thin films for different iron concentrations y (y = 0, 2, 4 and 6 at.%) are shown in Figure II.1. Undoped In_2O_3 thin layers exhibit a cubic structure with predominant (222) peak located at 30.58°. The peak positioned at 35.26° is attributed to the (400) reflection plan (JCPDS Card n°06-0416). When y increases until 4 at.%, we notice a simultaneous intensity increase of the diffraction peaks (222) and (400) which corresponds to an improvement of layer orientation and an enhancement of crystal quality. Doped indium oxide thin films crystallize into the cubic structure. However, when iron concentration reaches the value of 6 at.%, XRD spectra exhibit a strong (400) peak whereas the (222) diffraction peak decreases severely. Therefore, we can say that at this doping level, the atoms in the crystal lattice are rearranged. Indeed, a clear structural transition occurs from (222) to (400) plans for iron doping concentration equals to 6 at.% and the preferred orientation becomes (400) plan. Up to 4 at.%, some iron atoms substitute indium ones in the lattice and other iron atoms occupy interstitial sites [7]. But for y = 6 at.% the change of preferred orientation is followed by an increase of the strongest peak intensity may be due to occupancy of additional indium vacancy sites by iron atoms which are unoccupied previously. In fact, it is well known that In_2O_3 thin films present several types of defects levels, i.e: indium interstitial; oxygen and indium vacancies [8]. The same phenomena of change of preferred orien-tation from (222) to (400) have been observed after doping In_2O_3 thin films with molybdenum [7,9]. We can conclude that the doping element is the principal origin of change in the preferred orientation of crystal structure. However, the rate in which this modification is observed can be related to the nature of the doping element. Finally, XRD spectra suggest the presence of Fe_2O_3 phase. Indeed, there is a small peak (104) located at 2θ = 32.95 ° related to the presence of rhombohedra Fe_2O_3 phase [JCPDS Card n°88-2359].

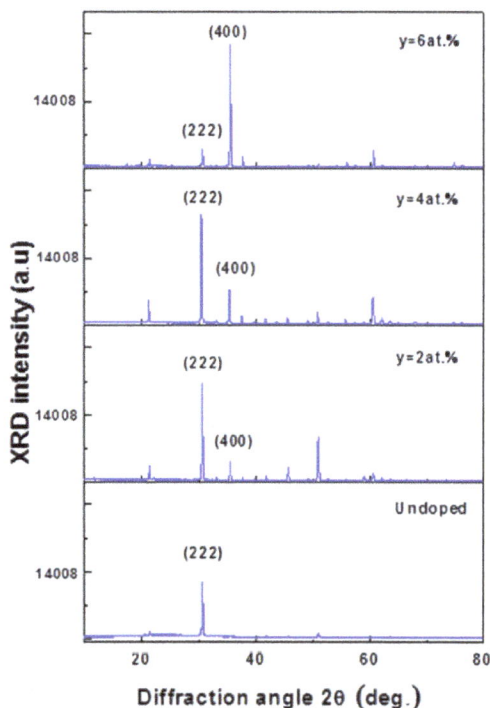

Figure II.1: XRD spectra of undoped and iron doped In_2O_3 thin films for different concentrations y = $([Fe^{2+}]/[In^{3+}])_{sol}$

2.1.2. Rietveld analysis

MAUD software [10] which is based on Rietveld analysis [10], was used to fit experimental XRD spectra. Some microstructural parameters can be estimated and possible phases can be identified after fitting process. The Rietveld refinements of undoped and iron doped indium oxide thin films are shown in Figure II.2. The deduced microstructural parameters (lattice parameters, microstrain and phases amounts) are listed in Table II.1. After fitting, two phases have been identified which are Fe_2O_3 and In_2O_3 (Figure II.2). In fact, secondary phases of rhombohedra Fe_2O_3 was observed for doped films. The percentage of Fe_2O_3 phases increases when iron concentration is increasing in the sprayed solution. It reaches 10.32% for 6 at.% iron doped In_2O_3 thin films. We note also an increase in the lattice parameter (a) after doping for bcc In_2O_3 thin films which indicate a clear expansion of the crystal lattice. The microstrain $<\sigma^2>^{1/2}$ increases after doping; this can be attributed to the incorporation of iron atoms in the crystal lattice. For doped thin films, the microstrain reaches a minimum value of 0.0301 % for y=6 at.% corresponding to the best crystallinity which is in good agreement with the XRD analysis.

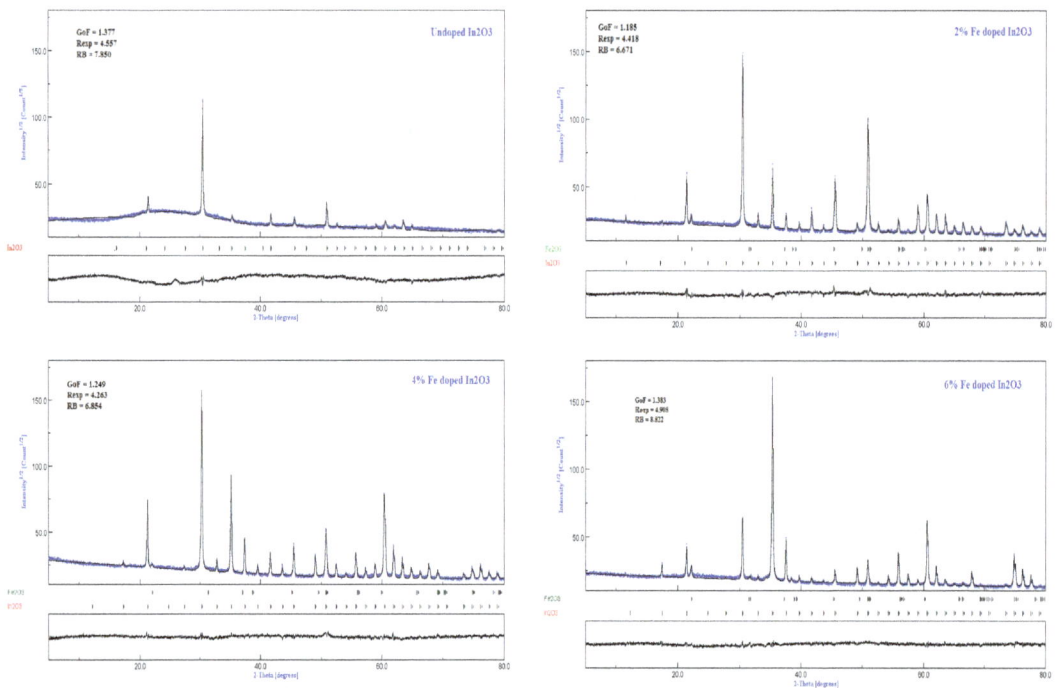

Figure II.2: Rietveld refinement for XRD patterns of undoped and iron doped indium oxide thin films. The difference between the experimental (dots) and the calculated (full line) patterns is given below.

Table II.1: Phases and structures parameter deduced from Rietveld Analysis for different doping concentrations y= ([Fe^{2+}]/[In^{3+}])$_{sol}$ in the sprayed solution.

	phases	Microstrain $<\sigma^2>^{1/2}$ (%)	Lattice parameter (Å)		Amount (%)
			a	**c**	
Undoped	In_2O_3	0.0208	10.12802	/	100
2 at.% Fe	In_2O_3	0.0412	10.13029	/	95.12
	Fe_2O_3	0.0302	5.6373	13.9537	4.88
4 at.% Fe	In_2O_3	0.0362	10.1311	/	93.53
	Fe_2O_3	0.0534	5.6537	13.8866	6.47
6 at.% Fe	In_2O_3	0.0301	10.1281	/	89.68
	Fe_2O_3	0.0432	5.6377	13.9474	10.32

2.1.3. Raman analysis

We have used Raman spectroscopy to confirm indium oxide's phase and to explore the vibrational modes. Raman spectra of undoped and 6 at.% iron doped In_2O_3 thin films are shown in Figure II.3. We observe two vibrational modes at 134 and 307 cm^{-1} for undoped indium oxide thin films which correspond to the modes of body centered cubic structure [11]. After doping, five other vibrational modes have been appeared at 109, 290, 338, 366 and 499 cm^{-1}. The peaks located at 109 and 366 cm^{-1} correspond also to the bbc In_2O_3 structure. The vibrational modes oberserved in 290 and 499 cm^{-1} are caused by the presence of undesirable rhombohedra Fe_2O_3 phase [12] as deduced from Rietvled analysis and XRD spectra. The presence of small peak in 338 cm^{-1} may be due to the incorporation of iron in the crystal lattice [11].

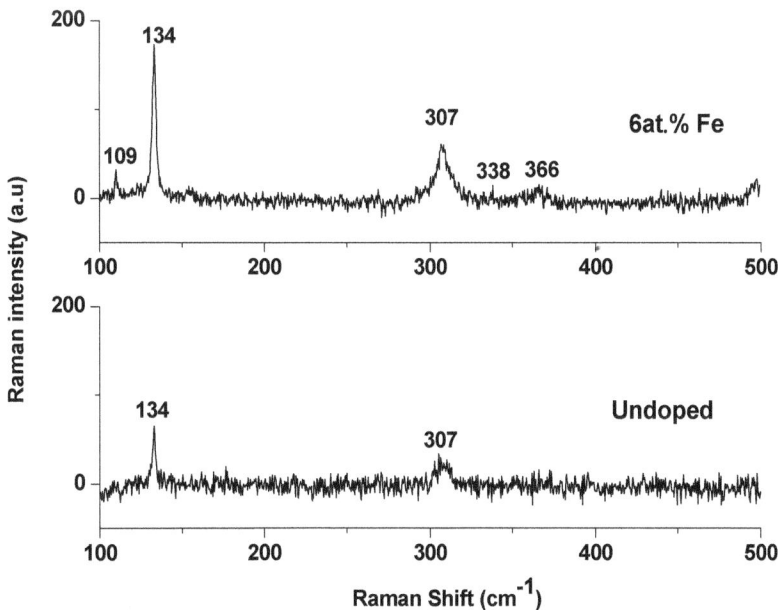

Figure II.3: Raman spectra measurements of In_2O_3:Fe for undoped and 6 at.% iron doped indium oxide thin films

2.1.4. **Calculation of grain size, dislocation density and number of crystallites:**

The grain size (d) of iron doped In_2O_3 thin films was calculated for different doping concentrations using the Debye-Sherrer formula applied to preferred orientations [13]:

$$d = \frac{0.94\lambda}{\sqrt{\beta^2 - \beta_0^2} \times cos\,\theta} \qquad (1)$$

Where λ is the X-ray wavelength of Cu Kα radiation ($\lambda = 1.5418$ Å), $\beta_0 = 0.125°$ is the width of the corresponding peak due to the instrumental expansion. β is the experimental full-width at half-maximum (FWHM) preferential diffracted peak measured in radians and θ is the Bragg's angle.

The dislocation density (δ_{disc}) and the number of crystallites per unit surface area (n_c) were calculated for the preferred orientation for each doping concentration y using the following expressions [13]:

$$\delta_{dis} = \frac{1}{d^2} \qquad (2)$$

$$n_c = \frac{t}{d^3} \qquad (3)$$

where t is the film thickness.

The thickness t of In_2O_3 thin films can be calculated using the weight difference method used by Yahmadi *et al.* [14]:

$$t = \frac{m}{D.S} \qquad (4)$$

Where m is the mass of the thin film expressed in gram, D is the density of the In_2O_3 in the bulk form and S (cm^2) is the effective area of the glass substrate on which the film was deposited.

The results are shown in **Figure II.4**. We observe a global growth of grain size after doping with iron. We notice also an increase in grain size when iron concentration increases to reach a maximum value of 98 nm for In_2O_3:Fe (6 at.%). However, dislocation density (δ_{dis}) and crystallites number (n_c) decrease after doping. Indeed, δ_{dis} decreases from 1.98 $\times 10^{10}$ cm^{-2} to 1.04 $\times 10^{10}$ cm^{-2} for respectively undoped and In_2O_3:Fe(6 at.%) thin films.

The lowest value of crystallite number n_c (in the order of 0.08 $\times 10^{12}$ cm^{-2}) is obtained for y= 6 at.%. The obtained value of (δ_{dis}) and (n_c) for In_2O_3:Fe(6 at.%) are in accordance with that y = 6 at.% correspond to the best crystallinity.

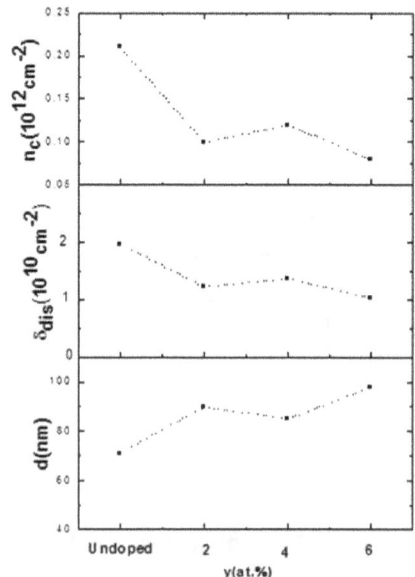

Figure II.4: Average grain size (d), dislocation density (δ_{dis}) and crystallites number (n_c) of undoped and iron doped In_2O_3 thin films for 2, 4 and 6 at.%

2.1.5. Morphological properties

AFM images of In_2O_3 thin films for different iron concentrations are presented in **Figure II.5**. Undoped films show a dense texture surface with pyramidal form. A noticeable increase of the grain size is observed after iron doping which may be attributed to incorporation of iron atoms in the lattice and the change in the preferred orientation particularly for y equal to 6 at.%. This observable grain size growth in AFM images deals clearly with calculation previously made by Debye-Sherrer formula. The surface morphology depends on iron doping concentration since the pyramidal grain has different density size and shape.

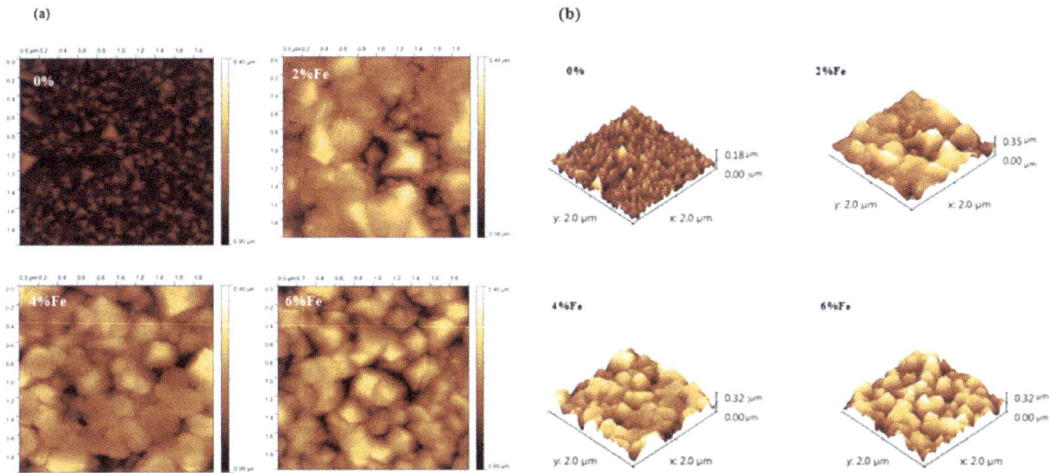

Figure II.5: AFM images (2D (a)) and (3D (b)) of undoped and iron doped In_2O_3 thin films for y concentration equals to 2, 4 and 6 at.%

2.2. Optical properties

Optical transmission and reflection spectra of iron doped indium oxide thin films are shown in **Figure II.6**. Interference fringes were observed in the visible and near infrared regions for undoped and iron doped In_2O_3 thin films. The observed oscillatory character indicates thickness uniformity and surface homogeneity of deposited material. It is seen from the figure 4 that all films are highly transparent over the visible and near-infrared regions (T≈80%). After doping with iron, we note a slight decrease of transmittance T which could be caused by photon scattering due to crystal defects created after doping. These experimental results are similar to those obtained after doping In_2O_3 thin films with zinc as described by Jothibas *et al.* [15]. It is noted a slight shift in the absorption edge when wavelength increases which indicates a decrease of the band gap energy for In_2O_3 after doping. The band gap energy (E_g) is calculated using Tauc equation [16]:

$$(\alpha h\nu) = A(h\nu - E_g)^n \qquad (5)$$

where (hν) is the photon energy, h is planck's constant, n is equal to (½) for direct band gap, A is a constant and α is the absorption coefficient which can be calculated using this formula [17]:

$$\alpha = -\frac{1}{t} \, Ln \left[\frac{T}{(1-R)^2} \right] \qquad (6)$$

where t, T and R, are film thickness, transmission and reflection coefficients, respectively.

The variation of $(\alpha h\nu)^2$ versus (hν) for In_2O_3:Fe thin films grown for different doping ratio y is shown in the **Figure II.7**. The straight line of the films over the wide range of photon energy indicates the direct transition type. The direct band gap energy was obtained by extrapolating the linear part of the Tauc plot curves to intercept the energy axis (at $\alpha h\nu$ = 0). The estimated values of E_g for iron doped indium oxide films decreases after doping with iron from 3.45 eV to 3.29 eV. This observed decrease of the band gap energy after doping may be due to an increase of the grain size [18] as shown in **Figure II.4**. The reduction of the band gap after doping can be also attributed to iron atoms located at donor level. The high rate of doped iron atoms may also reduce the difference between donor level and conduction band witch cause a decrease of the band gap. Similar results have been found by H. Khallef for boron doped cadminium sulphide deposited using CBD technique [19].

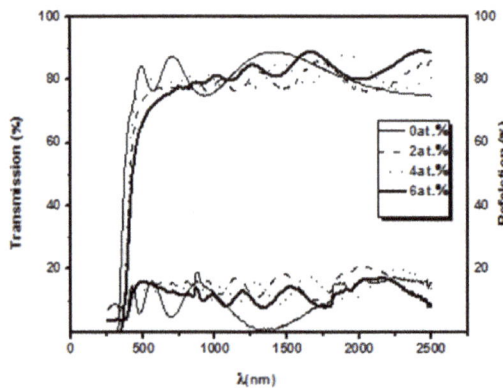

Figure II.6: Optical transmission and reflection spectra of undoped and iron doped In_2O_3 thin films for different concentration y= 0, 2, 4 and 6 at.%

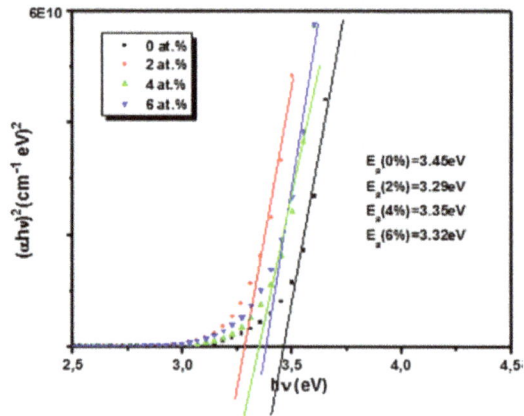

Figure II.7: Variation of $(\alpha h\nu)^2 = f(h\nu)$ of undoped and iron doped In_2O_3 thin films for different concentrations y = 0, 2, 4 and 6 at.%, and deduced values of E_g

The dispersion constants (E_d, E_0, λ_0, S_0) were determined using the Wemple model as detailed by Senthilkumar *et al.* [20]. The equation applied of this model is:

$$n^2 - 1 = \frac{E_d E_0}{E_0^2 - (h\nu)^2} \qquad (7)$$

E_d, E_0 are respectively the dispersion and the oscillator energy.

The constants S_0 and λ_0 can be deduced from the Wemple equation as:

$$\lambda_0 = \frac{hc}{E_0} \qquad (8)$$

$$S_0 = \frac{E_0 E_d}{(hc)^2} \qquad (9)$$

where S_0 and λ_0 are respectively the average strength and the wavelength of the oscillator.

The refractive index "n" was calculated in the transparency zone using the envelope method for $T_{max}(\lambda)$ and $T_{min}(\lambda)$ where n(λ) is given by [21]:

$$n(\lambda) = [N + (N^2 - n_0^2 n_{sub}^2)^{1/2}]^{1/2} \quad (10) \quad \text{and}$$

$$N = 2n_0 n_{sub} \left[\frac{(T_{max}-T_{min})}{(T_{max}T_{min})}\right] + \left(\frac{n_0^2 + n_{sub}^2}{2}\right).$$

n_0 is the refractive index of air and n_{sub} is the refractive index of glass substrate which is in the order of 1.52. In fact, as shown in Figure II.8, we have theoretically simulated the experimental spectra (T) using envelope method.

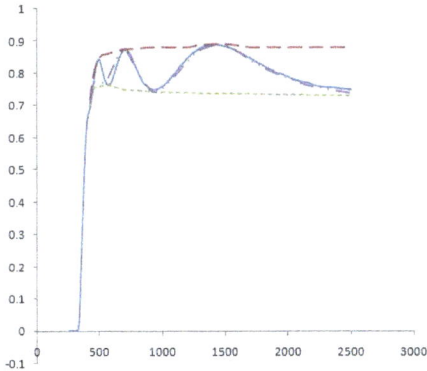

Figure II.8:
Calculation of transmission using envelope method for In$_2$O$_3$ thin films.

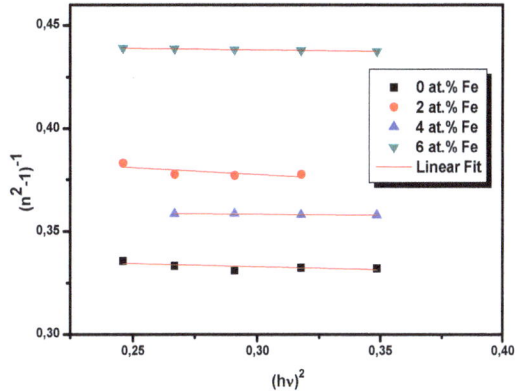

Figure II.9:
Plot of $(n^2-1)^{-1}$ as a function of $(hv)^2$ of undoped and iron doped In$_2$O$_3$ thin films for different ratio 0, 2, 4 and 6 at.%

Table II.2: Calculating dispersion and oscillator energies using Wemple model

y=[Fe^{2+}]/[In^{3+}] (at.%)	E$_0$ (eV)	E$_d$ (eV)	S$_0$ (10^{-5} nm^{-2})	λ_0 (nm)
0	5.05	15.02	4.93	245.54
2	7.49	18.86	9.18	165.55
4	6.14	17.01	6.79	201.95
6	7.71	17.55	8.80	160.83

2.3. Photoluminescence properties:

The photoluminescence (PL) spectra of iron doped indium oxide thin films with different iron concentrations y (y= 0, 2, 4 and 6 at.%) is shown in Figure II.10. We notice that PL characteristics of In$_2$O$_3$ thin layers depend clearly of iron atomic concentration. In fact, intensity of photoluminescence peaks increases after doping. For the undoped films we observe several emission peaks at 435 (2.85 eV), 485 (2.57 eV), 505 (2.45 eV) and 795 (1.56 eV) nm. After doping, photoluminescence spectra exhibit the same emission peaks with accentuated intensity unless a small shift for the first one from 435 nm to 417 nm (2.97 eV). Besides, we note after doping the appearance of a new emission peak at 527 nm (2.35 eV) which may be due to the presence of Fe$_2$O$_3$ defects. PL spectra illustrate also a large blue green emission band at 480-510 nm for all doped and undoped films. This photoluminescence emission could be due to lattice defects or oxygen deficiencies in the indium oxide thin films [22]. To explain this luminescence process Kaleemulla *et al.* [22] have reported that electrons positioned in donor level could be attracted by a hole present in acceptor level. Hence a trapped exciton is formed and the relaxation of the latter is radiative and leads to produce large blue green-emission. There are also a violet and infrared emissions centered respectively at 437 and 795 nm.

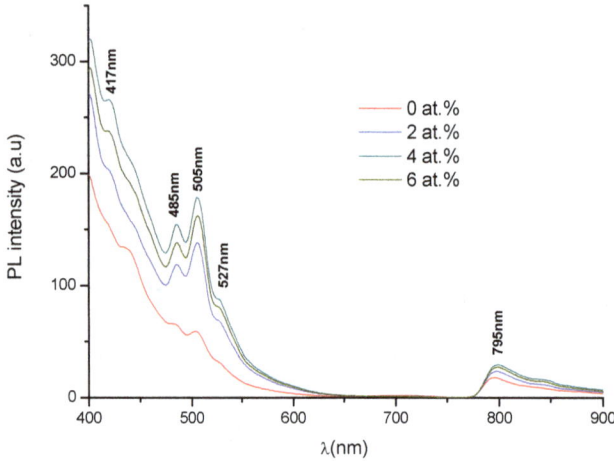

Figure II.10:
PL spectra of undoped and iron doped In_2O_3 thin films

2.4. Electrical properties

Measurements by Hall-effect of electrical parameters of In_2O_3:Fe thin films (electrical resistivity, carrier concentration and mobility) are made for different iron concentrations y (y = 0, 2, 4 and 6 at.%). The sign of the Hall Effect coefficient have confirmed the n-type of the indium oxide semiconductor. Table II.3 reveals a decrease of the resistivity from ρ=6502 x 10^{-4} Ω.cm to 197.90 x 10^{-4} Ω.cm for respectively undoped and In_2O_3:Fe (6 at.%). Such decrease can be attributed to an increase of the carrier concentration after doping. In fact, volume carrier concentration (N_v) was increased by two order of magnitude from 0.08 x 10^{20} cm^{-3} to 8.30 x 10^{20} cm^{-3} for respectively undoped In_2O_3 and 6 at.% doped ones. This can be attributed to the improvement of the crystallinity or may be to the change in the preferred orientation peak. Hall mobility (μ) was increased after doping process and it reaches a maximum value of 10.30 $cm^{-2}V^{-1}s^{-1}$ for y equals to 6 at.%.

Table II.3: Electrical parameters of undoped and iron doped indium oxide thin films

y(at.%)	ρ (Resistivity) $(\Omega.cm)*10^{-4}$	μ(mobility) $(cm^{-2}V^{-1}s^{-1})$	N_v (Volume carrier concentrations) $(cm^{-3})*10^{20}$	N_s (Surface carrier concentrations) $(cm^{-2})*10^{14}$
0	6502	1.17	0.08	2.45
2	5537	3.34	3.37	1.35
4	4194	2.29	6.50	2.60
6	197.90	10.30	8.30	12.30

In order to reduce more the resistivity, a heat treatment was applied on In_2O_3:Fe (6 at.%) thin films corresponding to optimized thin layers. The annealing was carried out under nitrogen atmosphere at 200, 300 and 400 °C. **Table II.4** shows electrical properties of annealed In_2O_3:Fe (at.6%). A slight increase of the resistivity is noted for annealed films at 200°C. A remarkable decrease is observed for films annealed at 300°C and ρ reaches a minimum value of 26.94×10^{-4} Ω.cm^{-1}. At T = 400 °C, electrical resistivity increases slightly keeping the same order of magnitude. It is also revealed an increase of the carrier concentration (N_v) from 8.30×10^{20} cm^{-3} to 10.04×10^{20} cm^{-3} for respectively as grown and annealed films at 300°C. This increase in free volume carrier concentrations N_v as a function of annealing temperature can be a good factor to improve the degenerate character of the films (N_v in the order of 10^{21} cm^{-3} for 300 and 400 °C annealing temperatures). This last result may contribute to a reduction of resistivity as explained by Yuan *et al.* [23]. The decrease of the resistivity can be also attributed to an increase of the grain size of the films [24] as illustrated previously in our structural analysis and to the reduction of structural defaults or grain boundaries.

Table II.4: Electrical parameters of as grown and annealed In_2O_3:Fe (6 at.%) at different temperatures

Annealing temperatures	ρ (**Resistivity**) (Ω.cm)$*10^{-4}$	μ (**mobility**) $(cm^{-2}V^{-1}s^{-1})$	N_v (**Volume carrier concentrations**) (cm^{-3}) **x** 10^{20}	N_s (**Surface carrier concentrations**) (cm^{-2}) **x** 10^{14}
As deposited	197.90	10.30	8.30	12.30
200 °C	346,50	0.40	4.49	135
300 °C	26.94	3.29	10.04	211
400 °C	64.89	1.27	10.57	227

2.5. Figure of Merit:

Optical transmittance and electrical resistivity are two important parameters to evaluate the quality of transparent conducting oxides. To correlate the properties of TCO films, we have used the figure of merit (φ) which is given by the relation [25]:

$$\phi = \frac{T^{10}}{R_s} \qquad (11)$$

where T is the average transmittance and R_s is the sheet resistance ($R_s = \rho/t$).

The variation of Φ of In_2O_3 as a function of iron doping concentration is shown in Figure II.11. We remark that Φ is increased from 0.09×10^{-4} Ω^{-1} to 1.73×10^{-4} Ω^{-1} when iron concentration is increasing. The optimum value of the figure of merit was obtained for y equal to 6 at.% which is in good agreement with that In_2O_3:Fe(6 at.%) thin films have best crystallinity, high transmittance and lowest electrical resistivity.

The Figure of Merit φ as a function of annealing temperature T_a for In_2O_3:Fe(6 at.%) are presented in the Figure II.12. We note an initial increase of φ. In fact, Figure of Merit increase from 1.73×10^{-4} Ω^{-1} to 24×10^{-4} Ω^{-1} respectively for the as grown films and for T_a

equal to 300 °C. Then ϕ decreases to $5 \times 10^{-4}\,\Omega^{-1}$ for annealing temperature equal to 400 °C. The Figure of Merit increases as the annealing temperature increases due to the reduction of electrical resistivity and enhancement of transmittance after annealing until 300 °C. Finally we can conclude that In_2O_3:Fe (6 at.%) annealed at 300 °C is the best TCO among our elaborated Indium oxide thin films.

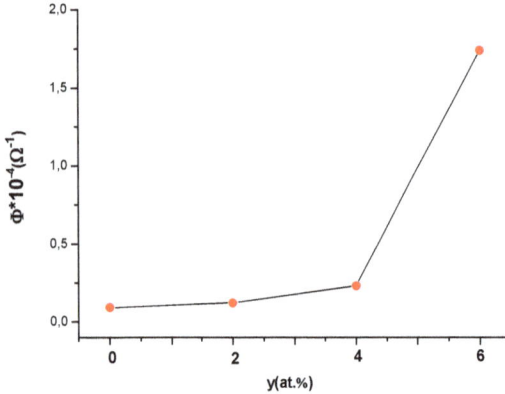

Figure II.11: Figure of Merit Φ as a function of iron atomic concentrations y

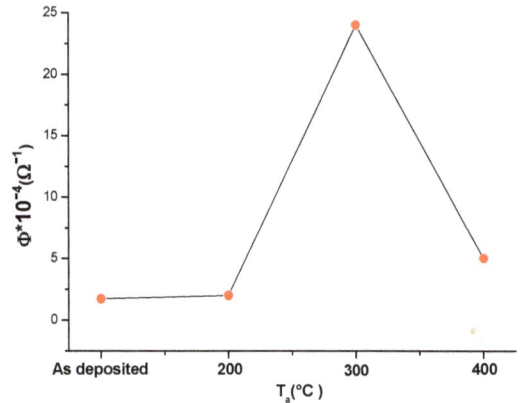

Figure II.12: Figure of Merit Φ as a function of annealing temperature for In_2O_3:Fe(6 at.%)

2.6. Conclusion

The effect of iron doped on the physical properties of In_2O_3 thin films elaborated by spray was studied in this chapter. Doping operation was carried out during the growth process by adding relative concentration of Fe^{2+} ions in the sprayed solutions. We have concluded that physical properties of In_2O_3:Fe have been globally improved after doping by iron. Structural analysis of XRD spectra shows a clear and strong change in the preferred orientation from (222) to (400) plan for iron 6%.at doped concentration which indicates an enhancement of the crystallinity. Calculation of grain size and results of Rietveld analysis corroborate this experimental obtained result. AFM images show that morphological properties have changed after doping and grain size becomes greater. Optical properties show a high transmittance around 80% in the visible and near infrared regions. Calculated band gap is in the range of [3.29 - 3.45] eV. Electrical resistivity has decreased after doping from $\rho = 6502 \times 10^{-4}\,\Omega.cm$ to $197.90 \times 10^{-4}\,\Omega.cm$ for respectively undoped and In_2O_3:Fe (6 at.%). A heat treatment has been performed to decrease more the resistivity and gives a value of $26.94 \times 10^{-4}\,\Omega.cm$ for annealing temperature equals to 300 °C. All these experimental results confirm the efficiency of iron doped indium oxide as a transparent conductive oxide.

2.7. References:

[1] C.Baratto, M.Ferroni, G.Faglia and G.Sberveglieri; Sens. Actuat. B 118 (2006) 221–225.
[2] J. Zhao, T. Yang, Y. Liu, Z. Wang, X. Li, Y. Sun, Y. Du, Y. Li and G. Lu; Sens. Actuat. B 191 (2014) 806–812.
[3] S. Kohiki, M. Sasaki, Y. Murakawa, K. Hori, K. Okada, H. Shimooka, T. Tajiri, H. Deguchi, S. Matsushima, M. Oku, T. Shishido, M. Arai, M. Mitome and Y. Bando; Thin Solid Films 505 (2006) 122–125.

[4] X. Peng-Fei, C. Yan-Xue, T. Min-Jian, Y. Shi-Shen, L. Guo-Lei, M. Liang-Mo and J. Jun; Chin. Phys. Lett. 27 (1) (2010) 117503.

[5] V.G. Myagkov, I.A. Tambasov, O.A. Bayukov, V.S. Zhigalov, L.E. Bykova, Yu.L. Mikhlin, M.N. Volochaev and G.N. Bondarenko; J.Alloy. Compd. 612 (2014) 189–194.

[6] N.B. Ibrahim, H. Baqiah and M.H. Abdullah; Sains Malaysiana 42 (7) (2013) 961–966.

[7] S. Parthiban, V. Gokulakrishnan, K. Ramamurthi, E. Elangovan, R. Martins, E. Fortunato, and R. Ganesan; Sol. Energy Mater. Sol.Cells 93 (2009) 92–97.

[8] Yang.-Ru. Lyu and Tsung.-Eong. Hsieh; Surf. Coat. Technol. 231 (2013) 219–223.

[9] S. Kaleemulla, N. Madhusudhana Rao, M. Girish Joshi, A. Sivasankar Reddy, S. Uthanna, P. Sreedhara Reddy; J. Alloy. Compd.504 (2010) 351–356.

[10] S. Louidi, F.Z. Bentayeb, W. Tebib, J.J. Suñol, A.M. Mercier, J.M. Grenèche, J. Non-Cryst. Solids 356 (2010) 1052–1056.

[11] Olivia M. Berengue, Ariano D. Rodrigues, Cleocir J. Dalmaschio, Alexandre J.C. Lanfredi, Edson R. Leite, Adenilson J. Chiquito,J. Phys. D Appl. Phys. 43 (2010) 045401 (4pp).

[12] C. Baratto, P.P. Lottici, D. Bersani, G. Antonioli, J. Sol–Gel Sci. Technol. 13 (1998) 667.

[13] A. Akkari, M. Reghima, C. Guasch, N. Kamoun-Turki, J. Mater. Sci. 47 (2012) 1365–1371.

[14] B. Yahmadi, N. Kamoun, C. Guasch, R. Bennaceur, Mater. Chem. Phys. 127 (2011) 239.

[15] M. Jothibas, C. Manoharan, S. Ramalingam, S. Dhanapandian, M. Bououdina, Spectrochim. Acta Part A Mol. Biomol. Spectrosc. 122 (2014) 171–178.

[16] D. Beena, K.J. Lethy, R. Vinodkumar, V.P. Mahadevan Pillai, V. Ganesan, D.M. Phase, S.K. Sudheer, Appl. Surf. Sci. 255 (2009) 8334–8342.

[17] S. Belgacem, R. Bennaceur, Revue. Phys. Appl. 25 (1990) 1245–1258.

[18] S. Benramache, B. Benhaoua, O. Belahssen, Optik 125 (2014) 5864–5868.

[19] H. Khallaf, G. Chai, O. Lupan, L. Chow, H. Heinrich, S. Park, A. Schulte, Phys. Status Solidi A 206 (2) (2009) 256–262.

[20] V. Senthilkumar, P. Vickraman, M. Jayachandran, C. Sanjeeviraj, Vacuum 84 (2010) 864.

[21] P. Prathap, Y.P.V. Subbaiah, M. Devika and K.T. Ramakrishna Reddy; Materials Chemistry and Physics 100 (2006) 375–379.

[22] S. Kaleemulla, A. Sivasankar Reddy, S. Uthanna and P. Sreedhara Reddy, Optoelectron. Adv. Mater. Rapid Commun. 2 (12) (2008) 782–787. December.

[23] Z. Yuan, X. Zhu, X. Wang, X. Cai, B. Zhang, D. Qiu and H. Wu; Thin Solid Films 519 (2011) 3254–3258.

[24] S. Erat, H. Metin, M. Ari, Mater. Chem. Phys. 111 (2008) 114–120.

[25] A.V. Moholkar, S.M. Pawar, K.Y. Rajpure, V. Ganesan, and C.H. Bhosale; J. Alloy. Compd. 464 (2008) 387–392.

3. Effects of zinc doping and heat treatment on the physical properties of indium oxide thin films

In this section, we will study the effect of zinc doping on the physical properties of In_2O_3 thin films. In the present chapter, we have synthesized zinc doped indium oxide thin films by spray pyrolysis technique. In our knowledge, only a report [1] has investigated about physical properties of In_2O_3:Zn thin films grown by spray. Jothibas *et al.* [1] have used quite different experimental conditions compared to us which have produced naturally other experimental results that enrich and enhance research studies about this material. In our case, In_2O_3:Zn have been synthesized with different doping concentration y ($y = (\frac{[Zn^{2+}]}{[In^{3+}]})_{sol} = 0, 1,$

2, 3, 4, 5 and 6 at.%) using other chemical precursors indium chloride ($InCl_3$) and Zinc acetate dehydrate (Zn ($CH_3COO)_2.2H_2O$)). In the next part of this chapter, we will study the influence of heat treatment on physical properties of optimized zinc doped indium oxide thin films which, in our knowledge, has not done before for sprayed In_2O_3:Zn. Indeed, zinc doped In_2O_3 thin layers are annealed in nitrogen atmosphere for 250°C and 450°C during 1 and 2 hours.

3.1. Zn-doping of In_2O_3 thin films

3.1.1. Structural analysis

XRD Analysis: XRD spectra of zinc doped indium oxide thin films for different zinc concentrations y are shown in Figure III.1. The diffraction peaks are very narrow indicating the good crystallinity of deposited films. All films crystallize into body centered cubic structure (JCPDS Card n°06-0416). XRD observations reveal that undoped films have a preferred orientation along (222) plan for $2\theta = 30.58°$. For y = 1 at.%, an enhancement of In_2O_3:Zn thin films crystallinity is observed. At this doping level, a change of preferred orientation from (222) to (400) occurs accompanied by an increase of the main diffraction peak intensity. The (400) peak is located at 35.29°. This change in the preferred orientation may be caused by the cation incorporation of zinc as shown by Parthiban *et al.* for In_2O_3:Mo thin films elaborated by spray [3]. In fact, this change could be due to the occupancy of additional indium vacancy sites by zinc elements which are unoccupied previously [4].

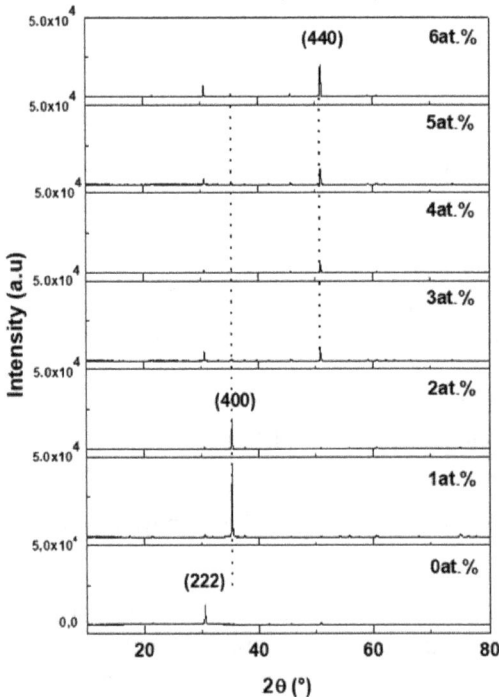

Figure III.1: XRD spectra of zinc doped In_2O_3 thin films for different zinc concentration y in the spray solution (y = $(\frac{[Zn^{2+}]}{[In^{3+}]})_{sol}$ = 0, 1, 2, 3, 4, 5 and 6 at.%)

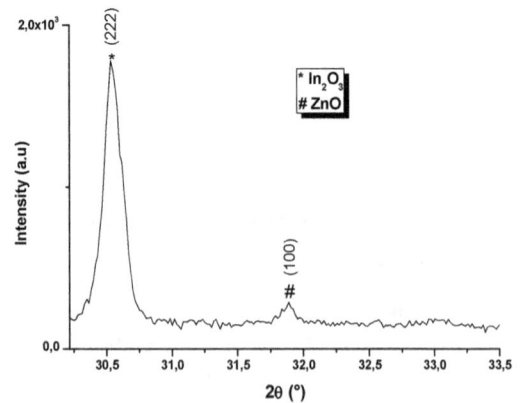

Figure III.2: Zoom of XRD spectrum for zinc concentration y equals to 1 at.%

Indeed, In_2O_3 thin layers have several types of defects levels like indium and oxygen vacancies and indium interstitial sites [5]. Moreover from y = 2 at.%, zinc atoms start to take place in the interstitial sites which probably lead to deterioration of films crystallinity and a complete change in the preferred orientation. In fact, for y = 2 at.%, the intensity of the main diffraction peak (400) decreases. From y = 3 at.%, an increase of zinc content in the sprayed solution leads to a noticeable degradation of the crystallinity of deposited thin layers. This deterioration of crystallinity is accompanied by a second transition of the preferred orientation from (400) to (440). We note also the appearance of secondary phase of ZnO due to the doping process. Indeed, the presence of (100) diffraction peak (Figure III.2) is located at 31.88° corresponds to the hexagonal ZnO phase (JCPDS Card n°89-1397). These experimental results are in good agreement with those observed elsewhere. In fact, Jothibas et al. [1] are founded ZnO phase at higher zinc doping concentration (11 at.%) characterized by (100) and (102) plans.

Rietveld Analysis: XRD patterns of zinc doped In_2O_3 thin films are fitted by MAUD software [2] which is based on Rietvled method [2]. Figure III.3 shows the Rietveld refinements of XRD patterns for different doping concentrations y. The structural parameters (lattice parameters, microstrain, crystallite sizes and phases amounts) given by fitting process are summarized in the Table III.1.

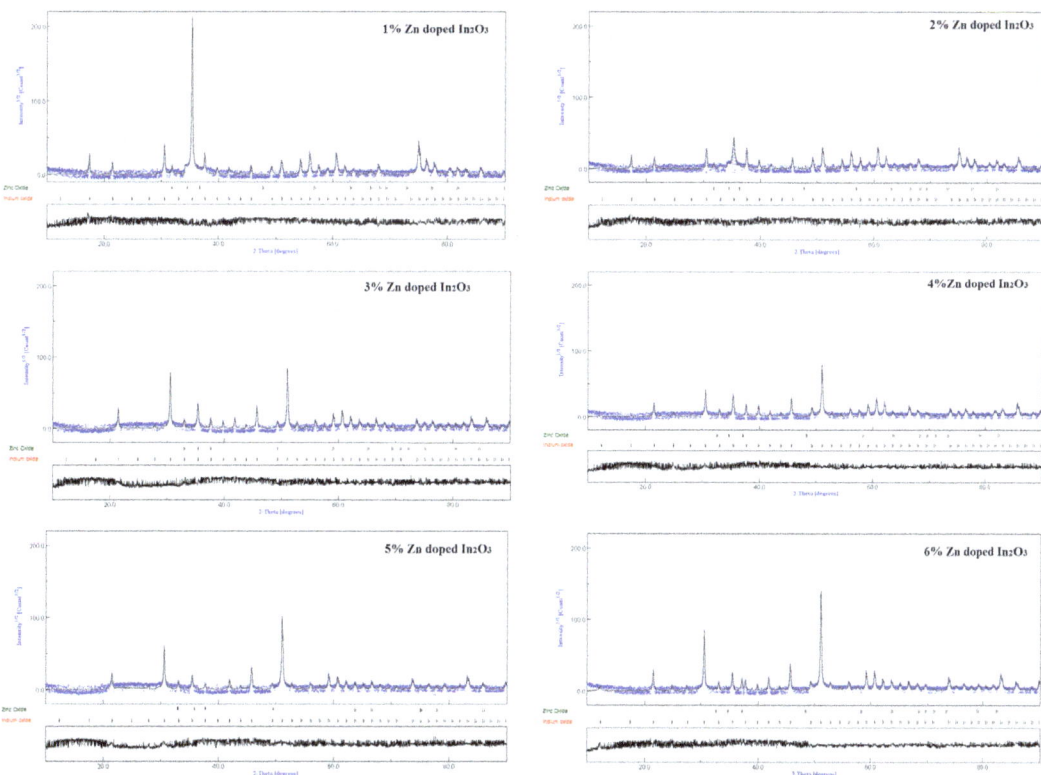

Figure III.3: Rietveld refinement of XRD spectrra of zinc doped indium oxide thin films for different doping concentation. The difference between the experimental (dots) and the calculated (full line) spectrum is given below

Table III.1: Phases and microstructure parameters determined by MAUD software for zinc doped In_2O_3 thin films (y $= \frac{[Zn^{2+}]}{[In^{3+}]}$)$_{sol}$ = 0, 1, 2, 3, 4, 5 and 6 at.%)

y(at.%)	Phases	Microstructure $<\sigma^2>^{1/2}$ (%) ($\pm$0.002)	Lattice parameters (Å) ($\pm$0.0001)		Amount (%) ($\pm$ 0.4)
			a	c	
0	In_2O_3	0.0208	10.1280	/	100
1	In_2O_3	0.041	10.1221	/	96.25
	ZnO	0.030	3.2417	5.1876	3.75
2	In_2O_3	0.054	10.1231	/	95.12
	ZnO	0.041	3.2393	5.1872	4.88
3	In_2O_3	0.051	10.1287	/	94.7
	ZnO	0.028	3.2385	5.1866	5.3
4	In_2O_3	0.060	10.1303	/	92.45
	ZnO	0.032	3.2317	5.1757	7.55
5	In_2O_3	0.052	10.1305	/	90.2
	ZnO	0.024	3.2313	5.1760	9.8
6	In_2O_3	0.031	10.1308	/	87.3
	ZnO	0.026	3.2309	5.1759	12.7

The Rietveld refinement reveals the presence of In_2O_3 and ZnO phase. The relative proportion of the zinc oxide phase increases when zinc concentration increases and reaches a maximum value of 12.7 % for 6 at.% (Table III.1). The lattice parameter (a) of the cubic centered In_2O_3 thin layers is in the order of 10.1280 and 10.1221Å for respectively undoped and In_2O_3:Zn (1 at.%). Furthermore, the lattice parameter (a) of the indium oxide In_2O_3 increases with zinc content and attains a maximum value of 10.1308Å for 6 at.% zinc concentration. Up to y equals to 2at.%, the crystal lattice is contracted after doping. However, beyond 3 at.% we note an expansion of the crystal lattice compared to the undoped ones.

The increase of the microstrain ($<\sigma^2>^{1/2}$) after doping may be caused by the incorporation of zinc atoms in the In_2O_3 crystal lattice. Figure III.4 exhibits the variation of the average grain size $<d>$ of zinc doped In_2O_3 thin films calculated by MAUD software as a function of doping concentration y. A maximum value of about 93.4 nm is obtained for y = 1 at.%. This result is in good agreement with the fact that In_2O_3:Zn (1 at.%) thin layers have the best crystallinity as found in XRD analysis.

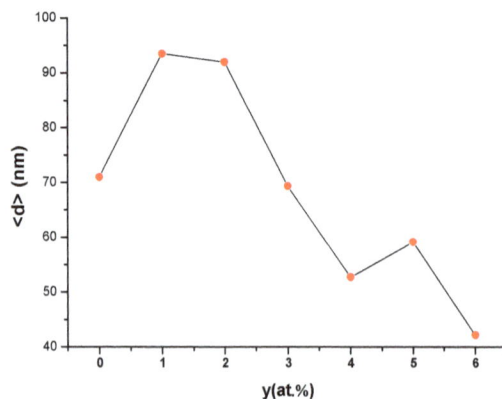

Figure III.4: Average grain size $<d>$ of zinc doped In_2O_3 thin films for different concentrations y $= \frac{[Zn^{2+}]}{[In^{3+}]}$)$_{sol}$ = 0, 1, 2, 3, 4, 5 and 6 at.%

3.1.2. Morphological properties

AFM images for undoped and zinc doped In_2O_3 thin films are presented in Figure III.5. It is noted an increase of the grain size for y = 1 at.% compared to the undoped one. This is in good accordance with the results obtained from the Rietveld refinements of the XRD patterns. When zinc concentration is increasing from 0 to 2 at.%, a clear change of grain form is observed. From y equal to 3 at.%, the micrographs reveal the presence of grains with different form and size in different spatial distribution. This last result may be attributed to the change of predominant peak when zinc concentration increases. The incorporation of zinc element and the variation of doping concentration affect extremely the surface morphology of elaborated films.

0 at.%

1 at.%

2 at.%

3 at.%

4 at.%

5 at.%

6 at.%

Figure III.5: AFM images of undoped and zinc doped In_2O_3 thin films for different zinc concentrations in the spray solution (y = $\frac{[Zn^{2+}]}{[In^{3+}]}$)$_{sol}$ = 0, 1, 2, 3, 4, 5 and 6 at.%)

RMS roughness values were determined using AFM images. All results are shown in Figure III.6. It is found that RMS roughness increases from 48 to 59 nm for respectively undoped and In_2O_3:Zn (1 at.%). The expansion of RMS values can be related to an enhancement of crystallite size as remarked by Liu *et al.* [6]. Beyond the level of 1 at.% doping, a decrease trend of RMS values is observed. In fact, some authors [7] have reported that the agglomeration can induces a modification in RMS roughness values.

3.1.3. Chemical compositions

Chemical elements of In_2O_3:Zn thin films are explored using EDS spectroscopy for different atomic concentration. The results are summarized in Table III.2

The [In]/[O] values in In_2O_3 thin films are greater than 0.76 % which is not closer to theoretical stochiometric value of indium oxide (0.67 %). It is clear that obtained layers are not stochiometric. This last result can be attributed to the presence

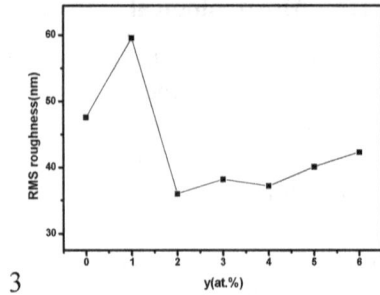

Figure III.6: RMS roughness of zinc doped In_2O_3 thin films for different concentrations $y = \frac{[Zn^{2+}]}{[In^{3+}]})_{sol} = 0, 1, 2, 3, 4, 5$ and 6 at.%

of oxygen vacancies. On the other hand, table 1 shows the presence of zinc atoms in doped films. The atomic percentages [Zn]/[In] present in the films are close to added one in the initial sprayed solution which means that spray technique permits practically the conservation of doping ratio after growth process.

Table III.2: Chemical compositions of undoped and zinc doped In_2O_3 thin films

y(at.%)	In(at.%)	O(at.%)	Zn (at.%)	[In/O](%)	[Zn/In](%)
0	45.34	54.66	0	0.91	0
1	44.40	55.18	0.42	0.80	0.94
2	44.14	54.92	0.94	0.80	2.13
3	42.78	55.87	1.35	0.76	3.15
4	42.77	55.52	1.71	0.77	3.99
5	43.06	55.01	1.93	0.78	4.48
6	44.08	53.52	2.40	0.82	5.44

3.1.4. Optical properties

The optical transmission and reflection spectra of zinc doped indium oxide thin films for different doping concentration y (y = 0, 1, 2, 3, 4, 5 and 6 at.%) are shown in Figure III.7. We note the presence of interference fringes in the transparency zone which are characteristics of uniform thickness and homogenous surface layers. This phenomenon is a consequence of interference of incident light beams at the interface between deposited layers and microscope slide substrate [8]. After doping, the transmission (T) decreases slightly but remains still high in the visible and infrared regions except for y = 5at.%. Indeed, the average transmission was around 80% for y equals to 1 at.% which corresponds to the best transmission of In_2O_3:Zn. Our results in terms of film transparency are promising and seem to be better than those reported in other works [1,9]. Indeed, some reports have found that doping atoms leads to a poor transmission. Pramod *et al.* [9] have investigated the effect of lithium doping on physical properties of In_2O_3 thin films used as gas sensors. They revealed that the transmission was drastically decreased after doping. Further, Jothibas et al. [1] have reported that zinc doped indium thin layers exhibit a poor percentage of transmission for higher doping level with the absence of interference fringes. The reduction of transmission after doping can be due to the enhanced scattering of photons caused by crystal defects produced by doping or to the absorption of photons by free carriers [1]. The band gap energy

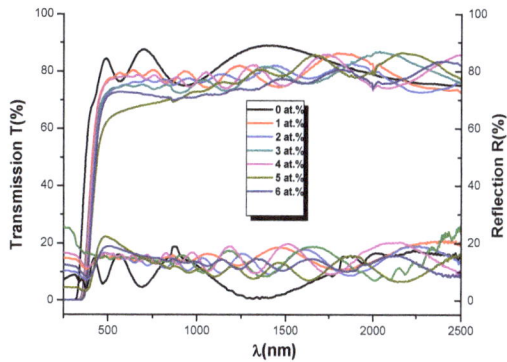

Figure III.7: Optical transmission and reflection spectra of undoped and zinc doped In_2O_3 thin films for different doping concentrations y

$$(y = \frac{[Zn^{2+}]}{[In^{3+}]})_{sol} = 0, 1, 2, 3, 4, 5 \text{ and } 6 \text{ at.\%})$$

Eg is calculated according to the Tauc equation as detailed in the previous chapter. The band gap values of zinc doped indium oxide thin films are presented in Figure III.8. The straight line of layers over the wide range of photon energy (hv) confirms the direct transition type. The direct band gap energy was determined by extrapolating the linear part of Tauc plot curves to intercept the energy axis (at αhv = 0). One observes that the band gap energy is in the range of [3.45-3.51]eV for y between 0 and 4 at.%. These values are closed to the theoretical band gap for TCO materials corresponding to 3.50 eV. However from 5at.%, E_g decreases slightly reaching 3.32 eV for y = 6 at.%. Our obtained results are in agreement with those reported by Jothibas *et al.* [1]. They found that band gap energy is in the range of [3.28; 3.61] eV. The decrease in the band gap for y = 5 at.% and 6 at.% may be due to the zinc elements positioned in donor level. In fact, The high doping concentration can reduce the difference between donor level and conduction band which leads to decrease of optical band gap energy [12].

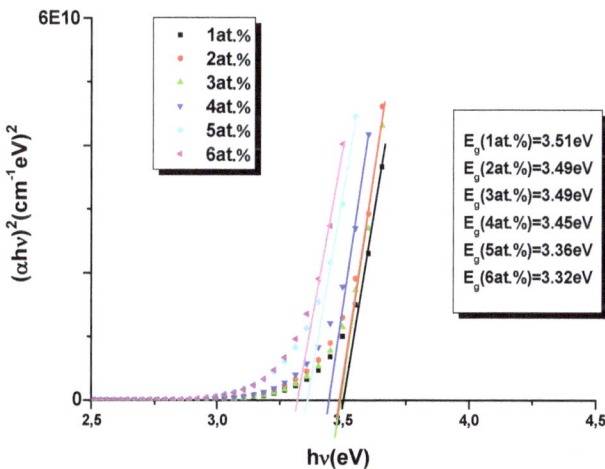

E_g(1at.%)=3.51eV
E_g(2at.%)=3.49eV
E_g(3at.%)=3.49eV
E_g(4at.%)=3.45eV
E_g(5at.%)=3.36eV
E_g(6at.%)=3.32eV

Figure III.8: $(\alpha hv)^2$ versus (hv) of zinc doped In_2O_3 thin films for different concentrations

$$y = (\frac{[Zn^{2+}]}{[In^{3+}]})_{sol} = 0, 1, 2, 3, 4, 5 \text{ and } 6 \text{ at.\%}$$

In order to determine the oscillator energy E_0, dispersion energy E_d, average strength S_0 and the wavelength λ_0, the Wemple model [13] have been applied to zinc doped indium oxide thin films. This model is explained in details in the previous chapter. The linear variations of $(n^2 - 1)^{-1}$ versus $(h\nu)^2$ for zinc doped indium oxide thin films (Figure III.9) permit us to determine E_0 and E_d from the slope $(E_d E_0)^{-1}$ and the offset (E_0/E_d) for each straight line.

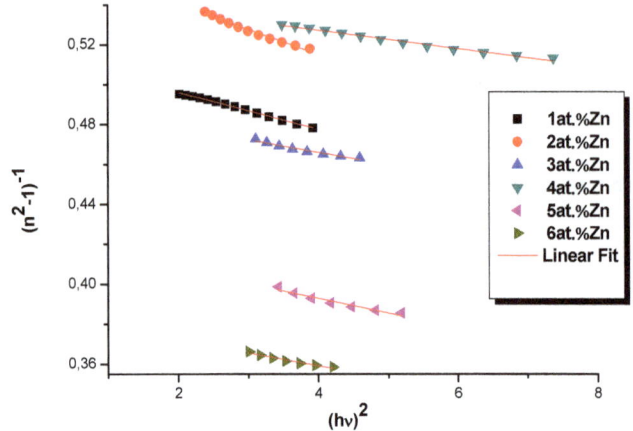

Figure III.9: Plot of $(n^2-1)^{-1}$ as a function of $(h\nu)^2$ of zinc doped In_2O_3 thin films for different ratios $y = \frac{[Zn^{2+}]}{[In^{3+}]})_{sol} = 0, 1, 2, 3, 4, 5$ and 6 at.%

All values of E_0, E_d, S_0 and λ_0, for zinc doped indium oxide thin films, are presented in Table III.3. It is seen that the oscillator energy (E_0) is the double of the optical band gap (E_g) for all zinc doped indium oxide thin films. However, $E_0 = 1.5 \times E_g$ for undoped In_2O_3 thin films [4]. This difference in the oscillator energy E_0 may be caused by the incorporation of doping element in the crystal lattice.

Table III.3: Calculating dispersion and oscillator energies using Wemple model for different doping concentrations y ($y = \frac{[Zn^{2+}]}{[In^{3+}]})_{sol} = 0, 1, 2, 3, 4, 5$ and 6 at.%)

y=[Zn²⁺]/[In³⁺]	E_0(eV)	E_d(eV)	λ_0(nm)	$S_0(10^{-5}nm^{-2})$
0	5.05	15.02	245.54	4.93
1	7.45	14.48	166.44	7.01
2	7.43	13.18	166.89	6.36
3	7.54	15.05	164.45	7.38
4	7.32	12.84	169.39	6.11
5	7.58	17.94	163.58	8.84
6	7.74	20.11	160.20	10.12

3.1.5. Photoluminescence analysis

Photoluminescence (PL) is a susceptible technique for the characterization of crystal structure or defect levels. Generally, the fluorescence of In_2O_3 compound is due to the presence of traps in indium oxide thin films. It is well known that In_2O_3 thin films have several types of defects levels such as indium interstitial, oxygen and indium vacancies [5]. Figure III.10 shows PL spectra of In$_2$O$_3$:Zn thin layers for different doping concentrations y. We note the presence of three principal emission peaks detected at 485, 505 and 795 nm corresponding to energy levels equal respectively to 2.56, 2.45 and 1.55 eV. This large blue-near green emission [485-505] may be related to lattice defects or oxygen deficiencies in the indium oxide thin films as explained by Kaleemulla *et al.* [14]. In fact, they have reported that photoluminescence process can be produced as follow: electrons located in donor lever can be attracted by a hole positioned in acceptor level. So a trapped exciton is formed leading to produce photoluminescence emis-sion [14]. A similar emission band is detected in our previous work for iron doped indium oxide thin films elaborated by spray [4]. Wu *et al.* [15] have reported that emission peaks observed in the visible range may be attributed to the recombination of carrier concentration that occurs between valence band and oxygen vacancies acting as donor levels. Further, the authors have revealed that PL peaks positions are different according to the literature. This fact may be related to the morphologies differences as shown in [15].

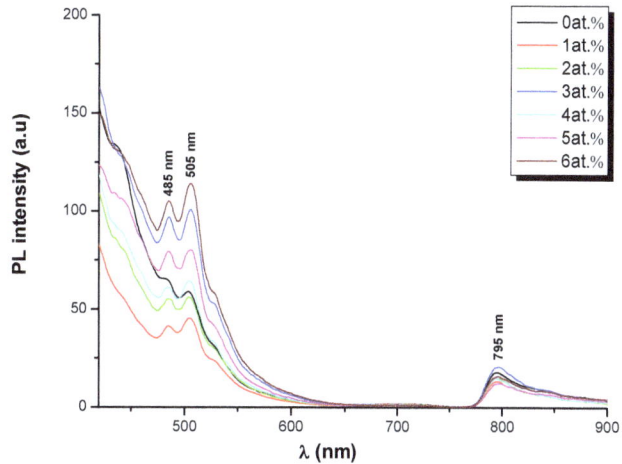

Figure III.10: Photoluminescence (PL) spectra of undoped and zinc doped In$_2$O$_3$ thin films for different concentrations $y = \frac{[Zn^{2+}]}{[In^{3+}]})_{sol}$

3.1.6. Electrical properties

The electrical parameters of undoped and zinc doped In$_2$O$_3$ thin films determined from the Hall Effect measurements are listed in Table III.4.

It is noted an initial decrease of the electrical resistivity (ρ) from 650.20 x 10^{-3} Ω.cm to 198.50 x 10^{-3} Ω.cm for respectively undoped and In$_2$O$_3$:Zn (1 at.%). This result is in accordance with the fact that In$_2$O$_3$:Zn (1 at.%) has the best crystallinity. Such decrease in ρ can be attributed to the enlargement of grain size due to the enhancement of the crystal quality as mentioned in the structural section. At this doping level (y = 1 at.%), one notes an increase of Hall mobility from 1.17 to 6.70 cm²/Vs. Except for y equals to 5 at.%, the volume carrier concentrations (N$_v$) decreases as a function of doping concentration y but it keep the same order of magnitude (about 10^{18} cm^{-3}).

Beyond y equals to 1 at.%, an increase of the electrical resistivity by an order of magnitude was noted. Such result can be attributed to the increase of crystal defaults and deterioration of crystallinity beyond 1 at.%.

Table III.4: Electrical parameters of undoped and zinc doped indium oxide thin films as a function of doping ratios $y = \frac{[Zn^{2+}]}{[In^{3+}]})_{sol} = 0, 1, 2, 3, 4, 5$ and 6 at.%

$y = [Zn^{2+}]/[In^{3+}]$ (at.%)	Resistivity ($\times 10^{-3}$ Ωcm)	Hall mobility (cm²/Vs)	Volume carrier concentration ($\times 10^{18}$cm^{-3})	Surface carrier concentration ($\times 10^{14}$cm^{-2})
0	650.20	1.17	8	2.45
1	198.50	6.70	4.70	1.41
2	4014	0.98	1.57	1.26
3	8801	4.28	0.17	0.06
4	1572	0.64	6.24	1.25
5	3431	0.18	10.20	2.05
6	4260	6.12	0.24	0.12

3.1.7. Figure of Merit

The quality of transparent conductive oxides (TCO) films is generally affected by optical and electrical properties. To have a relation linking these properties, we have used the figure of merit (ϕ) which correlates between optical transmission (T) and electrical resistivity (ρ). Figure III.11 shows the calculated ϕ versus zinc doping concentration.

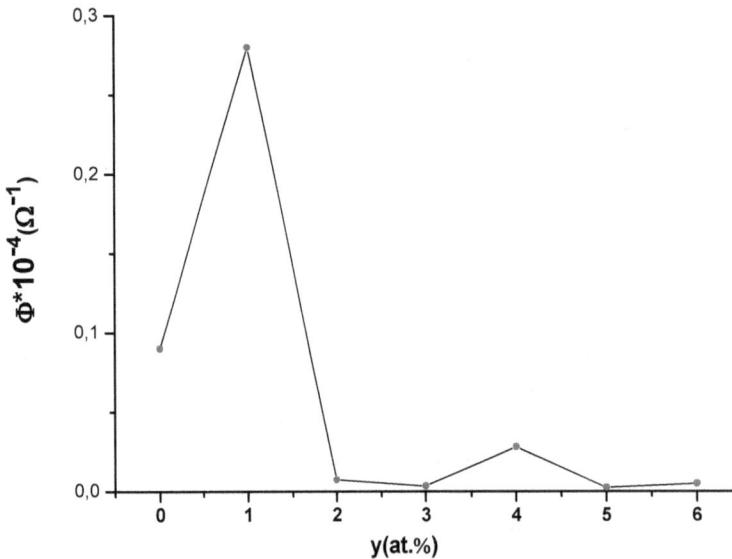

Figure III.11: Figure of Merit (Φ) of zinc doped indium oxide thin films for different atomic concentrations: $y = \frac{[Zn^{2+}]}{[In^{3+}]})_{sol} = 0, 1, 2, 3, 4, 5$ and 6 at.%

After doping, the figure of merit increases from 0.09 x 10^{-4} Ω^{-1} to 0.28 x 10^{-4} Ω^{-1} for respectively undoped and In$_2$O$_3$:Zn (1 at.%). Beyond y equals to 1 at.%, ϕ decreases significantly reaching a minimum value of 0.0023 x 10^{-4} Ω^{-1}. This result may be a consequence of the increase of resistivity by an order of magnitude for y greater than 1 at.%. The greater value of ϕ is obtained for In$_2$O$_3$:Zn (1 at.%) which correspond to the best crystalline quality films as shown is Figure III.1.

3.2. Effect of heat treatment on In_2O_3:Zn (1 at.%) thin films

In order to enhance the physical properties of optimized thin layers, a heat treatment was applied on In_2O_3:Zn (1 at.%) thin films. The annealing was carried out under nitrogen atmosphere at 250 and 450 °C during 1 and 2 hours.

3.2.1. Structural analysis

The XRD spectra of as deposited and annealed In_2O_3:Zn (1 at.%) thin films under nitrogen atmosphere at different temperatures (250 °C and 450 °C) during 1 and 2 hours are shown in Figure III.12. Reflection peaks were very narrow indicating a good crystallinity. The film diffractogram shows one main peak located at approximately $2\theta=35°$ assigned to (400) characteristic of body centered cubic structure. After annealing, there is no change of the preferred orientation and crystalline structure. An improvement of the crystallinity was observed for 250 and 450 °C during 2 hours which means that increase of annealing durations contributes to a reduction of grain boundaries leading to a compactness of the surface and then diffraction along (400) plan increases. Similar results have been observed for annealed aluminum doped indium sulphide thin films. The heat process was performed under nitrogen atmosphere on chemically deposited thin films for 200, 300 and 400 °C [17]. The best crystalline quality is attributed to film heated at 450 °C for 2 hours compared to all other layers. We can conclude that at higher annealing temperature (450 °C), the duration of heat treatment greatly influences on crystalline quality of thin layers. Further, one can conclude from XRD patterns that zinc doped indium oxide thin films is stable at higher annealing temperatures since there is no a structural change that occurs during heat treatment process.

On the other hand, the diffractograms do not show the appearance of secondary phases, it allows us concluding that annealing significantly influences the crystallinity without modification of chemical composition. The grain size (d) of annealed In_2O_3:Zn (1 at.%) thin films were determined from the Debye-Sherrer formula applied to (400) preferred orientations. The results are listed in Table III.5

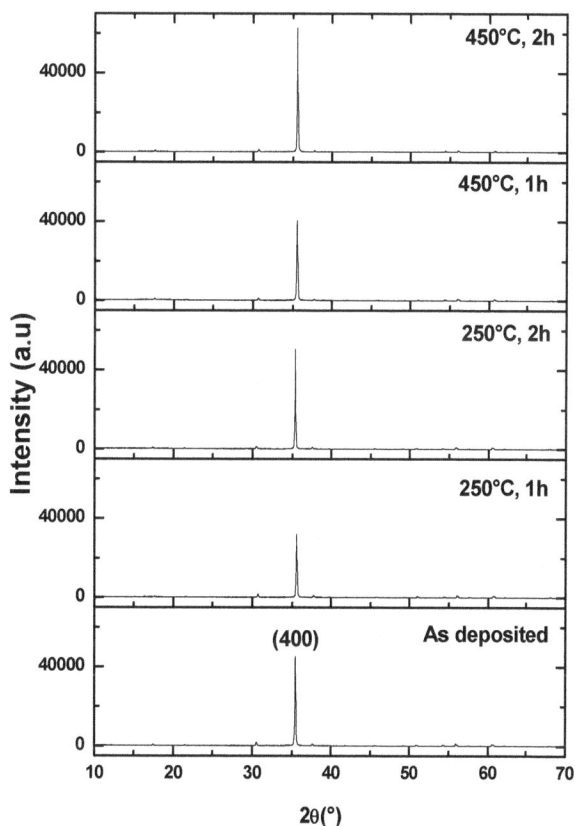

Figure III.12: XRD spectra of In_2O_3:Zn (1 at.%) annealed at different annealing conditions

Table III.5: Grain sizes of In_2O_3:Zn (1 at.%) for different annealing conditions

Annealing conditions	Grain size (nm)
As deposited	94
250°C, 1h	70
250°C, 2h	103
450°C, 1h	82
450°C, 2h	121

We found an initial decrease of the grain size from 94 to 70 nm for as deposited and films annealed at 250°C during 1 hour, respectively. Then d increases sharply to 103 nm for 250°C during 2 hours. By increasing the annealing temperature and duration to respectively 450 °C and 2 hours, the grain size reaches a maximum value of about 121 nm. This result is in good agreement with XRD analysis. Indeed, zinc doped indium thin films oxide annealed at 450°C for 2 hours have the best crystalline quality as mentioned in Figure III.12. This enlargement of the grain size may be due to coalescences of grains at higher annealing temperatures as mentioned by Sengupta *et al.* [18]. A similar result is obtained by Yuan *et al.* [26], where an enhancement of the In_2O_3 films crystallinity is followed by an increase of the grain size after annealing in vacuum.

3.2.2. Morphological properties

AFM images of as deposited and annealed indium oxide thin films are illustrated in the Figure III.13. The grain shapes and forms are changed as function of annealing temperatures. In fact, the surface topography is not the same for different annealing conditions. On the other hand, the surface of all films is dense. On notes that the grain size tends to be larger for films annealed at 250 and 450 °C during 2 hours compared to all other layers. This last funding support structural results. The variation of the RMS roughness as deposited of annealed zinc doped indium oxide thin films for y = 1 at.% is shown in Figure III.14. One notes that the RMS decreases for films annealed during 2 hours regardless of the annealing temperature. The RMS exhibits a minima value (35.2 nm) for 450 °C during 2 hours compared to all other films. For these annealing conditions, layers have the best crystallinity. A similar result was obtained for annealing nitrogen doped zinc oxide [19]. This significant decrease may be caused by the effect of suppression of agglomerates [20] or to the diffusion of little quantities of indium oxide compound to the inside of the film during the annealing process [21].

3.2.3. Optical properties:

The transmission and reflection spectra of as deposited and annealed In_2O_3:Zn (1 at.%) thin films are shown in Figure III.15. All layers present fringes interferences in transparency zone which are characteristics of uniform thickness and homogenous films. After heat treatment process, an improvement of transmission is noted in the visible region for T_a equals to 250 (2 hours) and 450 °C (2 hours). This improvement of transmission may be due to the enhancement of crystallinity [22] as mentioned in the structural analysis. In fact, the improvement of crystalline quality of In_2O_3 thin films is followed by a reduction of light scattering leading to an increase of transmission [22].

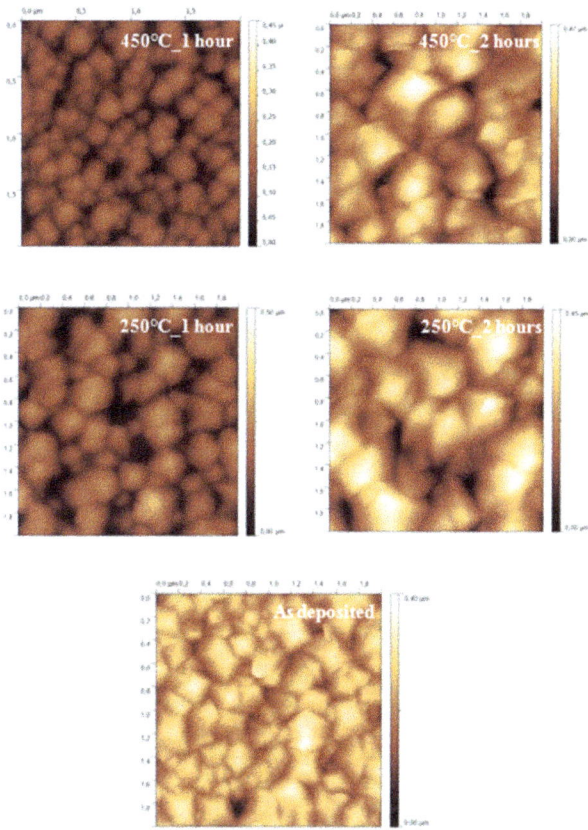

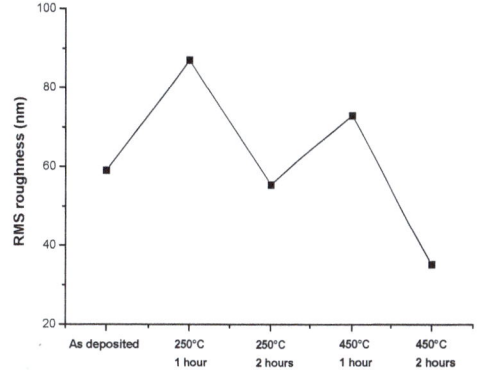

Figure III.13: AFM images of as deposited and annealed In_2O_3:Zn (1 at.%) thin films for different annealing conditions

Figure III.14: RMS roughness of zinc doped In_2O_3 thin films for different annealing conditions

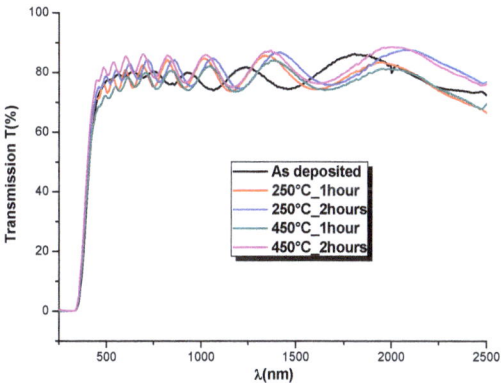

Figure III.15: Transmission spectra of annealed In_2O_3:Zn (1 at.%)

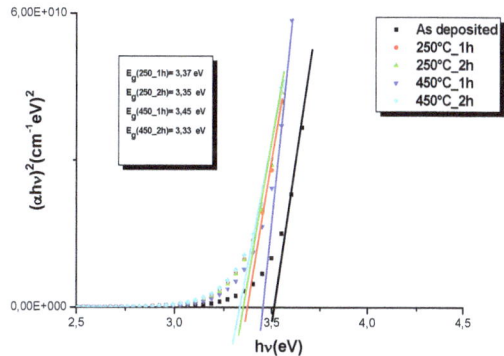

Figure III.16: $(\alpha h\nu)^2$ versus (hν) of In_2O_3:Zn (1 at.%) thin films for different annealing conditions

The variation of $(\alpha h \nu)^2$ versus photon energy (hν) is shown in Figure III.16. The straight line of the films over the wide range of photon energy indicates the direct transition type. The direct band gap energy was obtained by extrapolating the linear part of the Tauc plot curves to intercept the energy axis (at $\alpha h \nu = 0$). We found that the band gap energy tends to be lower after annealing. In fact, E_g is in the range of [3.33-3.45] eV after heated treatment. This decrease in E_g after heat treatment may be explained as fellow: The annealing of crystal defects leads to a decrease in donor concentration which may be inducing an increase of the band gap energy [23]. On the other hand, such decrease of band gap energy may be due to structural modification during heat treatment process [24].

3.2.4. Electrical properties:

Electrical properties were also enhanced by heat treatment. Table 6 shows electrical properties of annealed In_2O_3:Zn (1 at.%). It is observed a significant decrease of the electrical resistivity by two orders of magnitude from 198.50 x 10^{-3} Ω.cm to 2.82 x 10^{-3} Ω.cm for respectively undoped and layers annealed at 250 °C during 2 hours. At 450 °C and for different annealing durations, a gradual increase of the electrical resistivity is revealed followed by a reduction of the carrier concentration compared with films annealed at 250 °C. A sharp increase of the carrier concentration from 4.70 x 10^{18} cm^{-3} to 2.30 x 10^{20}cm^{-3} is shown for respectively as grown and films annealed at 250°C for 2 hours. This enhancement of N_v confirms that annealed films becomes highly degenerate (N_v in the order of 10^{20} cm^{-3}). Also, Yuan *et al.* [25] have explained that the reduction of resistivity after heat treatment may be attributed to an increase of volume carrier concentration. Similar results have been obtained for annealed In_2O_3 thin films in vacuum. In fact, Yuan *et al.* [25] have been shown that annealing In_2O_3 thin films at 500°C under vacuum for 1 hour leads to a decrease of electrical resistivity (1.05 x10^{-3} Ω.cm) and an increase of the carrier concentration (2.80 x 10^{20} cm^{-3}). Finally, the Hall effect measurements show that films annealed at 250 °C during 2 hours exhibit the best electrical properties.

Table III.6: Electrical parameters of as grown and annealed In_2O_3:Zn (1 at.%) at different temperatures and durations

Annealing temperatures	Resistivity (x10^{-3} Ωcm)	Hall mobility (cm²/Vs)	Volume carrier concentration (x10^{18}cm^{-3})
As grown	198.50	6.70	4.70
250°C, 1h	33.46	0.93	201
250°C, 2h	2.82	0.96	230
450°C, 1h	96.59	1.07	60.50
450°C, 2h	13.68	0.48	95.90

3.2.4. Figure of Merit:

The figure of merit ϕ versus annealing temperatures (T_a) and durations for y equals to 1 at.% is shown in Figure III.17. ϕ of annealed thin layers is greater than as deposited one. This result leads to confirm that the heat treatment process enhances the quality of elaborated TCO layers. In fact, ϕ increases sharply from 0.28 x 10^{-4} Ω^{-1} to 25.86 x 10^{-4} Ω^{-1} for respectively as deposited and films annealed at 250 °C during 2 hours. This result may be due to the significant decrease of the resistivity and enhance of transmission after annealing. Hence, In_2O_3:Zn (1 at.%) thin films heated at 250 °C for 2 hours are the most promising TCO among our elaborated In_2O_3 thin layers.

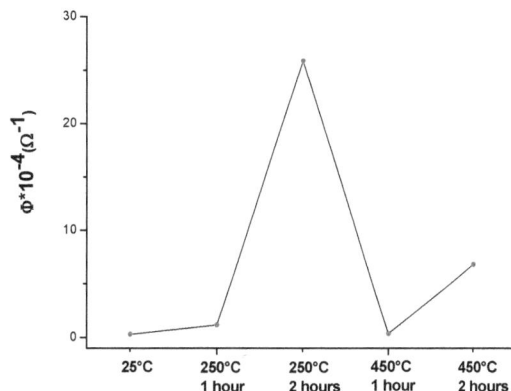

Figure III.17: Figure of Merit (Φ) of In_2O_3:Zn (1 at.%) as a function of annealing temperatures and for different durations.

3.2.5. Conclusion

Zinc doped indium oxide thin films have been prepared using spray pyrolysis. XRD analysis show that preferred orientation is changed after doping. The best crystallized In_2O_3:Zn were obtained for y equals to 1 at.%. The grain size was found to be maximum for In_2O_3:Zn (1 at.%) thin films reaching 93.4 nm. The maximum value of RMS roughness was obtained at this doping level. After doping a slight decrease in the transmission was observed but approximately all films are highly transparent (T in the order of 80 %). The band gap energy was around the theoretical TCO band gap material (3.5 eV) for films doped up to 4 at.%. Hall Effect measurements show that a minimum of electrical resistivity in the order of 198.50 x 10^{-3} Ω.cm was obtained for y equals to 1 at.%. According to the figure of merit results, the best physical properties was illustrated for a doping ratio equal to 1 at.%. Then a heat treatment was applied on this layer. An enhancement of crystallinity was obtained at higher annealing temperatures (450 °C for 2 hours). A further decrease of resistivity, followed by an increase of transmission, is shown after heat treatment at 250°C reaching a minimum value of 2.82 x 10^{-3} Ω.cm. The figure of merit of annealed thin layers is greater than as deposited one. This result leads to confirm that the heat treatment process enhances the quality of elaborated TCO layers. As a conclusion, the physical properties of In_2O_3 thin films are globally enhanced by zinc doping and heat treatment. All these experimental results leads to conclude that annealed In_2O_3:Zn (1 at.%) can be used as transparent conductive oxide material in many optoelectronic devices.

3.2.6. References

[1] M.Jothibas, C.Manoharan, S.Ramalingam, S.Dhanapandian, M.Bououdina; Spectrochimica Acta Part A: Molecular and Biomolecular Spectroscopy 122 (2014) 171–178
[2] S. Louidi, F.Z. Bentayeb, W. Tebib, J.J. Suñol, A.M. Mercier, J.M. Grenèche; Journal of Non-Crystalline Solids 356 (2010) 1052–1056.
[3] S. Parthiban, V.Gokulakrishnan, K.Ramamurthi, E.Elangovan, R.Martins, E. Fortunato, R.Ganesan; Solar Energy Materials & Solar Cells 93 (2009) 92–97.

[4] N. Beji, M. Souli, M. Ajili, S. Azzaza, S. Alleg and N. Kamoun Turki; Superlattices and Microstructures 81 (2015) 114–128.

[5] Y-R.Lyu, T-E.Hsieh; surface and coating technology 231 (2013) 219-223.

[6] Jiaxiang Liu, Da Wu, Shengnan Zeng; journal of materials processing technology 209 (2009) 3943–3948.

[7] H. Baqiah, N.B. Ibrahim, M.H. Abdi, S.A. Halim; Journal of Alloys and Compounds 575 (2013) 198–206.

[8] S.Cho; Microelectronic Engineering 89 (2012) 84-88.

[9] N.G. Pramod, S.N.Pandey; Ceramics International41(2015)527–532.

[10] D. Beena, K.J. Lethy, R. Vinodkumar, V.P. Mahadevan Pillai, V. Ganesan, D.M. Phase, S.K. Sudheer; Applied Surface Science 255 (2009) 8334–8342

[11] S. Belgacem et R. Bennaceur, Revue. Phys. Appl. 25, 1245-1258 (1990).

[12] H.Khallaf, G.Chai, O.Lupan, L.Chow, H.Heinrich, S. Park and A.Schulte; Phys. Status Solidi A 206, No. 2, 256 – 262 (2009).

[13] V. Senthilkumar, P. Vickraman, M. Jayachandran, C. Sanjeeviraj; Vacuum 84 (2010) 864–869.

[14] S. Kaleemulla, A.Sivasankar Reddy, S.Uthanna and P. Sreedhara Reddy; Optoelectronics and Advanced Materials – Rapid Communications Vol. 2, No. 12, December 2008, p. 782 – 787.

[15] P.Wu, Q.Li, C.X. Zhao, D.L. Zhang, L.F. Chi and T.Xiao; Applied Surface Science 255 (2008) 3201–3204.

[16] A.V. Moholkar, S.M. Pawar, K.Y. Rajpure, V. Ganesan and C.H. Bhosale, Journal of Alloys and Compounds 464 (2008) 387–392.

[17] "Etude de l'effet du dopage et du traitement thermique sur les propriétés optoélectroniques des couches minces d'In$_2$S$_3$ utilisées comme fenêtre optique dans les dispositifs photovoltaïques" Kilani Mouna; Thesis suuported at Montpellier University. 2013

[18] J. Sengupta, A. Ahmed and R. Labar; Materials Letters109 (2013) 265–268.

[19] S. Dhara, P.K. Giri; Thin Solid Films 520 (2012) 5000–5006.

[20]M.Devika, N. Koteeswaraeddy, K.Ramesh, KR.Gunasekhar, G.Esr, k.Ramaakrishnadd; Semiconductor science and technology, 21, 1125 (2006).

[21] H. Mahfoz Kotb, M.A. Dabban, A.Y. Abdel-latif, M.M. Hafiz; Journal of Alloys and Compounds 512, 115 (2012).

[22] V. Senthilkumar, P. Vickraman; Current Applied Physics 10 (2010) 880–885.

[23] O. Lupan, T. Pauporte, L. Chow, B. Viana, F. Pell, L.K. Ono, B. Roldan Cuenya, H. Heinrich; Applied Surface Science 256 (2010) 1895–1907.

[24] C. Liu, Z. Xu, Y. Zhang, J. Fu, S. Zang, Y. Zuo; Materials Letters139(2015)279–283.

[25] Z. Yuan, X. Zhu, X. Wang, X. Cai, B. Zhang, D. Qiu, H. Wu; Thin Solid Films 519 (2011) 3254–3258

4. Effects of molybdenum doping and heat treatment on physical properties of indium oxide thin films

As known, molybdenum doping and heat treatment process enhance significantly the physical properties of indium oxide thin films (IMO). Indeed, several papers have investigated physical properties of IMO thin layers [1-4]. In the chapter II and III, we have successfully synthesized promising TCO material with an acceptable resistivity ($\sim 10^{-3}$ Ω.cm) and optical transmission (~75%) but we have obtained a low Hall mobility compared to other authors. So, the aim of this chapter is to grow In_2O_3 thin films with a resistivity in the order of 10^{-4} $\Omega.cm$, high transmission and high hall mobility. As known, conventional ITO films have low resistivity value (10^{-4} $\Omega.cm$) and a high rate of transmission in the visible range. However, ITO has low transmission in the near infrared region. In fact, the significant decrease of the resistivity of ITO layers is attributed to the large density of carrier concentration ($10^{21}cm^{-3}$) which reduces abruptly the transmission of the ITO in NIR region [5]. Achieve combination between high transmissions with low resistivity still up to date a challenge for a promising TCO compound especially In_2O_3 films. In order to resolve this problem, various metals can be proposed as a doping element. Molybdenum atom appears to be one of the most and efficiency doping to achieve such challenge. In fact, due to the difference between valence band, Mo atoms lead to release three free electrons in crystal lattice [6-8] and therefore the resistivity decreases drastically. We can thus combine between high mobility and low resistivity and high transmission over the transparency zone. The same methodology used in previous chapter was adopted in the present one. To investigate the molybdenum doping effect on physical properties of indium oxide, the molybdenum concentration in the sprayed solution y was varied from 0 to 7 at.% (y = $[Mo^{6+}]$/ $[In^{3+}]$= 0, 1, 3, 5 et 7 at.%). Then optimized IMO thin films will be annealed under nitrogen atmosphere by varying annealing time and temperature. In this chapter, the physical properties of molybdenum doped In_2O_3 thin films using spray pyrolysis technique has been studied in part A. In the next part, the optimized IMO thin films will be annealed under nitrogen atmosphere by varying annealing time and temperature. So, the physical properties of heated molybdenum indium oxide thin films are investigated.

4.1. Molybdenum doping effect

4.1.1. Structural analysis

The XRD patterns of molybdenum doped indium oxide thin films for different concentrations y (y = $\frac{[Mo^{6+}]}{[In^{3+}]}$ = 0, 1, 3, 5 and 7 at.%) are shown in Figure IV.1. It is noted that doped In_2O_3 films crystallize into body centered cubic structure (bcc) as was the case of undoped layers (JCPDS Card n°06-0416). A main diffracted peak is located at approximately $2\theta = 30.50°$ corresponding to the (222) plan. It is also remarked a sharp increase of the strongest orientation (222) for y =3 at.% compared to undoped films. This result corresponds to an improvement of layer orientation and an enhancement of crystal quality by doping process. This result can be explained as follow: Up to 3 at.%, Mo^{6+} ions substitutes In^{3+} ions in the crystal lattice. A similar behavior was obtained by Kaleemulla et al. [7] for IMO films elaborated by an activated reactive evaporation method.

Beyond this level, a gradual decrease of (222) peak intensity is denoted compared to that observed for y = 3 at.%. So a deterioration of film's crystallinity is observed. Since films grown at molybdenum concentration equal to 7at.% are relatively more crystalline than those of undoped ones. This deterioration of crystallinity beyond 3 at.% may be attributed the formation of stresses caused by the difference in ion size between indium and molybdenum [9]. XRD results deal with those observed by Parthiban et al. [4]. In fact, they have revealed that doping with molybdenum enhances structural properties of indium oxide thin films. However, it should be mentioned that there is no change in the preferred orientation peak in our case in contrast with that noted in other papers [4,7]. The variation of diffracted angle 2θ

along (222) plan is exhibited in the inset of Figure IV.1. It is observed an increase of the diffracted angle for doped films compared to undoped ones. According to the Bragg equation (**2d sinθ=nλ (1)**), the increase in the diffracted angle is followed by a decrease of interplanar spacing (d) which lead to a contraction of crystal lattice. The grain size (d) of indium molybdenum oxide thin films was calculated for different doping concentrations using the Sherrer formula applied to (222) preferred orientation [10]. The dislocation density (δ_{disc}) and the number of crystallites per unit surface area (n_c) were calculated for the (222) preferred orientation for each doping concentration y [10].

The deduced parameters are summarized in Table IV.1. The grain size (d) tends to be larger after doping. In fact, d is in the range of [90 - 155] nm for doped layers. However, d is around 71 nm for undoped films. In_2O_3:Mo (3 at.%) thin films exhibit a maximum value of d which is equal to 155 nm. This results is in good accordance with that the best crystallinity is obtained for y equals to 3 at.%. The dislocation density (δ_{dis}) decreases after doping and it reaches a minimum value of about 0.41 x 10^{10}cm^{-2} for In_2O_3:Mo (3 at.%) compared to all other films. One notes that for this doping level, indium oxide thin films exhibit the lowest value of crystallite number n_c which is of around 1.88 x 10^{10}cm^{-2}. This last result is in agreement with that In_2O_3:Mo (3 at.%) films have the best crysllalline quality.

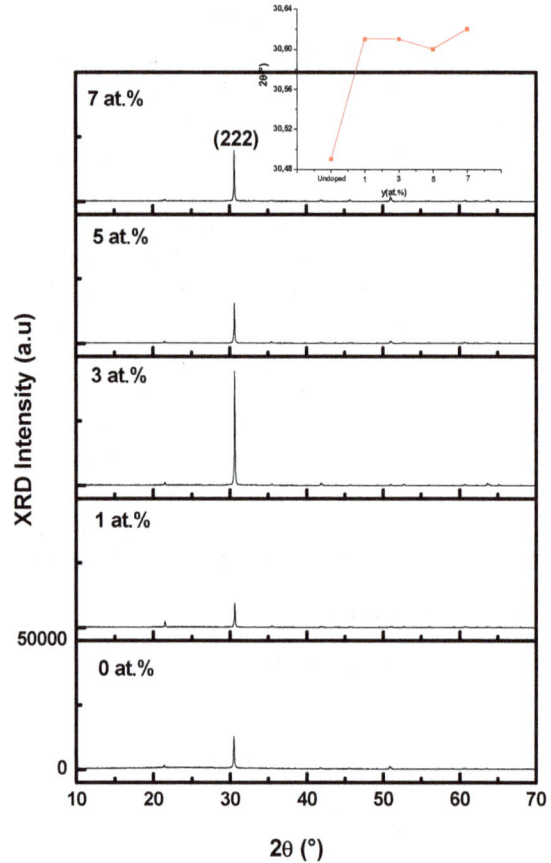

Fig. IV.1: XRD patterns of molybdenum doped indium oxide thin films for different molybdenum concentration y, **Figure inset**: Variation of diffracted angle 2θ along (222) plan as a function of molybdenum concentration y.

Table IV.1: Average grain size (d), dislocation density (δ_{dis}) and crystallites number (n_c) of molybdenum doped indium oxide thin films deduced from XRD spectra

y(at.%)	2θ(°)	FWHM(°)	d(nm)	δ_{dis} x 10^{10}(cm^{-2})	n_c x 10^{10} (cm^{-2})
0	30.58	0.1710	71	1.98	21
1	30.61	0.1550	90	1.23	9.60
3	30.61	0.1359	155	0.41	1.88
5	30.60	0.1440	115	0.76	4.60
7	30.62	0.1385	138	0.52	2.66

4.1.3. Rietveld Analysis:

XRD patterns of molybdenum doped In_2O_3 thin films were refined by MAUD program [11] which is based on Rietvled method [11]. Figure IV.2 shows Rietveld refinements of In_2O_3 thin films as function of atomic concentration y. The deduced results (lattice parameters a, grain size d, microstrain $< \sigma^2 >^{1/2}$ and phases amounts) are summarized Table IV.2. Fitted XRD spectra as shown in Figure IV.2 confirm the presence of two phases: In_2O_3 and MoO_3. The secondary phase MoO_3 has been detected after doping for $y \geq 3$ at.% with low amounts compared to In_2O_3 one. This phase crystallizes into orthorhombic structure. One can remark from Table 1 that molybdenum oxide phase increases as a function of doping concentration y and it reaches a maximum value of 8.25 % for 7 at.%. The absence of undesirable phase for 1 at.% can be explained as follow: Molybdenum element has been totally incorporated in In_2O_3 crystal lattice. On the other hand, the lattice parameter "a" corresponding to the cubic centered In_2O_3 structure is decreasing after doping revealing thus a contraction of the crystal lattice compared to undoped films. This result may be attributed to the fact that the ionic radius of Mo^{6+} (62 pm) is smaller that In^{3+} one (81 pm). The grain size d as a function of doping concentration y shows a maximum value in the order of 153 nm for y equals to 3 at.% revealing the improvement of crystalline quality. The grain size values deduced from Rietveld refinements match well with those calculated using Debye formula. The microstrain $< \sigma^2 >^{1/2}$ increases after doping except for y = 3 at.%. In fact, for this doping level, the microstrain reaches a minimum value of 0.0179 % corresponding to the best crystallinity which is in good agreement with XRD analysis.

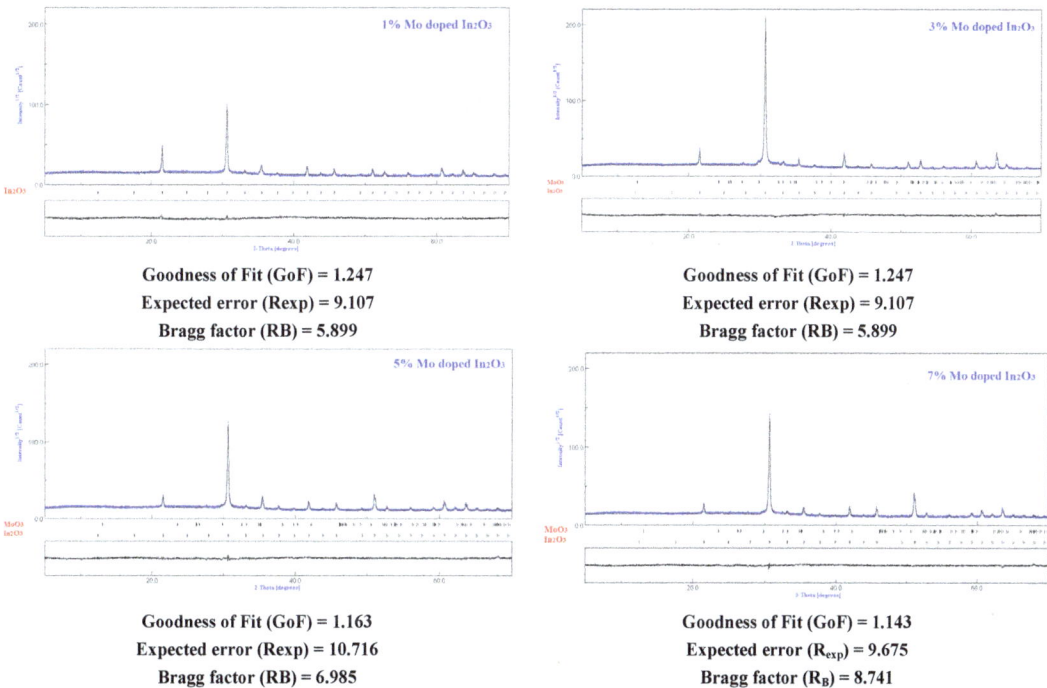

Goodness of Fit (GoF) = 1.247
Expected error (Rexp) = 9.107
Bragg factor (RB) = 5.899

Goodness of Fit (GoF) = 1.247
Expected error (Rexp) = 9.107
Bragg factor (RB) = 5.899

Goodness of Fit (GoF) = 1.163
Expected error (Rexp) = 10.716
Bragg factor (RB) = 6.985

Goodness of Fit (GoF) = 1.143
Expected error (R_{exp}) = 9.675
Bragg factor (R_B) = 8.741

Figure IV.2: Rietveld refinement of XRD spectra of molybdenum doped indium oxide thin films for different doping concentration y. The difference between the experimental (dots) and the calculated (full line) spectrum is given below

Table IV.2: Phases and microstructure parameters determined by MAUD software for molybdenum doped In_2O_3 thin films for $y = \frac{[Mo^{6+}]}{[In^{3+}]})_{sol} = 1, 3, 5$ and 7 at.%

y(at.%)	phase	d (nm) (±1.5)	$<\sigma^2>^{1/2}$ (%) (±0.01)	Lattice parameter (A°) (±0.0002)			Amount (%) (± 1)
				a	b	c	
0	In_2O_3	/	0.0208	10.1280			
1	In_2O_3	105	0.0214	10.1247	/	/	100
3	In_2O_3	153	0.0179	10.1250	/	/	96.31
	MoO_3	99.98	0.102	3.93822	13.9313	3.66085	3.69
5	In_2O_3	101	0.0254	10.1242	/	/	93.95
	MoO_3	99.99	0.143	3.94026	13.9459	3.66258	6.05
7	In_2O_3	103	0.0221	10.1241	/	/	91.75
	MoO_3	97.67	0.135	3.94025	13.9459	3.66257	8.25

4.1.4. Morphological analysis

The study of molybdenum doping on morphological properties of In_2O_3 thin films was investigated using atomic force microscopy (AFM) in tapping mode. The RMS roughness was then extracted. AFM images of IMO thin films are displayed in Figure IV.3. In the case of undoped films, the shape of the grains covering the sample was pyramidal forms. For doped films, the AFM images reveal the appearance of big agglomerates of crystallites over a dense texture. We also observe an increase of crystallite size after doping. This result is in good agreement with structural analysis where it was noted that grain size tends to be larger for doped films. It is also remarked that density and size of agglomerated grains differs for different molybdenum doping ratio. The RMS roughness values for different doping concentration y are shown in Figure IV.4. It is noted a remarkable decrease of the RMS after doping process. In fact, the RMS is in the order of [17.3 - 32] nm for doped layers. However, the Root mean square is around 48 nm for undoped ones. So, doping with molybdenum enhances the surface uniformity of elaborated layers.

4.1.5. Optical analysis

The transmission and reflection spectra of IMO thin films are presented in Figure IV.5. As observed, the molybdenum has an impact on optical properties of elaborated layers. Indeed, IMO thin films present a multiplied oscillation compared to undoped layers which corresponds to an enhancement of thickness uniformity and films homogeneity after doping. This fact can be attributed to the reduction of RMS roughness values after doping as seen previously. Although the transmission (T) has slightly declined after doping, it remains high of about 80% in visible and near infrared regions. This transmission decrease may be due to the enhanced scattering of photons caused by crystal defects produced by doping or to the absorption of photons by free carriers [12]. Other reporter have founded similar decrease of T by doping process for other doped In_2O_3 films such as In_2O_3:Mo [3], In_2O_3:Zn [12], and In_2O_3:Li [13]. On the other hand, the same phenomenon was observed in previous chapters for iron and zinc doped indium oxide thin films.

0 at.%

1 at.%

3 at.%

5 at.%

7 at.%

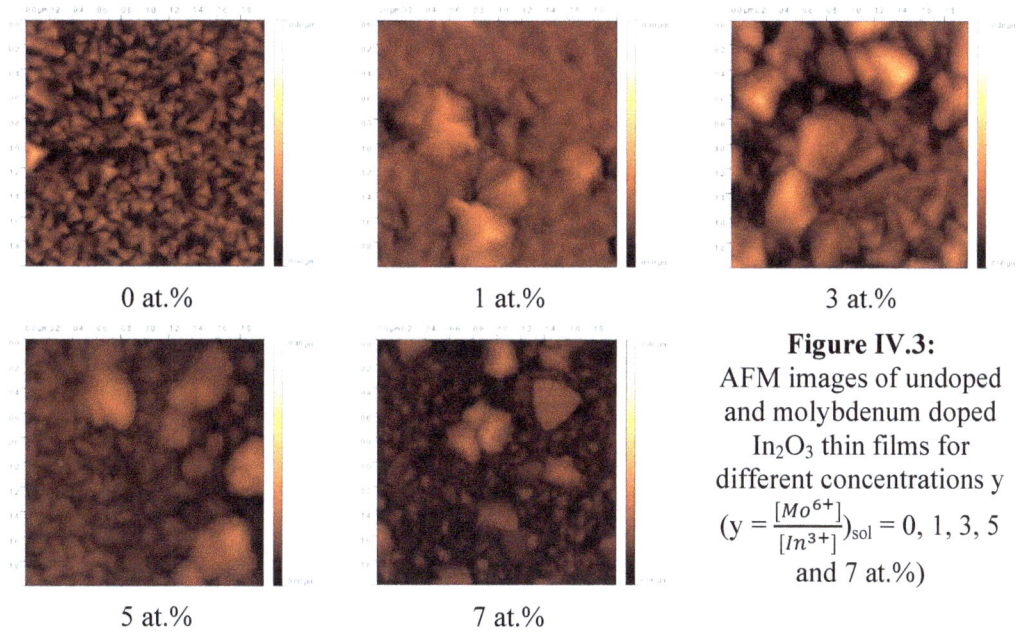

Figure IV.3:
AFM images of undoped and molybdenum doped In$_2$O$_3$ thin films for different concentrations y
$$\left(y = \frac{[Mo^{6+}]}{[In^{3+}]}\right)_{sol} = 0, 1, 3, 5 \text{ and } 7 \text{ at.\%}$$

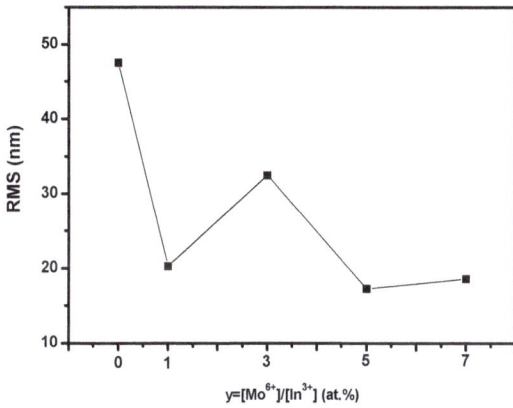

Figure IV.4: RMS roughness values of IMO thin films for different molybdenum concentrations, y

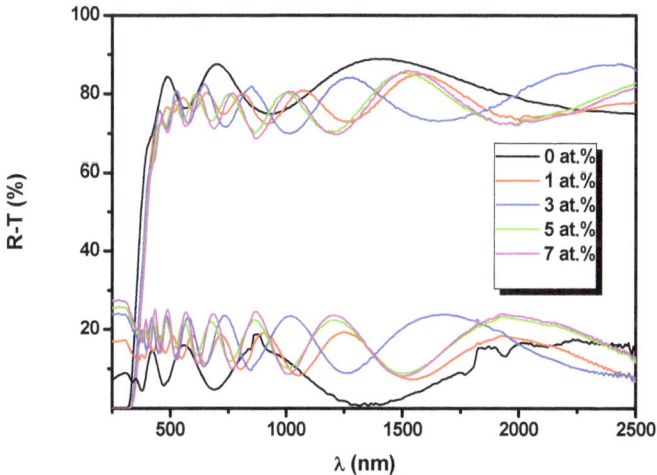

Fig IV.5: Transmission and reflection spectra of IMO thin films for different molybdenum concentration, y

The optical band gap of IMO thin films were calculated as a function of molybdenum doping concentration using Tauc equation [14]. The different results are summarized in Figure IV.7. The bang gap energy E_g is ranging between 3.52 and 3.56 eV for doped films. However, E_g is equals to 3.45 eV for undoped ones.

This expands of energy band gap after doping can be due to the increase of free carrier's concentrations in the films (Table IV.4). It can be also noted that the optical band gap variation is approximately proportional to that of carrier concentration [3].

The enlargement of optical band gap can be explained using Burstein-Moss effect as fellow [4]:

$$E_g = E_{g0} + \Delta E_g^{B-M} \quad (2)$$

where E_{g0} and ΔE_g^{B-M} are respectively intrinsic band energy and Burstein-Moss shift due to the filling of low levels in the conduction band.

The Burstein - Moss shift can be expressed as [4]: $\Delta E_g^{B-M} = \dfrac{h^2}{8m_{vc}^* \pi^2}(3n\pi^2)^{2/3} \quad (3)$

where n and m_{vc}^* are respectively the carrier concentration and reduced mass effective. It is therefore clearly seen that B-M shift is proportionnel to $n^{2/3}$.

On the other hand, Pramod *et al.* [13] have founded an increase of the band gap energy after doping In$_2$O$_3$ with lithium. In fact, they explained that an increase of free carrier concentration may be attributing to an enlargement of the band gap which is in good agreement with our obtained results (Table IV.4)

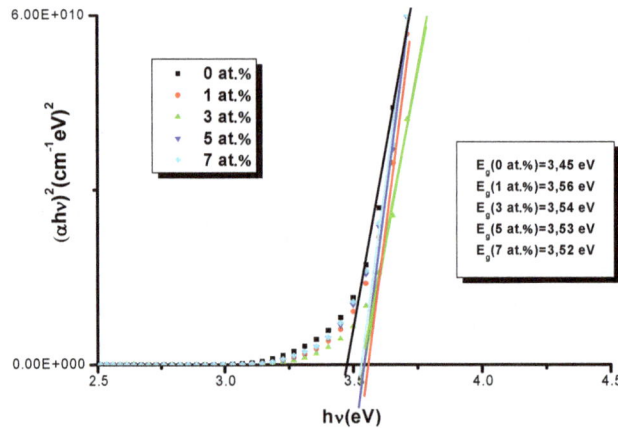

Figure IV.7: Optical band gap of IMO thin films for different molybdenum concentration y

Using the envelope method [15], the refractive index can be calculated for different molybdenum doping concentrations y. The results are given in Figure IV.8. The refractive index n is represented as a function of wavelength in the transparency region extending from visible to near infrared region. As shown, the refractive index increases after doping in the visible range. These n values are in accordance with that the transmission is decreased for doped layers. The highest value of refractive index is attributed for a doping ratio equals to 3 at.% reaching a value of about 1.93 at λ=400 nm compared to undoped films which is near to 1.82. The refractive index showed a Cauchy trend for photon energy smaller than band gap energy as it is revealed from the equation [16]:

$$n(\lambda) = A + \frac{B}{\lambda^2} \quad (4) \quad \text{where A and B are constants}$$

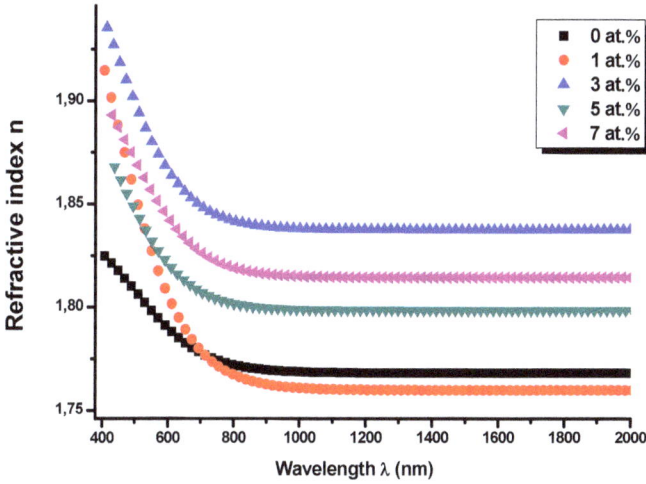

Figure IV.8: Refractive index n as a function of wavelength λ of IMO thin films for different molybdenum concentrations.

Based on refractive index values, the porosity and packing density (p) as a function of wavelength can be deduced for different molybdenum concentrations y according to the following equations [15,17]:

$$n^2 = \frac{(1-p)n_v^4+(1+p)n_v^2 n_s^2}{(1+p)n_v^2+(1-p)n_s^2} \qquad (5)$$

where n_v is the void refractive index (equal to one for air). n_s is the bulk value of refractive index.

$$Porosity = \left(1 - \frac{n^2-1}{n_d^2+1}\right) x\ 100\ \% \qquad (6)$$

where n_d is the refractive index of pore-free In_2O_3 which is equals to 2 [18] and n is the refractive index of the porous In_2O_3 thin films.

All results are shown in Figure IV.9. It is observed that the packing density (p) (Figure IV.9a) is in range of [0.86 - 0.97]. The packing density values are in the same range with those obtained by Senthilkumar *et al.* [17] for ITO films. On the other hand, p exhibits in the visible range highest values for doped layers compared to undoped ones which supports the refractive index studies. One notes that p is maximum for 3 at.% which correspond to the best crystallinity. As seen from Figure IV.9b, the porosity decreases in the visible range for doped films. IMO thin films grown at 3 at.% are the less porous layers which can be the synonym of homogeneous film as revealed from transmission spectra. The oscillator energy E_0 and the dispersion energy E_d were determined using the Wemple model as detailed by Senthilkumar *et al.* [17]. E_0 and E_d were determined from the slope $(E_d E_0)^{-1}$ and the intercept at the origin (E_0/E_d) for each dopig ratio. Figure IV.10 shows the linear variations of $(n^2 - 1)^{-1}$ versus $(hv)^2$ for molybdenum doped indium oxide thin films. The deduced Wemple are listed in Table IV.3. It is revealed from this table that $E_0=2\ x\ E_g$ for molybdenum doped films. However, $E_0 = 1.5\ x\ E_g$ for undoped films. The oscillator energy E_0 increases after doping which is in goods agreement with the expand band gap energy E_g for doped films.

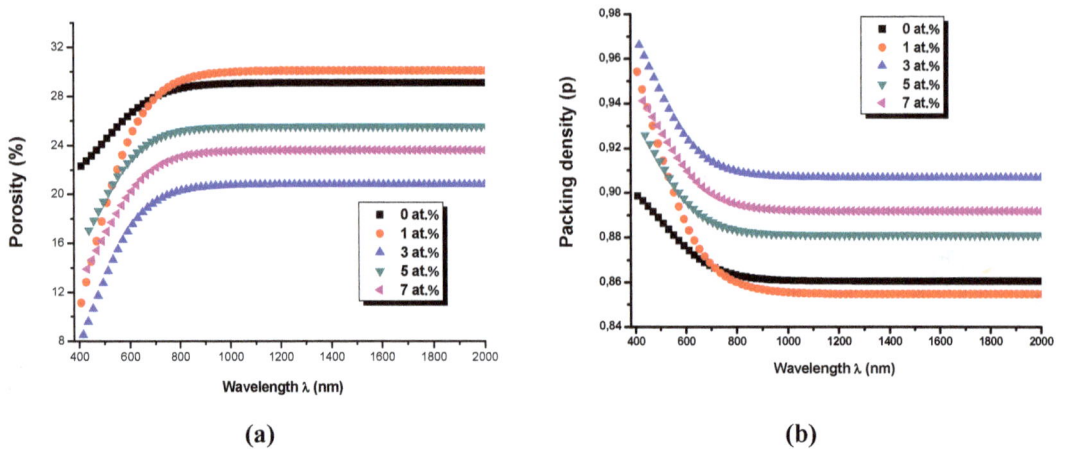

(a)   (b)

Figure IV.9: Porosity and packing density as a function of wavelength λ of IMO thin films at different molybdenum concentration y

Table IV.3: Optical dispersion constants derived from Wemple model of molybdenum doped indium oxide thin films as function of doping concentration y

y (at.%)	E_0 (eV)	E_d (eV)	S_0 (10^{-5} nm^{-2})	λ_0 (nm)
0	5.05	15.02	4.93	245.54
1	6.87	14.73	6.58	180.50
3	6.92	15.70	7.06	179.20
5	7.16	15.21	7.08	173.20
7	7.00	15.30	6.97	177.14

4.1.6. Photoluminescence analysis

The photoluminescence (PL) performed on IMO thin layers for different molybdenum concentrations are displayed on Figure IV.11. Three peaks were detected at wavelength of about 485; 505 and 795 nm. The blue green emission band located in the range of [485-505] nm for doped and undoped films could be due to lattice defects or oxygen deficiencies [19]. On the other hand, Wu *et al.* [20] have reported that PL emissions observed in the visible range may be attributed to the recombination of carrier concentration that occurs between valence band and oxygen vacancies acting as donor levels. Moreover, we observe the same peaks for undoped and molybdenum doped films whatever the doping ratio. It can be so deduced that molybdenum atoms are not responsible of the observed recombination. Then we can assume that molybdenum atoms do not lead to the appearance of new emission peaks but it affects photoluminescence peak intensities.

4.1.7. Electrical properties

Electrical parameters of IMO thin films derived from Hall Effect measurements are summarized in Table IV.4. A considerable decrease of resistivity by two orders of magnitude is observed for a doping ratio equals to 3 at.% reaching a lowest value of about 2.03×10^{-3} $\Omega.cm$ compared to undoped films (650.20×10^{-3} $\Omega.cm$). The decrease of the resistivity can be attributed to the difference between electronic valences. In fact, the introduction of Mo^{6+} ions into indium oxide matrix leads to the release of three electrons in the In_2O_3 lattice which induces the decrease of the resistivity. This reduction may be also attributed to the increase of mobility (μ) and carrier concentration (N_v) by an order of magnitude. Indeed, it is revealed for y = 3 at.% an increase of Hall mobility from 1.17 cm^2/Vs to 34.67 cm^2/Vs compared to as deposited films. On the same way, volume carrier concentration is increased from 0.8 to 8.85 $\times 10^{19} cm^{-3}$. Such increase of N_v may be related to the improvement of crystallinity and thus contribute to the reduction of trapped sites located at grain boundaries [21]. These obtained values means that molybdenum doping enhances electrical properties of indium oxide thin films. The measured electrical parameters for doped films at y = 3 at.% are in the same range to that reported previously by Parthiban *et al.* [3], i.e. where they mentioned that mobility, resistivity and carrier concentration are 77 cm^2/Vs, 1.8×10^{-3} $\Omega.cm$ and $4.6 \times 10^{19} cm^{-3}$ for IMO films, respectively.

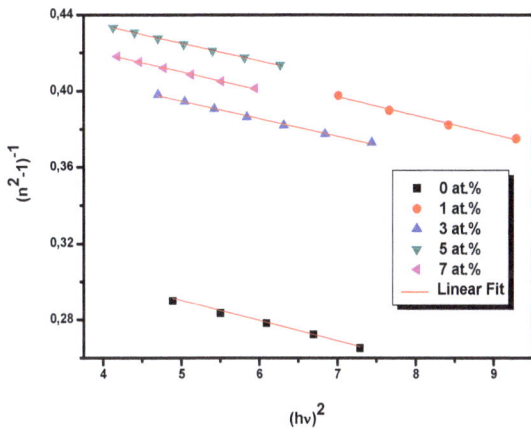

Figure IV.10: Variation of $(n^2-1)^{-1}$ as a function of $(h\upsilon)^2$ of IMO thin films at different molybdenum concentration y

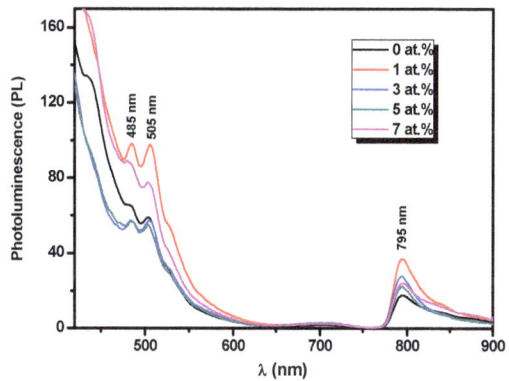

Figure IV.11: Photoluminescence spectra of IMO thin films at different molybdenum concentration y

Table IV.4: Electrical parameters derived from Hall Effect measurements of IMO thin films at different molybdenum concentration y.

$y=[Mo^{6+}]/[In^{3+}]$ (at.%)	Resistivity $\times 10^{-3}(\Omega.cm)$	Hall mobility (cm^2/Vs)	Volume carrier concentration $(\times 10^{19} cm^{-3})$	Surface carrier concentration $(\times 10^{15} cm^{-2})$
0	650.20×10^{-3}	1.17	0.80	0.24
1	19.10×10^{-3}	10.79	3.03	1.06
3	2.03×10^{-3}	34.67	8.85	3.10
5	195×10^{-3}	14.62	2.19	0.78
7	210×10^{-3}	22.11	1.35	4.73

The mobility increase as observed at y=3 at.% can be due to the introduction of little amount of molybdenum doping atoms into the sprayed solution leading thus to the reduction of neutral impurity scattering effects as explained Parthiban *et al.* [3]. On the other hand, Prathap *et al.* [22]. showed that carriers mobility increases as the result of the improvement of crystallite size. Beyond 3 at.%, a gradual increase of resistivity followed by a decrease in mobility achieving the same orders of magnitude to that of undoped films. Parthiban *et al.* [3], have found similar behavior where increasing concentrations of Mo, the doping atoms cannot occupy appropriate sites into In_2O_3 lattice. Considering the ionic radius difference between indium and molybdenum atoms, it can be produced a crystal deformation since the molybdenum may occupy interstitial sites for increasing molybdenum concentration. Therefore, defects such as grain boundaries, crystal defects and scattering of ionized impurities are enlarged for increasing doping ratio and hence the electron mobility is attenuated.

4.1.8. Figure of Merit

The Figure of Merit of IMO thin films for different molybdenum ratio are shown in Figure IV.12. It is revealed an increase of Figure of Merit value by three orders of magnitude for a doping ratio equals to 3 at.% reaching a maximum value of about 32.65×10^{-4} Ω^{-1} compared to undoped layers (0.09×10^{-4} Ω^{-1}). After that, the Figure of Merit decreases drastically and attains similar values to undoped films. It is thus concluded from the present part, the optimized concentration of molybdenum concentration needed for the growth of promising IMO thin films is equals to 3 at.%. The heat treatment affects significantly the structural, morphological, optical and electrical properties of indium oxide thin films. So, a heat treatment was performed on In_2O_3:Mo (3 at.%) under nitrogen gas for 1 and 2 hours at $T_a = 250$ and $450°C$.

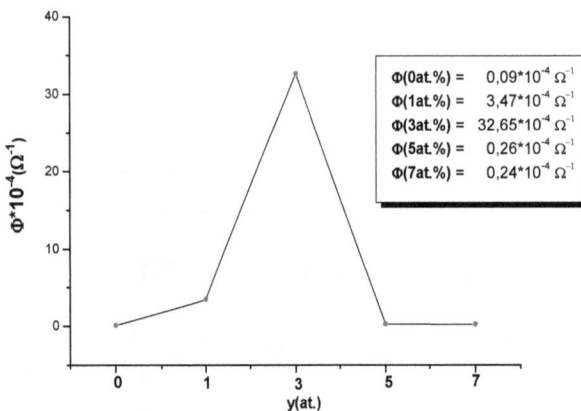

$\Phi(0 at.\%) = 0.09*10^{-4} \Omega^{-1}$
$\Phi(1 at.\%) = 3.47*10^{-4} \Omega^{-1}$
$\Phi(3 at.\%) = 32.65*10^{-4} \Omega^{-1}$
$\Phi(5 at.\%) = 0.26*10^{-4} \Omega^{-1}$
$\Phi(7 at.\%) = 0.24*10^{-4} \Omega^{-1}$

Figure IV.12: Plots of Figure of Merit versus molybdenum concentration y for IMO tin layers

4.2. Effect of annealing on physical properties of optimized IMO thin films

4.2.1. Structural properties

XRD patterns of annealed In_2O_3:Mo (3 at.%) thin films at different conditions are presented in Figure IV.13. The diffraction peaks of annealed films matched well with those of body centered cubic structure as was the case for as deposited films. It is noted a decrease of peak intensity of the preferred orientation along (222) plan for sample annealed at 250°C for 1 hour. For increasing annealing time and keeping annealing temperature constant at 250°C, a gradual increase of peak intensity is noted.

This trend increase is observed at 450°C for 1 hour exhibiting the best crystalline quality after heating. In fact, for this annealing condition, the intensity of the strongest peak is close to as deposited one. This may be explained as the result of rearrangement of atoms in crystal lattice along (222) plan at 450°C for 1 hour. It can be deduced that heat treatment does not affect the preferential direction but it assists the grain growth. After that, a crystalline deterioration is observed although the films are crystalline. The extend of annealing time at higher annealing temperatures (450°C) may contribute to the creation of atoms disorder which may contribute to the orientation of grains into different plan orientation and then the atoms do not grow in the same direction leading to a poor crystallinity. Zhu *et al.* [23] have founded a similar XRD behavior in term of intensity for ZnO annealed in air at 150, 300, 450°C. On the other hand, an annealing treatment has been performed by Shin *et al.* [7] on IMO thin films for annealing temperatures equal to 400, 500 and 600°C under vacuum. The authors reveal the enhancement of crystal quality with heat treatment and the films are grown with (222) texture even after annealing process.

Using Debye Scherer Formula, the grain size of annealed IMO thin films are calculated and the results are summarized in Table IV.5. It is noted a decrease of grain size after heat treatment. In fact, heated films exhibit grain size in the range of [119-139] nm. However, d is in the order of 155 nm for as deposited ones. This last finding may be attributed to the fact that the crystallinity is deteriorated by heating. A similar result was obtained by *Kim et al.* [24] for annealing ZnO:Al in N_2 gas. In fact, they found a degradation of film's crystallinity after heating followed by a decrease of grain size.

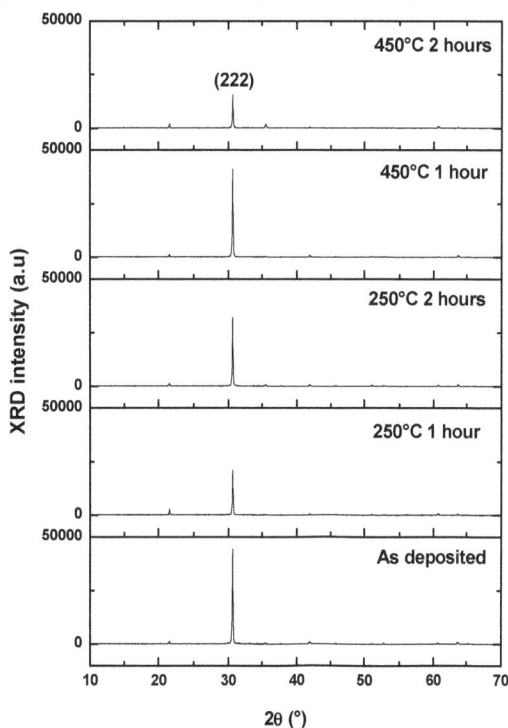

Figure IV.13: XRD spectra of annealed IMO thin films (3 at%) at different annealing conditions.

Table IV.5: Grain size of annealed IMO thin films for y = 3 at%

Annealing conditions	Position of (222) peak [2θ(°)]	FWHM(°)	Grain size (nm)
As deposited	30.61	0.1359	155
250°C, 1h	30.62	0.1392	134
250°C, 2h	30.62	0.1390	135
450°C, 1h	30.64	0.1383	139
450°C, 2h	30.64	0.1427	119

4.2.2. Morphological properties

The surface topography of In_2O_3:Mo (3 at.%) thin films annealed at different conditions is presented in Figure IV.14. The annealing temperature and annealing times affect significantly the surface topography. Indeed, a modification of grain form and size is noted for annealed films since the films exhibit different textures. It can be remarked that the topography surface was disturbed after heat treatment by clusters. The as deposited films show a compact structure with bigger grain size. The grain size is decreased for annealed films. Such result was confirmed previously in XRD analysis. IMO film's surface annealed at 250°C for 1 hour have a non-uniform distribution of grains leading to the non-homogeneity of the surface. For increasing annealing temperature 450°C (1 hour), a noticeable change in surface topography has been occurred. This texture may be due to the diffusion of atoms inside films through grains boundaries which induces a rearrangement of atoms and thus promotes a best growth as observed in XRD spectra.

The RMS roughness was extracted from AFM images and the results are recapitulated **Table IV.6**. It can be remarked that the highest RMS value is obtained for annealed films at 250°C during 2 hours. It is also revealed the increase of roughness values for increasing annealing times. Indeed, the RMS roughness reaches 52 and 27.5 nm instead of 46 and 12 for respectively annealed films at 250 and 450°C. Whereas, for increasing annealing temperature, the films looks smoother and more uniform and hence the RMS decreases significantly for example for annealed films during 1 hour, RMS decreases from 46 to 12 nm for respectively 250 and 450°C.

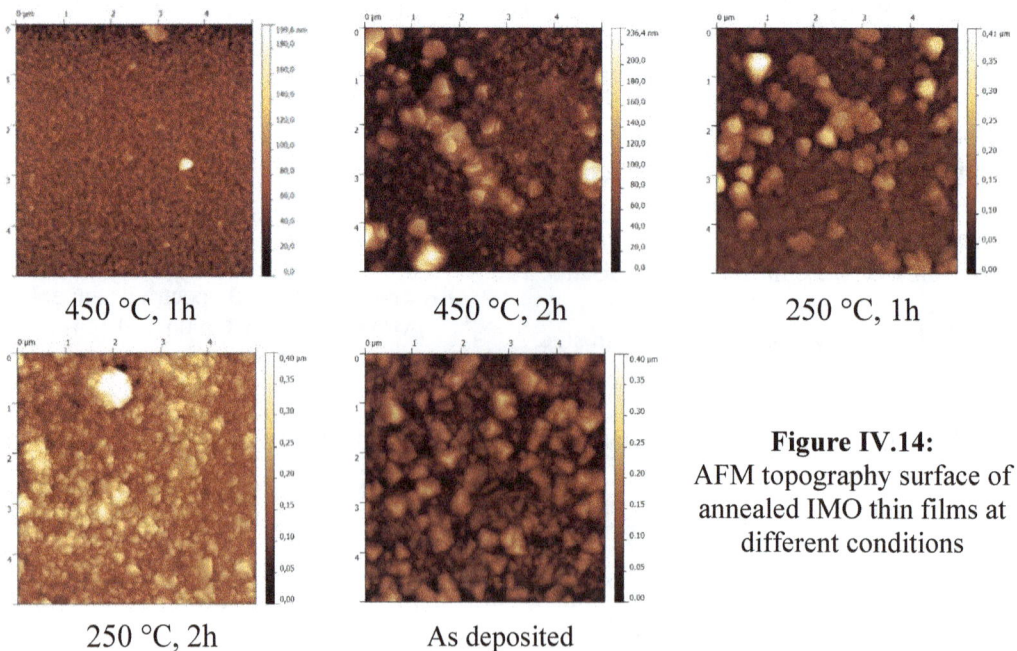

450 °C, 1h

450 °C, 2h

250 °C, 1h

250 °C, 2h

As deposited

Figure IV.14:
AFM topography surface of annealed IMO thin films at different conditions

Table IV.6: RMS roughness values of as deposited and annealed IMO thin films for 3 at%

Annealing Conditions	As deposited	250°C, 1h	250°C, 2h	450°C, 1h	450°C, 2h
RMS (nm)	20	46	52	12	27.5

4.2.3. Optical properties:

Transmission and reflection spectra of annealed IMO thin films for y = 3 at.% are displayed in Figure IV.15. It is noted the presence of interference fringes for all films which justifies the homogeneity and uniformity surface of the films after heat treatment. The average transmission of annealed IMO thin films is of about 75%. It also reaches higher value for films annealed at 450°C for 1 hour corresponding to the lower value of roughness and to the best cristallinity as observed from XRD analysis. This last result can be due to the low carrier concentration as mentioned in the next section (electrical properties). Annealed IMO thin films exhibit higher transmission compared to the conventional ITO thin films which led to conclude that the latter material contain higher carrier concentration than that of annealed IMO thin films. For films annealed at 250°C during 2 hours, it is noted an attenuation of interference fringes in the visible region this can be due to the high surface roughness as observed from AFM images leading to the enhanced of photons scattering caused by crystal defects.

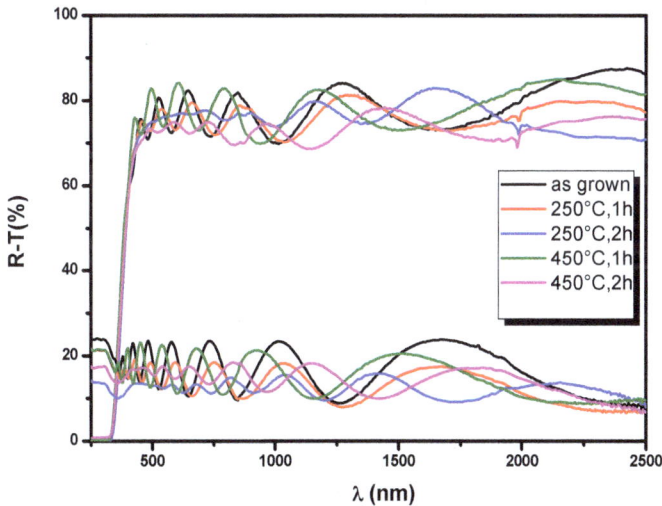

Figure IV.15: Transmission and reflection spectra of annealed IMO thin films (3 at%) at different annealing conditions

The calculated band gap was estimated using Tauc equation and the obtained results are given in table Table IV.7.

Table IV.7: Optical band gap of annealed IMO thin films (3 at.%)

Annealing Conditions	As deposited	250°C, 1h	250°C, 2h	450°C, 1h	450°C, 2h
E_g (eV)	3.54	3.64	3.59	3.65	3.56

An increase of the optical band gap energy has been detected for annealed films compared to as deposited ones. Such decrease in E_g may be related to smaller grain size obtained after annealing [24]. However, by maintaining constant the annealing temperature, E_g does not practically changes as function of annealing times.

4.2.4. Electrical properties

Electrical properties of annealed IMO (3 at.%) are determined using Hall effect measurements. All results are summarized in Table IV.8. It can be observed that the resistivity of annealed IMO thin films at 250°C for 1 hour decreases by an order of magnitude compared to as deposited ones from 2.03 x 10^{-3} Ω.cm to 9.58 x 10^{-4} Ω.cm. The expand in annealing time by maintaining annealing temperature constant at 250°C leads to further decrease of ρ and reach a minimum value of about 5.55 x 10^{-4} Ω.cm. At this annealing condition a high mobility value in the order of 42.69 cm²/Vs has been recorded which can be ascribed to the lower scattering frequency as reported by Cho *et al.* [25]. According to the above equation, the lower values of resistivity can be provided for higher carrier concentration [25]:

$$\mu = \frac{1}{ne\rho} = \frac{e}{fm^*} \quad (7)$$

where: n: carrier concentration; e: charge of the electron; ρ: resistivity; f: free electron scattering frequency and m^*: effective mass of the electron

For increasing annealing temperature (T = 450 °C), the resistivity rises again. This may be related to the growth of oxygen vacancies during heat treatment process [16], which probably led to the scattering of ionized impurities [16].

Table IV.8: Electrical parameters of annealed IMO thin films for 3 at.% at different annealing conditions.

Annealing conditions	Resistivity (Ω.cm)	Hall mobility (cm²/VS)	Volume carrier concentration (cm⁻³)	Surface carrier concentration (x10¹⁵cm⁻²)
As deposited	2.03 x 10^{-3}	34.67	8.85 x 10^{19}	3.10
250°C, 1h	9.58 x 10^{-4}	42.90	1.52 x 10^{20}	5.31
250°C, 2h	5.55 x 10^{-4}	42.69	1.63 x 10^{20}	5.69
450°C, 1h	4.34 x 10^{-3}	20.66	6.96 x 10^{19}	2.43
450°C, 2h	1.41 x 10^{-3}	22.38	1.98 x 10^{20}	6.92

It is well known that carrier mobility is related to the free mean path which is calculated using the above equation [8]:

$$l = (3\pi^2)^{1/3} x \left(\frac{h}{2\pi e^2}\right) x \frac{1}{\rho} x (n^{-\frac{2}{3}}) \quad (8)$$

where: l: free mean path; h: the Planck's constant; e: electron charge; ρ: Resistivity and n: carrier concentration. The calculated free mean path of annealed IMO thin films (3 at.%) for different annealing conditions is presented in Figure IV.16. It can be remarked that free mean path increases for increasing annealing duration however it decreases for annealing temperatures. The longest l is found to be in the order of 4.89 nm for annealing temperature at 250°C for 2 hours corresponding to films which present the best electrical properties. The obtained values of free mean path are much smaller than that of the measured grain size from XRD analysis. Shin *et al.* [8] have found similar results for annealed IMO thin films grown by hetero target sputtering. They reported that the effect of scattering at dislocations and grain boundaries may be neglected when the mean free path is smaller than grain size. Thus, they conclude that scattering of conduction electron is overall changed by ionized and neutral impurities [8].

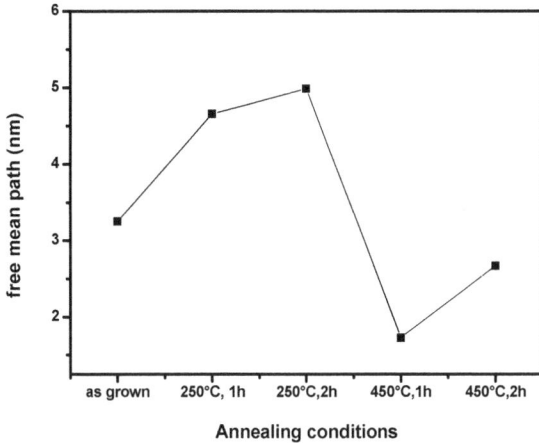

Figure IV.16:

Free mean path of an electron in IMO thin

4.2.5. Figure of Merit

In order to identify to most suitable TCO material among our heated films, we have calculated the Figure of Merit Φ for different annealing conditions. All values are summarized in Fig IV.17. It is seen that the Figure of Merit increases for 250°C compared to as deposited films. In fact, Φ increases from 35.65×10^{-4} Ω^{-1} to 49.80×10^{-4} Ω^{-1} for respectively as deposited and films annealed at 250°C during 1 hour. By increasing the annealing duration to 2 hours, the Figure of Merit exhibits a maximum of about 81.82×10^{-4} Ω^{-1}. On notes that for 450°C, Φ decreases. The calculated Φ values permit us to conclude that In_2O_3:Mo (3 at.%) annealed at 250°C during 2 hours has the best optoelectronic properties compared to other elaborated films.

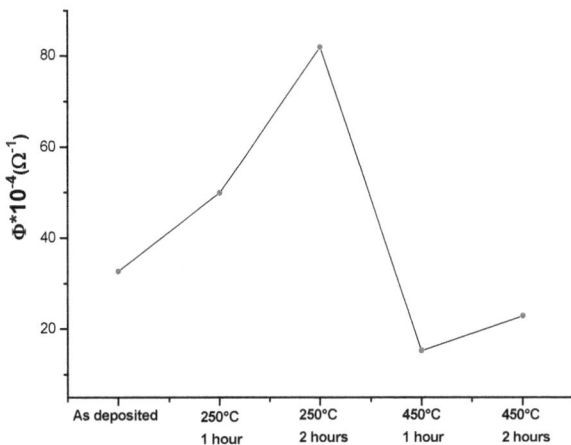

Fig IV-17:
Plots of Figure of Merits of In_2O_3:Mo (3 at.%) for different annealing conditions

4.2.6. Conclusion

Indium oxide thin films doped with different molybdenum concentrations have been elaborated by spray pyrolysis technique. Doping operation was carried out during the growth process by adding relative concentration of Mo^{6+} ions in the sprayed solutions. The physical properties of In_2O_3:Mo have been globally improved after doping by molybdenum. Structural analysis of XRD spectra shows an enhancement of the crystallinity for a doping ratio equals to 3 at.% with a grain size of about 155 nm. Optical properties show a high transmittance around 80% in the visible and near infrared regions with the presence of interference fringes showing the homogeneity surface of the obtained films. Calculated band gap is in the range of [3.45 - 3.56] eV. Electrical resistivity has decreased after doping from $\rho = 650.20 \times 10^{-3}$ Ω.cm to 2.03×10^{-3} Ω.cm for respectively undoped and In_2O_3:Mo (3 at.%). According to Figure of Merit values In_2O_3:Mo (3 at.%) exhibits the best optoelectronic properties. A heat treatment has been applied then on the optimized IMO films (3 at.%). XRD spectra of heated films show polycrystalline character with (222) as preferred orientation characteristic of bcc structure. The band gap energy increases after heat treatment. The resistivity decreases to 5.55×10^{-4} Ω.cm for annealing temperature equals to 250°C during 2 hours. All these experimental results confirm the efficiency of molybdenum doped indium oxide as a transparent conductive oxide since the introduction of molybdenum (3 at.%) followed by heat treatment at during 2 hours at 250°C leads to combine between low resistivity, high mobility and high average transmission in the visible and near infrared region.

4.2.7. References

[1] S. Parthiban, K. Ramamurthi, E. Elangovan, R. Martins, E. Fortunato and R. Ganesan; Materials Letters, 62 (2008) 3217-3219

[2] E. Elangovan, A. Marques, A. Pimentel, R. Martins and E. Fortunato; Vacuum 82 (2008) 1489-1494

[3] S. Parthiban, V. Gokulakrishnan, K. Ramamurthi, E. Elangovan, R. Martins, E. Fortunato and R. Ganesan; Solar Energy Materials and Solar Cells, 93 (2009) 92-97

[4] S. Parthiban, E. Elangovan, K. Ramamurthi, R. Martins and E. Fortunato; Solar Energy Materials and Solar Cells, 94 (2010) 406-412

[5] Y. Sawada, C. Kobayashi, S. Seki and H. Funakubo; Thin Solid Films 409 (2002) 46–50

[6] F. Nanni, F.R. Lamastra, F. Franceschetti, F. Biccari and I. Cacciotti; Ceramics International, 40 (2014) 1851-1858

[7] S. Kaleemulla, N. Madhusudhana Rao, M. Girish Joshi, A. Sivasankar Reddy, S. Uthanna and P. Sreedhara Reddy; Journal of Alloys and Compounds 504 (2010) 351-356

[8] Y.H. Shin, S.B. Kang, S. Lee, J.J. Kim and H.K. Kim; Organic Electronics, 14 (2013) 926.

[9] M. Ajili, M. Castagné and N. Kamoun Turki; Superlattices and Microstructures 53 (2013) 213–222

[10] A. Akkari, M. Reghima, C. Guasch and N. Kamoun-Turki; J Mater Sci (2012), 47:1365.

[11] S. Louidi, F.Z. Bentayeb, W. Tebib, J.J. Suñol, A.M. Mercier, J.M. Grenèche; Journal of Non-Crystalline Solids 356 (2010) 1052–1056

[12] M. Jothibas, C. Manoharan, S. Ramalingam, S. Dhanapandian and M. Bououdina; Spectrochimica Acta Part A: Molecular and Biomolecular Spectroscopy 122 (2014) 171–178

[13] N.G. Pramod and S.N. Pandey; Ceramics International 41 (2015) 527 – 532

[14] J. Tauc and A. Menth, Journal of Non-Crystalline Solids 8-10 (1972) 569-585 North-

Holland Publishing Co

[15] P. Prathap, Y.P.V. Subbaiah, M. Devika and K.T. Ramakrishna Reddy; Materials Chemistry and Physics 100 (2006) 375–379

[16] A. Barhoumi, G. Leroy, B. Duponchel, J. Gest, L. Yang, N. Waldhoff, S. Guermazi; Superlattices and Microstructures 82 (2015) 483–498

[17] V.Senthilkumar, P.Vickraman, M.Jayachandran and C.Sanjeeviraj; *Vacuum* 84, 2010, 864

[18] V. Senthilkumar and P. Vickraman; Current Applied Physics 10 (2010) 880–885

[19] S.Kaleemulla, A.Sivasankar Reddy, S.Uthanna and P.Sreedhara Reddy; Optoelectronics and Advanced Materials Rapid Communications Vol. 2, No. 12, December 2008, p. 782-787

[20] P. Wu, Q. Li, C.X. Zhao, D.L. Zhang, L.F. Chi and T. Xiao; Applied Surface Science 255 (2008) 3201 – 3204

[21] E. Benamar, M. Rami, C. Messaoudi, D. Sayah and A. Ennaoui; Solar Energy Materials and Solar Cells 56 (1999) 125-139

[22] P. Prathap, N. Revathi, K.T. Ramakrishna Reddy and R.W. Miles; Thin Solid Films 518 (2009) 1271-1274

[23] B.L. Zhu, X.Z. Zhao, F.H. Su, G.H. Li, X.G. Wu, J. Wu and R. Wu; Vacuum 84 (2010) 1280–1286

[24] D.K. Kim and H.B. Kim; Current Applied Physics 13 (2013) 2001 - 2004

[25] S. Cho; Microelectronic Engineering 89 (2012) 84-88

5. Effects of europium doping and annealing on physical properties of indium oxide thin films

Intensive studies have been done to improve physical properties of indium oxide thin films. Various atoms have been used as dopant element to be incorporated into In_2O_3 lattice such as molybdenum [1], fluor [2], iron [3] and zinc [4]. However, few reports have focused on doping In_2O_3 with rare earth atom like erbium (Er), europium (Eu), lanthanum (La), and ytterbium (Yb). For example, Niu *et al.* [5] have been reported the physical properties of Eu^{3+}, Gd^{3+} (gadolimium) and Ho^{3+} (holmium) doped In_2O_3 thin films grown by sol gel method. Further, Dakhel [6] has prepared ytterbium doped In_2O_3 by evaporation technique. The authors indicate that doping with Yb^{3+} introduces stress into In_2O_3 crystal lattice with a change in optical and electrical properties. On the other hand, Kim *et al.* [7] have been illustrated that the structure of indium oxide is the same to that of rare earth oxide which facilitate hence the introduction of rare earth elements into the In_2O_3 lattice. In this report, the authors have studied the effect of europium doping on the physical properties of ITO films prepared using sol gel technique [7].

In this part, indium oxide doped with europium (Eu^{3+}) is presented, using the optimized conditions and the same methodology investigated in previous chapter. In the part A, we have studied the effect of europium doping on the structural, morphological, optical and electrical properties of indium oxide thin layers. However, in the part B, the physical properties of annealed europium doped indium oxide thin films will be investigated.

5.1. Effect of europium doping

5.1.1. Structural properties:

XRD analysis: XRD patterns of europium doped indium oxide thin films are shown in Figure V.1 for different ratio y (y = $(\frac{[Eu^{3+}]}{[In^{3+}]})_{sol}$ = 0; 0.1; 0.3 and 0.5 at.%). Undoped In_2O_3

thin films present a sharp peak located at approximately $2\theta = 30.51°$ corresponding to the (222) plan characteristic of body centered cubic structure (JCPDS Card n°06-0416). It is observed after doping, a reduction of the (222) peak intensity followed by an increase of the (400) peak intensity which is positioned at 35.57°.

One notes that doped films crystallize also into bcc structure. When the europium amount in the sprayed solution increases, the peak intensity of (400) plan increases compared to (222) one indicating that doped films are preferentially orientated along (400) plan. This change in the preferred orientation may be attributed to the incorporation of europium element as mentioned by Parthiban *et al.* for molybdenum doped In_2O_3 thin films [8]. A similar change in the preferred orientation is obtained in our previous study for iron doped indium oxide thin films grown by spray [3] and for zinc doped indium oxide thin films. It is also revealed from Figure V.1 that the best crystallinity is obtained for y = 0.3 at.% which indicates an enhancement of films crystallinity after doping with this content of europium. Beyond this rate, a small decrease of preferential peak intensity is remarked but the films remain crystalline.

One notes that no secondary phases such as Eu_2O_3 have been detected from XRD spectra. This last observation may be due to the low amount of europium used during the doping process. Moreover, Ting *et al.* [9] have revealed for europium doped zinc oxide thin films the absence of Eu_2O_3 phase which can be formed even for high concentration of added Eu^{3+}. It can be thus assumed that europium doping elements have been totally incorporated into In_2O_3 lattice and may occupy indium sites as reported by Kim [7]. In fact, they mentioned that Eu^{3+} occupied In^{3+} sites for Eu doped ITO films prepared by sol gel technique. Since the ionic radius of Eu^{3+} is near to that of In^{3+}, so we can consider that Eu^{3+} can be totally incorporated into In_2O_3 lattice [6]. The grain size (d) was calculated using the Debye Scherrer formula [10] as described earlier along the preferred orientation for each doping ratio. The dislocation density (δ_{disc}) and the number of crystallites per unit surface area (n_c) were calculated using expressions as detailed in the second chapter. The obtained results are summarized in Table V.1.

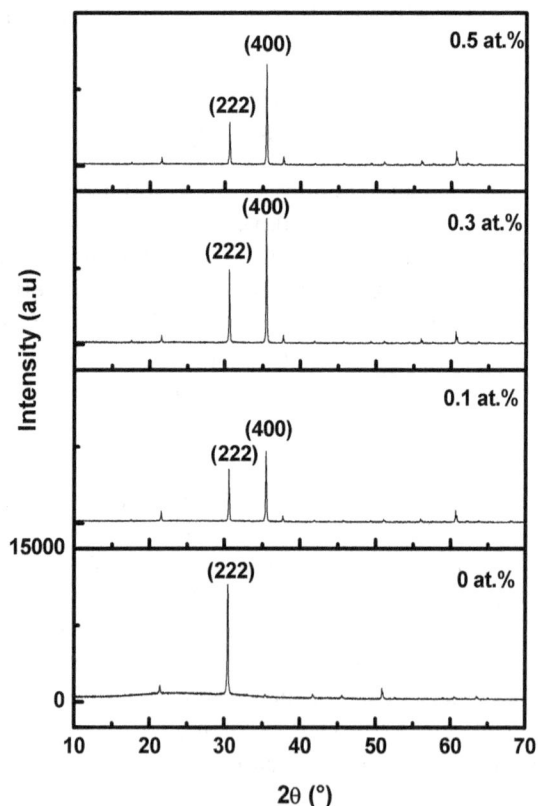

Figure V.1: XRD spectra of undoped and europium doped indium oxide thin films for different doping ratio $y = (\frac{[Eu^{2+}]}{[In^{3+}]})_{sol}$

Table V.1: Average grain size (d), dislocation density (δ_{dis}) and crystallites number (n_c) of europium doped indium oxide thin films deduced from XRD spectra

$y=[Eu^{3+}]/[In^{3+}]$ (at.%)	d (nm)	δ_{dis} (x 10^{10} cm^{-2})	n_c (x 10^{10} cm^{-2})
0	71	1.98	21
0.1	106.7	0.88	5.76
0.3	142.4	0.50	2.42
0.5	105.3	0.90	6

It is shown a noticeable increase of the grain size after doping from 71 to 142.4 nm for respectively undoped and In_2O_3:Eu (0.3 at.%). This result may be related to the fact that indium oxide thin films have the best crystallinity for y = 0.3 at.%. Such result may contribute to the reduction of grain boundaries which play a considerable role in electrical conductivity. On notes that, the dislocation density δ_{dis} and the number of crystallites n_c decrease after doping reaching minimum values of about respectively 0.50 x 10^{10} cm^{-2} and 2.42 x 10^{10} cm^{-2} for a doping ratio equals to 0.3 at.%.

Rietveld analysis: XRD patterns of europium doped indium oxide thin films are refined using MAUD software [11] which is based on the Rietveld method [11]. The Rietveld refinement of elaborated thin films for atomic concentration $y=(\frac{[Eu^{3+}]}{[In^{3+}]})_{sol} = 0.1$, 0.3 and 0.5 at.% are shown in Figure V.2. The structural and micro-structural parameters are presented in Table V.2. As evident, there is no secondary phase detected after doping process. This result is in good accordance with XRD analysis which confirms that europium element is totally incorporated in the crystal lattice due to the fact that we have doped with low amount. One notes that "a" increases after doping with increasing Eu content in the sprayed solution. Indeed, an increase from 10.1280 to 10.1298 Å for respectively undoped and In_2O_3:Eu (0.3 at.%) indicating an expansion of crystal lattice parameters. This can be the result of replacement of indium atoms by rare earth ones [12]. Such result can be correlated with ionic radii of the atoms. Indeed, indium atoms exhibit an ionic radius of about 80 pm which is smaller than that of europium ones (95 pm). In accordance with other report [10], doping indium oxide with ytterbium atoms leads to a dilatation of crystal lattice which can be due to the fact that ionic radius of In^{3+} is smaller that Yb^{3+}. Similar results were also obtained for lanthanide and erbium doped In_2O_3 compounds [12]. The microstrain $<\sigma^2>^{1/2}$ increases slightly after doping for 0.1 at% then it decreases reaching a minimum value of around 0.0203% for y = 0.3 at.%. Such result is in accordance with that In_2O_3:Eu (0.3 at.%) thin films have the best crystallinity.

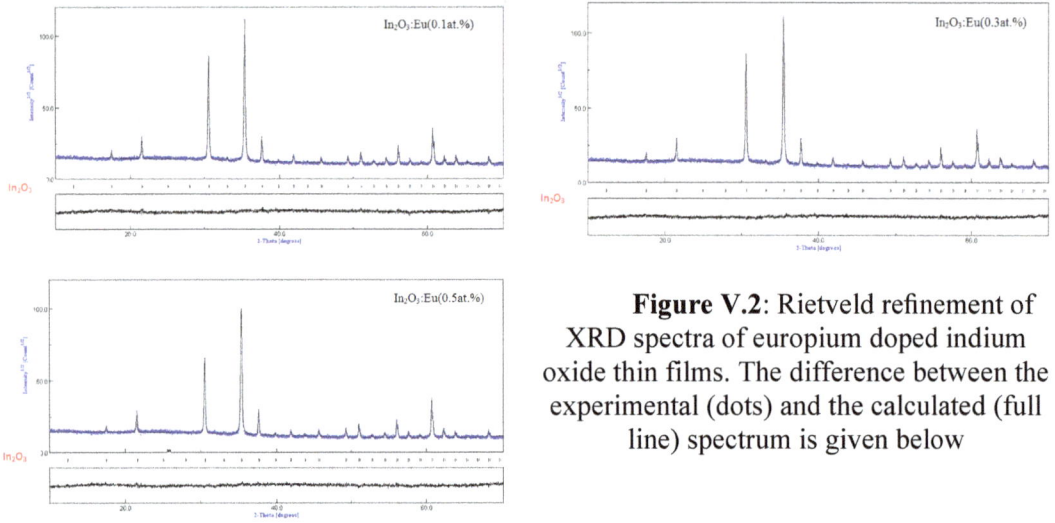

Figure V.2: Rietveld refinement of XRD spectra of europium doped indium oxide thin films. The difference between the experimental (dots) and the calculated (full line) spectrum is given below

Table V.2: Phases and structural parameters deduced from Rietveld analysis of europium doped indium oxide thin films for different doping concentrations $y = (\frac{[Eu^{2+}]}{[In^{3+}]})_{sol}$

y(at.%)	Phase	$<\sigma^2>^{1/2}$ (%) (±0.001)	a (Å) (±0.0001)
0	In_2O_3	0.0208	10.1280
0.1	In_2O_3	0.0373	10.1285
0.3	In_2O_3	0.0203	10.1298
0.5	In_2O_3	0.0402	10.1310

5.1.2. Morphological properties:

The respective 2D and 3D AFM images of In_2O_3:Eu thin films for different doping ratio $(y=\frac{[Eu^{3+}]}{[In^{3+}]})_{sol}$ =0; 0.1; 0.3 and 0.5 at.%) are presented in **Fig V.3**. The films were scanned in contact mode over an area of 2μm x 2μm. It is revealed that the introduction of europium affects extremely the surface topography of the films. Then, the incorporation of europium into the crystal lattice has a noticeable impact on grain size and shape. Indeed, it is well remarked that grain size tends to be larger after doping with a uniform distribution especially for 0.3 at.%. Such increase of grain size deduced from AFM images confirms our previous XRD results.

The RMS roughness values were determined from AFM analysis. All the results are presented in Figure V.4. It can be seen that the RMS roughness is reduced from 48 to 25.6 nm for respectively undoped and In_2O_3:Eu (0.1 at.%). After that, an increasing trend was observed and the obtained values became greater than those for undoped one. Cabello *et al.* [13] have also reported that europium doped ZrO_2 thin films have a non-uniform RMS.

Fig V.3: AFM images of europium doped indium oxide thin films for different doping concentrations $y = (\frac{[Eu^{2+}]}{[In^{3+}]})_{sol}$ and deduced RMS roughness values.

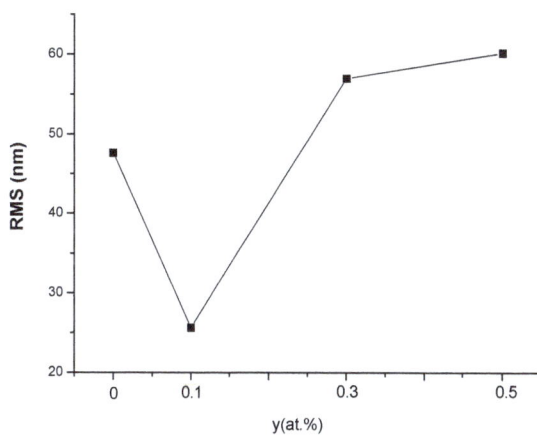

Figure V.4: RMS roughness of europium doped indium oxide thin films for different doping ratio $y = (\frac{[Eu^{2+}]}{[In^{3+}]})_{sol}$

5.1.3. Optical properties:

The transmission and reflection spectra of In_2O_3:Eu thin films at different doping ratio y are shown in Figure V.5. The presence of interference fringes after doping indicates good thickness uniformity, as was the case of undoped indium oxide films. Beyond y equals to 0.1 at.%, the interference fringes intensities are attenuated. The doped films show a lower transmission in the visible and near infrared range with increasing doping ratio compared to undoped ones. The reduction of transmission after doping can be due to the enhanced of photons scattering caused by crystal defects and produced after doping or to the absorption of photons by free

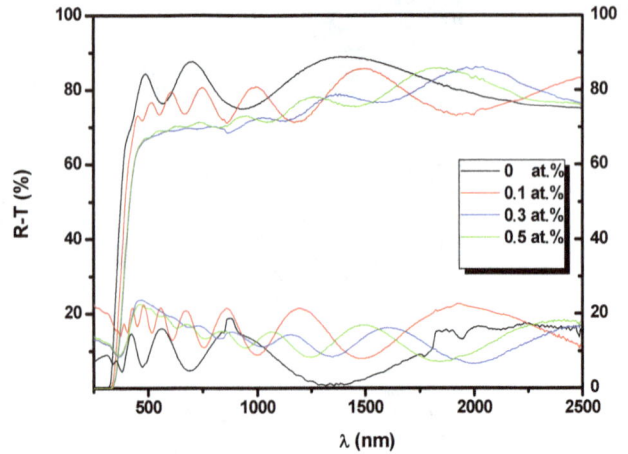

Figure V.5: Transmission and reflection spectra of undoped and europium doped indium oxide thin films for different doping ratio y = $(\frac{[Eu^{2+}]}{[In^{3+}]})_{sol}$

carriers [4]. Unlike for y = 0.3 and 0.5 at.%, europium doped layers exhibit a transmission decrease in the visible region reaching 70%. However, an enhancement of film transparency is noted in the infrared region. Moreover, the decrease of the transmission for doped films may be due to the decreases of the porosity after doping as seen in next part (Figure V.8_b). A similar behavior was also observed in the chapter II, III and VI by doping In_2O_3 thin films with iron zinc and molybdenum. In the literature, many authors have founded a decrease of transmission by doping process like for In_2O_3:Zn [4], In_2O_3:Mo [8] and In_2O_3:Li [14].

The band gap energy E_g was determined for all films from their reflection and transmission spectra based on Tauc equation as reported earlier. All results are summarized in Figure V.6. The optical band gap was found to be ranging between 3.43 and 3.51 eV indicating large band gap energy.

The refractive index "n" was calculated in the transparency zone using the envelope method for $T_{max}(\lambda)$ and $T_{min}(\lambda)$ [15]. The variation of refractive index "n" as a function of wavelength λ for different europium doping concentration is shown in Fig V.7. Doped films exhibit the highest value of refractive index "n" compared to undoped films. This result may be attributed to the enlargement of grain size and decrease of the porosity after doping (Fig V.8_b). Senthilkumar *et al.* [16] have obtained for indium tin oxide thin films a decrease of the refractive index after doping with tin. The authors have mentioned that the reduction of refractive index after doping may be the result of lower grain size value or to the increase of porosity by doping process [16]. Since for europium doping ratio equal to 0.3 and 0.5 at.%, the transmission spectra have a similar trend which leads to the superposition of their refractive index curves. On the other hand, the variation of the refractive index with doping ratio is in accordance with transmission spectra. In fact, undoped films which exhibit the highest transmission have the lowest refractive index. Moreover, it is well known that there is correlation between the refractive index and the packing density (p) of grains in layers. In fact, from the refractive index "n", we can determine the packing density "p" as mentioned by Harris *et al.* [15]. As shown in Figure V.8_a, the packing density "p" is increased after doping. This result may be attributed to the enlargement of grain size after doping.

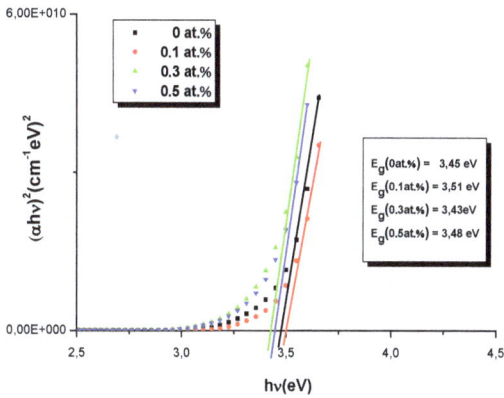

Figure V.6: Variation of $(\alpha h v)^2$ versus (hv) of undoped and europium doped In$_2$O$_3$ thin films for different doping ratio $y = (\frac{[Eu^{2+}]}{[In^{3+}]})_{sol}$

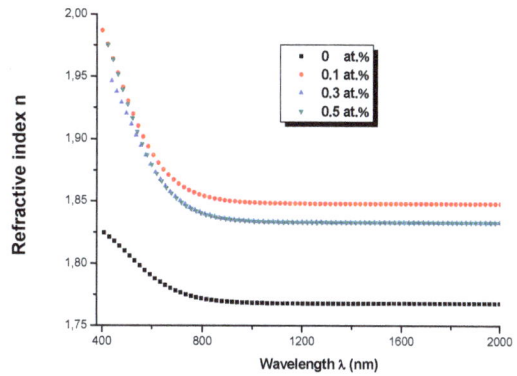

Figure V.7: Refractive index "n" of europium doped indium oxide thin films as a function of wavelength for different doping ratio $y = (\frac{[Eu^{2+}]}{[In^{3+}]})_{sol}$

The porosity of all films is calculated using equation which is described in the chapter IV.[16] It is seen from Figure V.8_**b** where we present the porosity of In$_2$O$_3$:Eu, that the lowest porosity value was observed for y=0.1 at.%. This last finding can be in agreement with morphological analysis where the topography surface obtained at this doping ratio exhibits a surface without pinholes. On the other hand, undoped layers are the most porous films. Indeed, the doping process minimizes significantly the porosity of the films. This decrease of the porosity may be due to the significant enlargement of grain size as seen from Table 1 due to the incorporation of europium element in In$_2$O$_3$ lattice.

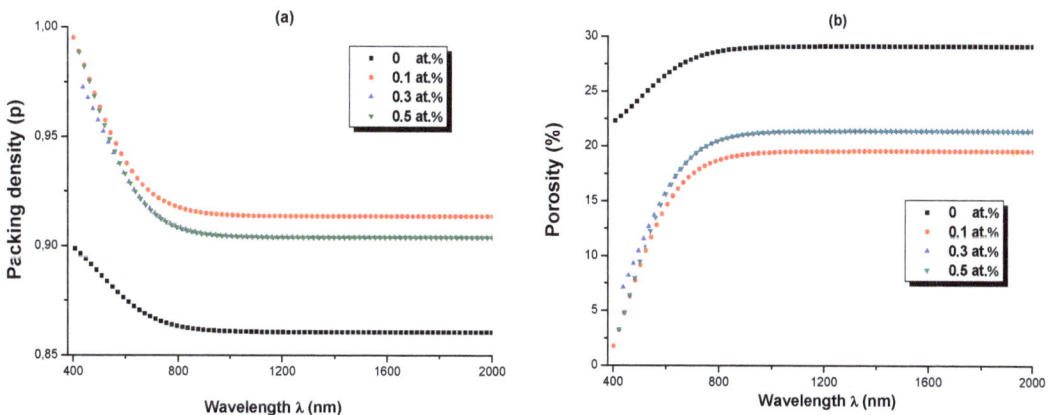

Figure V.8: Packing density p (a) and Porosity (b) of europium doped indium oxide thin films for different doping ratio $y = (\frac{[Eu^{2+}]}{[In^{3+}]})_{sol}$

The refractive index dispersion (n) was studied using Wemple model as detailed in the chapter II. The linear variations of $(n^2 - 1)^{-1}$ versus $(hv)^2$ for europium doped indium oxide thin films are shown in Figure V.9. E_0 and E_d were determined from the slope $(E_d E_0)^{-1}$ and the intercept of the origin (E_0/E_d) for each straight line. The average strength S_0 and the wavelength λ_0 of the oscillator are also determined.

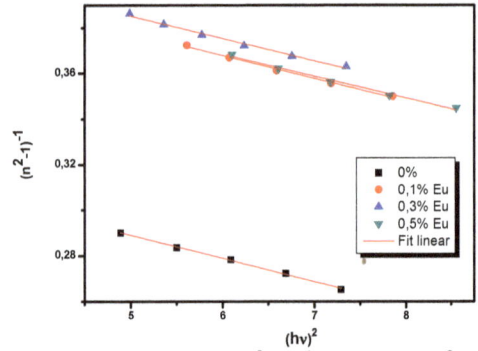

Figure V.9: Plot of $(n^2-1)^{-1}$ versus $(hv)^2$ of europium doped indium oxide thin films

The deduced values of E_0, E_d, S_0 and λ_0, for europium doped indium oxide thin films, are summarized in Table V.3. It has seen that the oscillator energy (E_0) is the double of the optical band gap (E_g) for all europium doped indium oxide thin films. However, $E_0 = 1.5 \times E_g$ for undoped In_2O_3 thin films. This result is similar to those obtained for In_2O_3:Zn, In_2O_3:Fe and In_2O_3:Mo.

Table V.3: The oscillator energy (E_0), dispersion energy (E_d), average strength (S_0) and wavelength (λ_0) of the oscillator extracted from wemple model.

y(at.%)	E_0 (eV)	E_d (eV)	S_0 (10^{-5} nm^{-2})	λ_0 (nm)
0	5.05	15.02	4.93	245.54
0.1	6.91	15.48	7.41	179.45
0.3	6.63	15.26	6.58	187.03
0.5	6.66	15.64	6.77	186.19

5.1.4. Photoluminescence analysis

The photoluminescence (PL) spectra of europium doped indium oxide thin films with different concentrations y (y= 0; 0.1; 0.3 and 0.5 at.%) are shown in Figure V.10. We note the presence of three emission peaks located at 485 (2.57 eV), 505 (2.45 eV) and 795 nm (1.56 eV). PL spectra illustrate a large blue green emission band at [485-505] nm for doped and undoped films. This emission could be due to lattice defects or oxygen deficiencies in the layers [17]. Wu *et al.* [18] have reported that emission peaks observed in the visible range may be attributed to the recombination of carrier concentration that occurs between

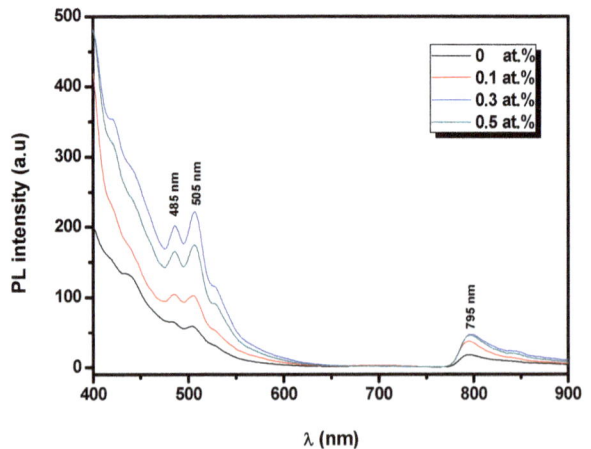

Figure V.10: PL spectra of undoped and europium doped indium oxide thin films for different doping ratio $y = \left(\dfrac{[Eu^{2+}]}{[In^{3+}]}\right)_{sol}$

valence band and oxygen vacancies acting as donor levels. Different peaks positions from PL spectra have been founded in the literature. Indeed, Jothibas *et al.* [4] detect PL peaks at 418, 440 and 674 nm for zinc doped In_2O_3 thin films. These differences between PL results may be caused by the difference in film morphologies as revealed in [18].

5.1.5. Electrical properties

The electrical parameters of In_2O_3:Eu are deduced from Hall effect measurements. All results are shown in Table V.4. It is observed a decrease in resistivity values for increasing europium content until $y = \frac{[Eu^{3+}]}{[In^{3+}]})_{sol} = 0.3$ at.%. In fact, the electrical resistivity ρ is decreased from 650.20×10^{-3} Ω.cm to 30.70×10^{-3} Ω.cm for respectively as deposited and In_2O_3:Eu (0.3 at.%). Then a slight increase of resistivity was obtained for y = 0.5at.% keeping the same order of magnitude. Similar results were reported by J.K. Kim for europium doped ITO films [7]. Moreover, an enhancement of Hall mobility is shown for increasing Eu content in the sprayed solution reaching a maximum value of 12.01 cm²/Vs for y = 0.5at.%.

Table V.4: Hall Effect measurements of europium doped indium oxide thin films at different europium concentration

$y=[Eu^{3+}]/[In^{3+}]$ (at.%)	Resistivity x 10^{-3} (Ω.cm)	Hall mobility (cm²/Vs)	Volume carrier concentration ($\times 10^{19}$cm^{-3})	Surface carrier concentration ($\times 10^{15}$cm^{-2})
0	650.20	1.17	0.80	0.24
0.1	209.30	5.25	0.56	0.19
0.3	30.71	6.07	3.35	1.20
0.5	42.92	12.01	1.20	0.42

However, the highest volume carrier concentration N_v is obtained for doping ratio equal to 0.3 at.% reaching 3.35×10^{19} cm^{-3}. The decrease of resistivity can be assigned to the increase of carrier concentration as observed in the Table V.4. Such increase of N_v may be related to the improvement of crystallinity and thus contribute to the reduction of trapped sites presents at grain boundaries [19]. However, for y = 0.5 at.%, free carrier density decreases to 1.20×10^{19} cm^{-3}. This fact can be attributed to the decrease of oxygen vacancies as reported by Ting *et al.* [9]. However, the excess of europium atoms into In_2O_3 matrix leads to an increase of structural defects acting as trap levels and hence the resistivity increases. Similar findings have been obtained for sprayed Al doped ZnO thin films [20]. The report deals with the role of gallium in electrical properties of ZnO thin films. In fact, Al atoms contribute in the first time to the enhancement of electrical conductivity. Whereas, the excess of Al atoms into the sprayed solution leads to the creation of trapping level and thus the reduction of electrical conductivity.

5.1.6. Figure of Merit:

It is well known that optical transmission and electrical resistivity have an important impact on the TCO material quality. So, in order to summarize optoelectronic properties of elaborated films, we have calculated the Figure of Merit for all layers using Haacke's equation [21]. The variation of the figure of merit ϕ as function of doping concentration y is shown is the Figure V.11. It is seen that the Figure of Merit is increasing after doping. In fact, ϕ increases from $0.09 \times 10^{-4} \ \Omega^{-1}$ to $0.48 \times 10^{-4} \ \Omega^{-1}$ for respectively undoped films and In$_2$O$_3$:Eu (0.3 at.%). These ϕ results leads to conclude that doping with europium can ameliorate the optoelectronic properties of indium oxide thin films.

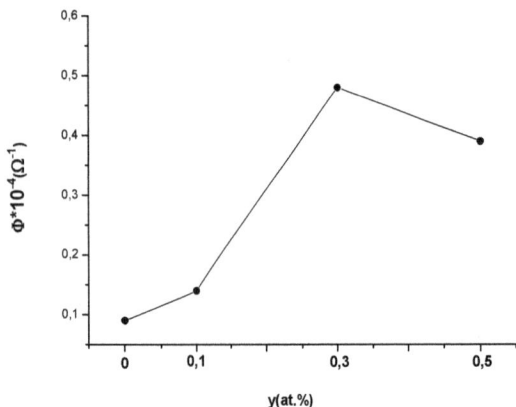

Figure V.11: Figure of Merit (Φ) of europium doped indium oxide thin films for different atomic concentrations $y = \frac{[Eu^{2+}]}{[In^{3+}]})_{sol}$

5.2. Effect of heat treatment on In$_2$O$_3$:Eu (0.3 at.%) thin films

In the part A, we have globally improved the physical properties of indium oxide thin layers by doping with europium. Indeed, the crystallinity of elaborated layers is enhanced and the electrical resistivity is decreased but the obtained values of ρ in greater than 10^{-3} Ω.cm. Unfortunately, the transmission is decreased after doping. So a heat treatment is necessary to reduce more ρ and to enhance optical transmission. In this section, In$_2$O$_3$:Eu (0.3 at.%) thin films are annealed in nitrogen atmosphere for 250 and 450°C during 1 and 2 hours. The investigation on physical properties of heated In$_2$O$_3$:Eu (0.3 at.%) thin films will be the aim of the part B of this chapter.

5.2.1. Structural properties

The X-ray diffraction patterns of annealed In$_2$O$_3$:Eu (0.3 at.%) thin layers are shown in **Figure V.12**. All heated films are polycrystalline and crystallize in to the body centered cubic structure as was the case of as deposited ones. One notes that there is no change in the predominant peaks after heat treatment. In fact, we note the presence of two main peaks (222) and (400) located respectively at 30.54 and 35.58°. The intensity of the preferred orientation (400) does not practically changes as function of annealing temperatures which may be means that the crystallization is competed after doping process. These last results lead to conclude that the crystalline structure is stable during heat treatment.
A slight enhancement of crystallinity degree is observed for 450 °C during 1 hour. A similar result is obtained by Kato *et al.* [22] for In$_2$O$_3$ thin films elaborated by sputtering and

annealed in air. Indeed, they found that the crystallinity is slightly improved by annealing process. So there is no significant change of the preferred orientation intensity. However, In_2O_3:Sn film's crystallinity is enhanced by annealing in N_2 atmosphere for 250 and 350°C [23]. Moreover, Fang *et al.* [24] have founded for another TCO material which is ZnO that annealing in air for a temperature over to 600 °C leads to decrease the preferred peak's intensity. It can be concluded that heat treatment and deposition conditions as well as TCO nature's material affect extremely the crystalline quality of TCO material.

The grain sizes (d) for different annealing conditions are determined by the Debye Scherrer formula along the (400) orientation. As seen from Table V.5, d varies slightly compared to as deposited one. Indeed, d is in the range of [133-150] nm for annealed films whereas d is around 142 nm for as deposited layers.

The calculated grain sizes are in good agreement with XRD spectra as function of annealing conditions. Indeed, as seen previously the film's crystallinity does not influenced by heat treatment One notes that films heated at 450°C during 1 hour exhibit the maximum value of grain size which is 150 nm. This last results is in accordance with that for these annealing conditions In_2O_3:Eu (0.3 at.%) has the best crystallinity. Lin *et al.* [23] have also revealed for annealed In_2O_3:Sn in N_2, that the enlargement of grain size may be attributes to the improvement of crystallinity.

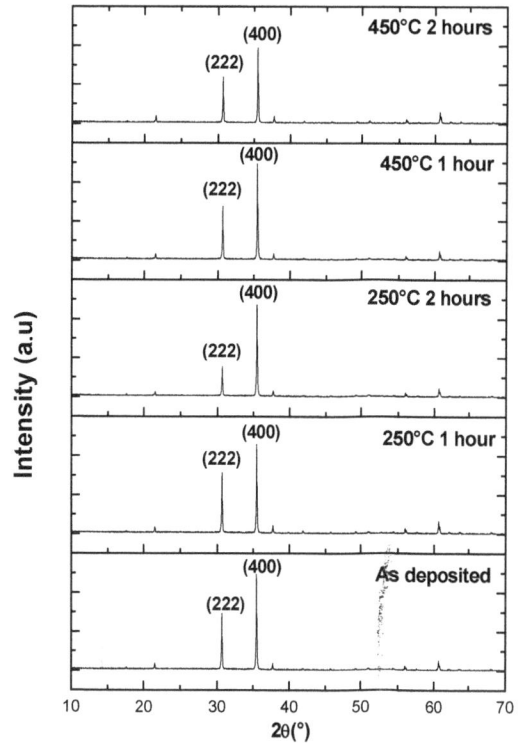

Figure V.12: XRD spectra of as deposited and heated In_2O_3:Eu (0.3 at.%) for different annealing conditions

Table V.5: Full-Width at Half-Maximum (FWHM), position of (400) peak, grain size (d) as function of annealing conditions

Annealing conditions	FWHM(°)	Position of (400) peak [2θ(°)]	d(nm)
As deposited	0.1380	35.47	142
250°C_1 hour	0.1384	35.48	140
250°C_2 hours	0.1397	35.49	134
450°C_1 hour	0.1368	35.49	150
450°C_2 hours	0.1398	35.50	133

5.2.2. Morphological properties

AFM images of In$_2$O$_3$:Eu (0.3 at.%) thin films for different annealing conditions are shown in Figure V.13. The variation of grain sizes observed from AFM images matched well with those calculated using Debye Scherrer formula. A significant change of grain forms is observed after heat treatment. In addition, the AFM graphs of heated films reveal the presence of uniform and dense grains without any cracks for 250°C and 450°C (1 hour). One notes that there are no scratch defects in the surface after heat treatment which reflects the uniformity of surface layers. Our results seem to be better than those obtained by Lin *et al.* [23] for In$_2$O$_3$:Sn annealed in N$_2$ atmosphere. In fact, other authors have founded the presence of some defects in films surface.

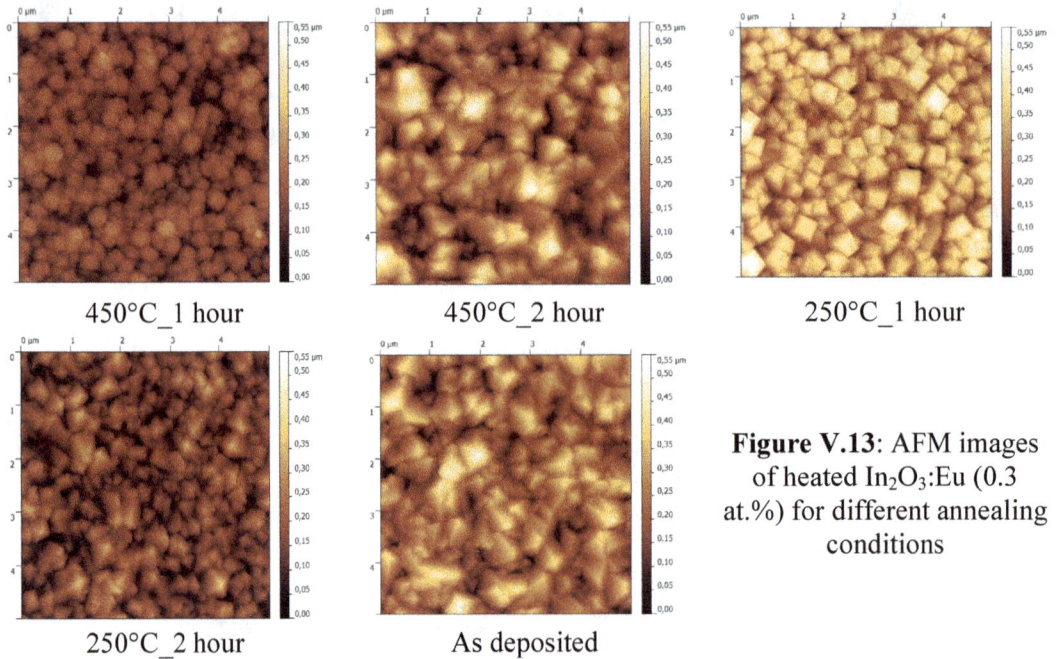

450°C_1 hour

450°C_2 hour

250°C_1 hour

250°C_2 hour

As deposited

Figure V.13: AFM images of heated In$_2$O$_3$:Eu (0.3 at.%) for different annealing conditions

The RMS values of heated In$_2$O$_3$:Eu (0.3%) thin films are presented Figure V.14. RMS roughness varies as function of heating conditions. In fact, it initially increases from 57 nm to 78 nm for respectively as deposited and layers annealed at 250 °C during 1 hour. Then, RMS decreases reaching a lower value of about 47 nm for 450°C (1 hour). This difference in RMS with annealing conditions may be related to the change in surface morphologies as shown by AFM images.

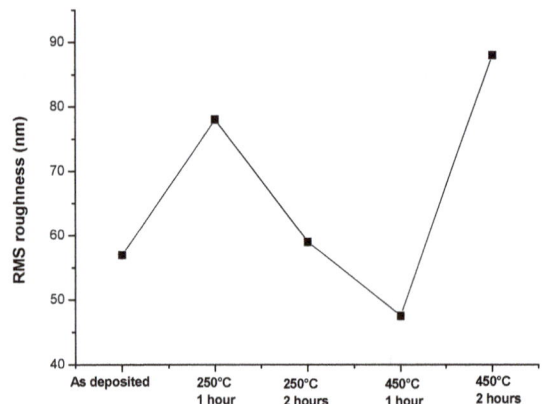

Figure V.14: RMS roughness of as deposited and annealed In$_2$O$_3$:Eu (0.3 at.%) for different annealing conditions

5.2.3. Optical properties

Transmission and reflection spectra of heated indium oxide thin films for different annealing conditions are shown in Figure V.15. The transmission increases in the visible range after heat treatment for 250 (2 hours) and 450°C (1 hour) leading to an enhancement of thin films transparency. In fact, T increases from 68 to an average value of 75 % in the visible range. Moreover, for these annealing conditions, new interferences fringes have been appeared compared to as deposited ones. These results reflect the efficiency of annealing in N_2 atmosphere to obtain smooth and homogeneous surface layers. However, T deceases for 250°C (1 hour) and 450°C (2 hours). Optical results in term of transmission and reflection lead to conclude that annealing temperatures and durations play an important role in the quality of films transparency. Sengupta *et al.* [25] have obtained a decrease in T after heat treatment for Mg doped ZnO thin films. However, Senthikulmar *et al.* [26] have reported that an annealing of In_2O_3 thin films contribute an enhancement of films transparency. The band gap energy for heated layers is calculated using TAUC equation as well as the case of as deposited ones. As shown in Figure V.16, there is no significant change in the band gap energy after heat treatment. In fact E_g is in the order of [3.41-3.49] eV.

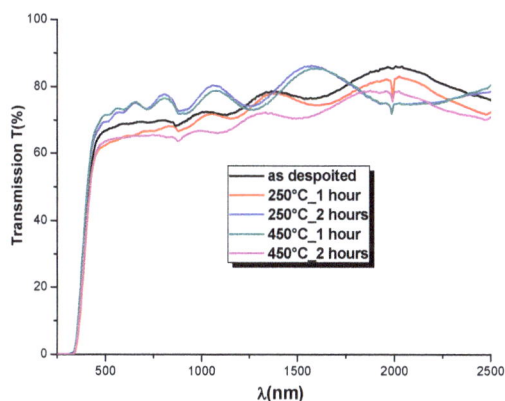

Figure V.15: Transmission spectra of annealed In_2O_3:Eu (0.3 at.%)

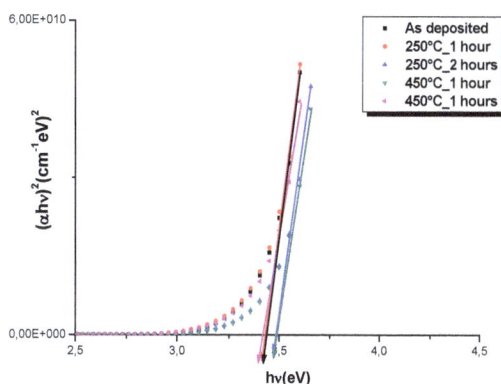

Figure V.16: Variation of $(\alpha h\nu)^2$ versus $(h\nu)$ of as deposited and annealed In_2O_3:Eu (0.3 at.%)

5.2.4. Electrical properties:

The electrical properties of annealed In_2O_3:Eu (0.3 at.%) thin films were explored by Hall Effect measurements. All results are presented in Table V.6. The resistivity decreases from 30.70 x 10^{-3} Ω.cm to 4.32 x 10^{-3} Ω.cm for respectively as deposited and annealed films at 250°C during 2 hours. Then, the resistivity increases reaching 98.67 x 10^{-3} Ω.cm for 450°C during 1 hour. A similar result is obtained for annealed indium tin oxide thin films [27]. Indeed, the authors revealed that electrical resistivity is decreased for lower annealed temperature (300°C) then ρ increases by increasing annealing temperature. On the other hand, an increase of Hall mobility is observed from 6.07 to 11.85 cm²/Vs is observed after annealing for 250°C (1 hour). One notes that volume free carrier concentration (N_v) are enhanced due to heat treatment keeping the same order of magnitude. In fact, N_v increases from 3.35 x10^{19} to 8.16 x10^{19} cm^{-3} after annealing at 250°C during 2 hours. The decrease of the resistivity after annealing may be due to the increase of volume carrier concentration and

Hall mobility [26]. This last funding is also consistent with that obtained by Yuan *et al.* for annealed In_2O_3 thin films in vacuum [28]. Moreover, Lin *et al.* [23] have revealed that a further decrease of ρ is followed by an increase of free carrier. In fact, they found that the annealing of In_2O_3:Sn in N_2 gas is marked by an enhancement of N_v from 1.76 $x10^{20}$ cm^{-3} to 11.10 $x10^{20}$ cm^{-3} which leads to a decrease in ρ from 14.8 to 3.99 10^{-3} Ω.cm. One notes that in our case, annealing at lower temperature (250°C) is more suitable to obtain promising electrical resistivity and volume carrier concentration. In addition, the annealing time affect extremely the electrical properties of the material. Xu *et al.* [27] have also mentioned the dependence of annealing temperatures and times the electrical properties of ITO films annealed in air. It can be deduced that heat treatment of europium doped indium oxide thin layers at lower temperature is efficient to obtain a low resistivity (in the order of 10^{-3} Ω.cm) and to enhance at the same time Hall mobility and free carrier concentration.

Table V.6: Hall Effect measurements of heated europium In_2O_3:Eu (0.3 at.%) at different annealing conditions.

Annealing conditions	Resistivity x 10^{-3}(Ω.cm)	Hall mobility (cm²/Vs)	Volume carrier concentration ($x10^{19}cm^{-3}$)	Surface carrier concentration ($x10^{15}cm^{-2}$)
As deposited	30.71	6.07	3.35	1.20
250°C, 1h	12.14	11.85	4.34	1.52
250°C, 2h	4.32	9.97	8.16	2.85
450°C, 1h	98.67	3.46	1.82	0.64
450°C, 2h	38.56	5.61	2.88	1.01

5.2.5. Figure of Merit

In order to find to most suitable annealing conditions for improving the optoelectronic properties of our elaborated layers, we have calculated the figure of merit after heating. Figure V.17 shows the variation of the Figure of Merit for In_2O_3:Eu (0.3 at.%) with different annealing conditions. As seen, ϕ increases after annealing at 250°C and reaches a maximum value of about 6.1 x 10^{-4} Ω^{-1} for 250°C during 2 hours. This obtained value of ϕ is in agreement with that for 250°C during 2 hours, indium oxide exhibit the lowest value of resistivity. As a deduction, In_2O_3:Eu (0.3 at.%) heated at 250 during 2 hours is the most promising TCO among our elaborated layers to be used in many potential applications like optoelectronic devices.

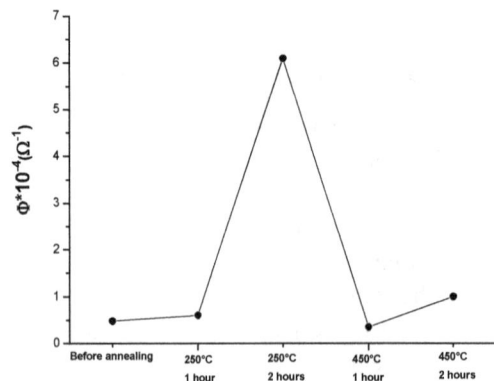

Figure V.17: Figure of Merit (Φ) of In_2O_3:Eu (0.3 at.%) as a function of annealing temperatures and for different durations.

5.2.6. Conclusion

Influence of europium doping as well as heat treatment on sprayed In_2O_3 thin films have been studied. Indium oxide is doped with low amount of europium ($y = \frac{[Eu^{3+}]}{[In^{3+}]})_{sol} = 0.1$; 0.3 and 0.5at.%). XRD analysis show a polycrystalline nature with body centered cubic structure for doped In_2O_3 thin films. An enhancement of the crystallinity is obtained for y = 0.3 at.% followed by a change in the preferred orientation from (222) to (400) plan. The grain size seems to be larger after doping reaching a maximum value of 142.4 nm. AFM images show a dense texture characterized with pyramidal form. The transmission decreases due to doping process. The band gap energy is in the range of [3.43-3.51] eV. The electrical resistivity is reduced from 650.20 x 10^{-3} Ω.cm to 30.70 x 10^{-3} Ω.cm for respectively undoped films and In_2O_3:Eu (0.3 at.%). The best physical properties for europium doped In_2O_3 thin films were recorded for a ratio equal to 0.3 at.% according to the Figure of Merit values. A heat treatment in nitrogen atmosphere was carried out to further enhance optical and electrical properties of optimized In_2O_3:Eu (0.3 at.%). An enhancement of films transparency is founded after annealing for 250 (2 hours) and 450°C (1 hour). Electrical resistivity is decreased by an order of magnitude reaching 4.32 x 10^{-3} Ω.cm at 250°C during 2 hours. The Figure of Merit results calculated after heat treatment permit us to conclude that In_2O_3:Eu (0.3 at.%) thin layers heated at 250°C during 2 hours exhibit the best optoelectronic properties among our elaborated films. It can be seen that doping with low amount of rare earth europium atom followed by heat treatment at low temperature can be sufficient to obtain a good TCO material without undesirable phases. So, few quantities of rare earth europium element can replace high rate of other metal transition dopant elements to produce a promising transparent conductive oxide.

5.2.7. References

[1] S. Parthiban, K. Ramamurthi, E. Elangovan, R. Martins, E. Fortunato and R. Ganesann; Materials Letters 62 (2008) 3217–3219

[2] N. Fellahi, M.addou, Z. Sofiani, M. El Jouad, K. Bahedi, S. Bayoud, M. Haouti, B. Sahraoui and J.C. Bernede, Journal Of Optoelectronics and Advanced Materials, Vol. 12, No. 5, May 2010, p. 1087 – 1091.

[3] N. Beji, M. Souli, M. Ajili, S. Azzaza, S. Alleg, N. Kamoun Turki; Superlattices and Microstructures 81 (2015) 114–128

[4] M. Jothibas, C. Manoharan, S. Ramalingam, S. Dhanapandian, M. Bououdina; Spectrochimica Acta Part A: Molecular and Biomolecular Spectroscopy 122 (2014) 171–178

[5] X. Niu, H. Zhong and X. Wang, K. Jiang; Sensors and Actuators B 115 (2006) 434–438

[6] A.A. Dakhel; Microelectronics Reliability 50 (2010) 211–216

[7] J.K. Kim and Y.G. Choi; Thin Solid Films 517 (2009) 5084–50864

[8] S. Parthiban, V. Gokulakrishnan, K. Ramamurthi, E. Elangovan, R. Martins, E. Fortunato, R. Ganesan; Solar Energy Materials & Solar Cells 93 (2009) 92–974

[9] C.C. Ting, W.Y. Li, C. Wang and H.E. Yong; Thin Solid Films 562 (2014) 625–631

[10] A. Akkari, M. Reghima, C. Guasch, N. Kamoun-Turki; J Mater Sci (2012), 47:1365–71

[11] S. Louidi, F.Z. Bentayeb, W. Tebib, J.J. Suñol, A.M. Mercier, J.M. Grenèche; Journal of Non-Crystalline Solids 356 (2010) 1052–1056

[26] T. Zhang, F. Gu, D. Han, Z. Wang and G. Guo; Sensors and Actuators B 177 (2013), p1180

[13] G. Cabello, L. Lillo, C. Caro, G.E. Buono-Core, B. Chornik and M.A. Soto; Journal of Non-Crystalline Solids 354 (2008) 3919–3928

[14] N.G. Pramod, S.N.Pandey; Ceramics International 41 (2015)527–532

[15] P. Prathap, Y.P.V. Subbaiah, M. Devika, K.T. Ramakrishna Reddy; Materials Chemistry and Physics 100 (2006) 375–379

[16] V. Senthilkumar, P. Vickraman, M. Jayachandran, C. Sanjeeviraj; Vacuum 84 (2010) 864. [17] S. Kaleemulla, A. Sivasankar Reddy, S. Uthanna and P. Sreedhara Reddy; Optoelectronics and Advanced Materials – Rapid Communications Vol. 2, No. 12, December 2008, p. 782 – 787

[18] P. Wu, Q. Li, C.X. Zhao, D.L. Zhang, L.F. Chi and T. Xiao; Applied Surface Science 255 (2008) 3201–3204.

[19] E. Benamar, M. Rami, C. Messaoudi, D. Sayah and A. Ennaoui; Solar Energy Materials and Solar Cells 56 (1999) 125-139

[20] K. Mahmoud and S. Bin Park; Electronic MaterialsLetters; 9(2013), 161

[21] M. Ait Aouaj, R. Diaz, A. Belayachi, F. Rueda and M. Abd-Lefdi; Materials Research Bulletin 44 (2009) 1458–1461

[22] Kazuhiro Kato, Hideo Omoto, Takao Tomioka and Atsushi Takamatsu; Thin Solid Films 520 (2011) 110–116

[23] Limei Lin, Fachun Lai, Yan Qu, Rongquan Gai and Zhigao Huang; Materials Science and Engineering B 138 (2007) 166–171

[24] Z.B. Fang, Z.J. Yan, Y.S. Tan, X.Q. Liu and Y.Y. Wang; Applied Surface Science 241 (2005) 303–308

[25] J. Sengupta, A. Ahmed and R. Labar; Materials Letters109 (2013) 265–268.

[26] V. Senthilkumar and P. Vickraman; Current Applied Physics 10 (2010) 880–885

[27] Z. Xu, P. Chen, Z. Wu, F. Xu, G. Yang, B. Liu, C. Tan, L. Zhang, R. Zhang and Y. Zheng; Materials Science in Semiconductor Processing 26 (2014) 588–592

[28] Z. Yuan, X. Zhu, X. Wang, X. Cai, B. Zhang, D. Qiu, H. Wu; Thin Solid Films 519 (2011) 3254–3258

6. General Conclusion

To meet the current need of the optoelectronics devices, transparent conductive oxide (TCO) compounds must combine high optical transmission with low electrical resistivity. Among them, doped and undoped SnO_2, ZnO and In_2O_3 thin films have been widely used in optoelectronics devices. However, little attention has been paid to the growth of doped In_2O_3 with transition metals and rare earth element. Therefore, the focus of this chapter was the effect of doping In_2O_3 thin films with iron, zinc, molybdenum, and europium.

Doped indium oxide thin films have been successfully prepared using chemical spray pyrolysis. This technique is one of the most reproducible and efficient technique for the growth of thin films due to simplicity, low process cost, large production scale and homogeneity of the films. For the elaboration of undoped indium oxide thin films, we have dissolved indium chloride ($InCl_3$) in bi-distilled water. Doping operation was carried out during the growth process by adding relative concentration of doping atoms in the sprayed solutions.

To further enhance the physical properties of doped indium oxide layers, especially the electrical properties, an annealing process under nitrogen atmosphere (N_2) was performed on the optimized layers for each doping element. For such investigation, different techniques of characterization have been applied. Among them, X-Ray diffraction, Raman spectroscopy, Atomic force microscopy (AFM), Spectrophotometer, Fluorescence spectroscopy and Hall Effect measurements. In addition, Maud software which is based on Rietveld Method has been used to determine some structural and micro-structural parameters like: microstrain, lattice parameters and phase amounts.

Indium oxide thin films were doped with iron for different iron concentrations y (y = $(\frac{[Fe^{2+}]}{[In^{3+}]})_{sol}$ = 0, 2, 4 and 6 at.%). The structural properties were enhanced compared to undoped layers. The best crystalline was founded for y = 6 at.% with a maximum grain size of about 98 nm. Optical properties show a high transmittance of about 80% in the visible and near infrared regions. The band gap energy is in the range of [3.29 - 3.45] eV. A significant decrease of the electrical resistivity ρ is also founded for this doping level. In fact, ρ deceased from 6502 x 10^{-4} Ω.cm to 197.9 x 10^{-4} Ω.cm for undoped and In_2O_3:Fe (6 at.%), respectively The resistivity continues to diminish by annealing in nitrogen gas for 300 °C during 45 min reaching 26.94 10^{-4} Ω.cm.

For doping with zinc (y = $(\frac{[Zn^{2+}]}{[In^{3+}]})_{sol}$ = 0, 1, 2, 3, 4, 5 and 6 at.%), it is revealed that the best crystallized In_2O_3 films were obtained for a doping ratio equals to 1 at.% with an average of grain size of about 93.4 nm. After doping a slight decrease in the transmission was observed but approximately all films are highly transparent (T in the order of 80 %). Hall Effect measurements show that a minimum of electrical resistivity in the order of 198.50 x 10^{-3} Ω.cm was obtained for this doping ratio. A further decrease of resistivity (ρ) followed by an increase in transmission is shown for optimized In_2O_3:Zn (1 at.%) after heat treatment at 250 °C for 2 hours, reaching a minimum value of about 2.82 x 10^{-3} Ω.cm.

In the case of indium oxide thin films doped with molybdenum y = $(\frac{[Mo^{6+}]}{[In^{3+}]})_{sol}$ = 0, 1, 2, 3 and 4 at.%), it was revealed that physical properties of In_2O_3:Mo have been globally improved after doping by molybdenum. Indeed, structural analysis shows an enhancement of the crystallinity for a doping ratio equals to 3 at.%. For this doping level, the grain size shows a maximum value of about 155 nm. The transmission is in the order of 80%. Electrical resistivity decreased after doping 650.20 x 10^{-3} Ω.cm to 2.03 x 10^{-3} Ω.cm for undoped and In_2O_3:Mo (3 at.%), respectively. Then, the electrical resistivity decreased to 5.55 x $10^{-4}\Omega$.cm after annealing under nitrogen atmosphere at 250°C for 2 hours.

Doping the indium oxide with rare earth element was carried out using low amount of europium (y = $(\frac{[Eu^{3+}]}{[In^{3+}]})_{sol}$ = 0.1; 0.3 and 0.5 at.%). The best physical properties of In_2O_3:Eu thin films were recorded for a ratio y equal to 0.3 at.% according to the Figure of Merit values. At this doping ratio, the grain size was estimated to be about 142.4 nm. The electrical resistivity is reduced up to 30.70 x 10^{-3} Ω.cm for In_2O_3:Eu (0.3 at.%) compared to the value of undoped In_2O_3 layers. A heat treatment process at 250°C for 2 hours leads to a further decrease of ρ by an order of magnitude to about 4.32 x 10^{-3} Ω.cm.

Maud software applied on doped indium oxide thin films reveal the appearance of secondary phases with low amounts for doping with iron, zinc and molybdenum. However, no undesirable phases were detected in the case of europium doped In_2O_3 thin films. Thus, it can be seen that doping with low amount of rare earth europium atom followed by heat treatment at low temperature can be sufficient to obtain a good TCO material without undesirable phases.

Irrespective of the doping atoms, the optical band gap of doped indium oxide thin films was close to the theoretical band gap for TCO compound which is equal to 3.5 eV. Besides, the optical transmission was found to be in the range of 80 % for doped thin films.

Due to the presence of interference fringes for all doped which in turn reflects the homogeneity of the obtained films, we have applied the envelope method in order to determine some optical parameters like refractive index (n), porosity, packing density (p), oscillator energy (E_0) and dispersive energy (E_d) for doped In_2O_3 thin layers

The photoluminescence (PL) performed on doped In_2O_3 thin layers at room temperature for different concentration revealed the presence of three main peaks in the blue green and near infrared region located at 485, 505 and 795 nm. The observed peaks in the visible range were attributed to the recombination of carrier concentration that occurs between valence band and oxygen vacancies acting as donor levels. Whatever the doping atoms used, we observe practically the same peaks for undoped and doped films, such behavior can be explained by the fact that doping atoms do not contribute to the appearance of new emission peaks. However, the PL intensities are sensitive to doping process. In fact, the intensity of the photoluminescence peaks varies as function of atomic concentration of Eu.

Finally, we can conclude that doping with certain atoms followed by an adequate heat treatment leads to an overall enhancement in structural and optoelectronic properties of indium oxide thin films. So, In_2O_3 thin layers presented in this chapter can be a good candidate to be used in various optoelectronic devices and especially as an optical window or transparent electrode in solar cells as follows:

SnO_2:F/**In_2O_3:Fe**/In_2S_3/CZTS or CIGS or SnS or CIS/Au

SnO_2:F/**In_2O_3:Eu**/In_2S_3/CZTS or CIGS or SnS or CIS/Au

SnO_2:F/**In_2O_3:Mo**/In_2S_3/CZTS or CIGS or SnS or CIS/Au

SnO_2:F/**In_2O_3:Zn**/In_2S_3/CZTS or CIGS or SnS or CIS/Au

In_2O_3:Mo/ZnO:Al/In_2S_3/CZTS or CIGS or SnS or CIS/Au

where the growth of CIS [1], SnO_2:F [2], CZTS [3], CIGS [4], In_2S_3[5] and SnS [6] thin films are well controlled in our laboratory.

References:

[1] Z. Seboui, M. Ajili, N. Jebbari, and N. Kamoun-Turki; ur. Phys. J. Appl. Phys. (2013) 62: 30302

[2] N. Jebbari, B. Ouertani, M. Ramonda, C. Guasch, N. Kamoun-Turki and R. Bennaceur; Energy Procedia 2 (2010) 79

[3] Z. Seboui, A. Gassoumi and N. Kamoun-Turki; Materials Science in Semiconductor Processing 26 (2014) 360–366

[4] M. Ajili, M. Castagné and N. Kamoun-Turki, J. Lumin. 150, 1–7 (2014)

[5] M. Ajili and N. Kamoun- Turki; J Mater Sci: Mater Electron (2014) 25:3840–3845

[6] M. Reghima, A. Akkari, C. Guasch and N. Kamoun; Journal of Electronic Materials 44 (2015) 4392

Synthesis by Chemical Spray Pyrolysis and Characterization of TiO$_2$ Thin Films

Wafa Naffouti, Tarek Ben Nasr and Najoua Kamoun Tuki

Université Tunis El Manar, Faculté des Sciences de Tunis, Département de Physique, LR99ES13
Laboratoire de Physique de la Matière Condensée (LPMC), 2092 Tunis Tunisie, Tunisia

Abstract: Photovoltaic devices are based on numerous compounds. A lot of them are unsafe to handle and costly to obtain. Therefore, the need of alternative compounds that are abundant in the earth's crust and non-toxic has become an essential challenge. Taking all these considerations into account, titanium dioxide semiconductor can be a promising candidate for industrial photovoltaic applications due to its perfect chemical stability, oxidative ability, wide indirect band gap, n-type conductivity and non-toxicity. In fact, titanium dioxide (TiO$_2$) has been studied extensively for optoelectronic applications including supercapacitors, biosensors, photocatalysts and photovoltaic applications. It is a promising candidate for many technological applications due to its long-term stability, low cost, low biological toxicity, large electron mobility and exciton binding energy. Moreover, this compound possesses many interesting structural characteristics, as well as optical properties such as high optical transmission in the visible spectral range and good refractive index.

Table of Contents

1. Introduction

Despite the global development of energy consumption, reserves of fossil fuels such as oil, coal and natural gas, will be exhausted if alternatives are not developed. In fact, the energy demand and consumption increase drastically with world's growing population as well as technological development (Figure 1). Thus, this drastic increase in energy requirement is one of the biggest challenges faced by our generation. Therefore, both scientific and political point of views are focused on renewable and sustainable energy such as geothermal, wind, biomass, marine flow and photovoltaic solar cells as attractive alternatives to preserve the comfort of human beings while respecting nature. Indeed, renewable energy does not generate waste or polluting emissions. It participates on the fight against the greenhouse effects and CO_2 emissions in the atmosphere resulting from human activity, such as fossil fuel burning and deforestation.

Solar energy is the most abundant renewable and sustainable energy source. It is inexhaustible, respectful to the environment and it has several advantages. Figure 2 depicts the average global insolation all over the world. The total amount of solar energy received at the ground for a week exceeds the energy produced by the world's reserves of oil, coal, gas and uranium. The photovoltaic effect, discovered by the French physicist Alexandre-Edmond Becquerel in 1839, involves the direct conversion of sunlight into electrical energy. Currently, photovoltaic devices are based on a huge number of compounds. A lot of them are unsafe to handle and costly to obtain. Therefore, the need of alternative compounds that are abundant in the earth's crust and nontoxic becomes an essential challenge. Taking all these considerations into account, titanium dioxide semiconductor can be a promising candidate for industrial photovoltaic applications due to its perfect chemical stability, oxidative ability, wide indirect band gap, n-type conductivity and non-toxicity.

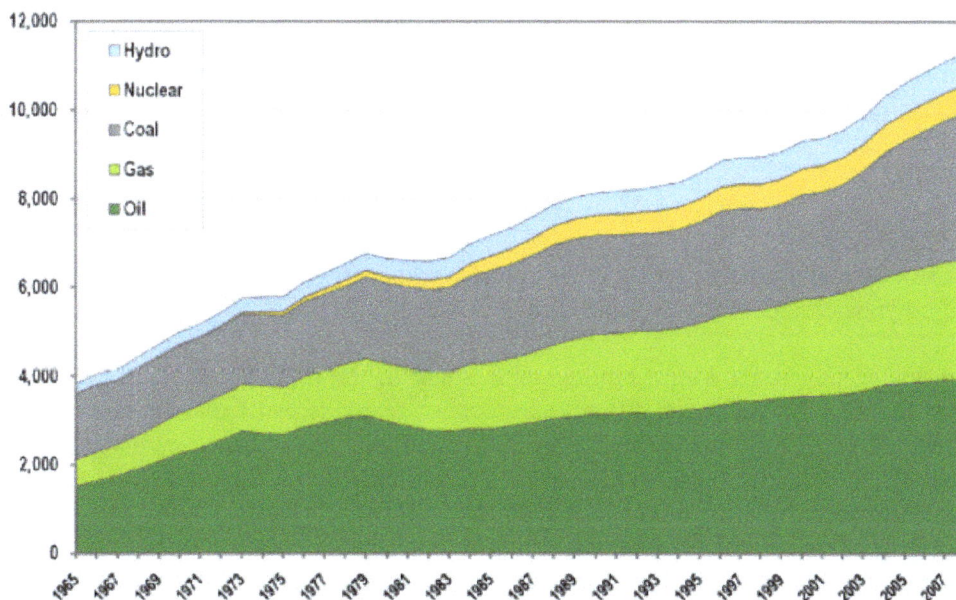

Figure 1. Worldwide energy consumption from 1965 to 2008. The units are in million tons of oil equivalent [1].

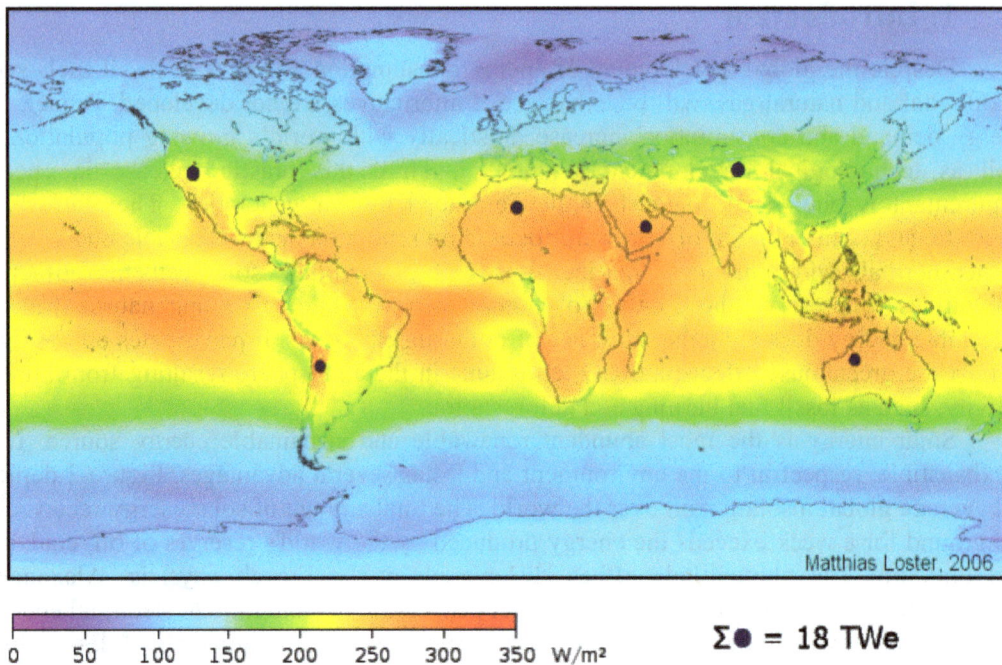

Figure 2. Map of average global insolation. The black dots represent possible locations of large solar power plants that in total could provide the world's total energy demand [2].

Therefore, this chapter details the use of spray pyrolysis technique to deposit TiO$_2$ thin films for potential use in several optoelectronic devices. It is worth noting that spray pyrolysis technique is not frequently employed for the growth of TiO$_2$ thin films compared to the sol gel technique. Thus, this study has shown the desirable ability to control the effect of growth parameters as well as doping process on the structural and optical properties of TiO$_2$ thin films. Furthermore, an investigation of thermal and transport properties of anatase TiO$_2$ compound using WIEN2k package has been achieved, which are not discussed previously.

References
[1] British Petroleum (BP). Statistical review of world energy. Technical report, 2009.
[2] http://www.ez2c.de/ml/solar_land_area/

This section deals with a bibliographic study of the physical properties of titanium dioxide thin films as well as its different preparation methods. Particular attention is paid to describe the TiO$_2$ thin films elaboration technique "chemical spray pyrolysis" and the deposition process, in the second part of the section. Structural, morphological and optical characterization will be briefly described after. These characterizations include X-ray diffraction, Atomic Force Microscopy, Scanning Electron Microscopy, spectrophotometry and spectrofluorimetry.

I.1. Applications of TiO₂

Titanium dioxide (TiO_2) compound with a wide indirect band gap and n-type conductivity has been studied extensively for optoelectronic applications including supercapacitors [1], biosensors [2], photocatalysts [3,4,5], light emitting diodes [6,7] and photovoltaic applications [8,9,10]. It is a promising candidate for many technological applications due to its long-term stability, low cost, low biological toxicity, large electron mobility and exciton binding energy. Moreover, this compound possesses many interesting structural characteristics, as well as optical properties such as high optical transmission in the visible spectral range and good refractive index.

I.2. Elaboration techniques

Currently, various synthetic routes have been employed to grow TiO_2 thin films including: chemical bath deposition [11,12], sol-gel method [5,13], electron beam evaporation [3,14], polymer assisted hydrothermal deposition (PAHD) method [15], pulsed laser deposition (PLD) [16], high energy reactive magnetron sputtering (HE RMS) [17], DC-magnetron sputtering deposition [18] and spray pyrolysis technique [19,20,21]. Among them, spray pyrolysis technique is cheap, easy-to-achieve and suitable for large area deposition. Compared to the others, it offers opportunities to precise control of growth parameters.

I. 3. Structural properties of TiO₂ thin films

Titanium dioxide (TiO_2), also known as titania, exhibits a rich phase diagram including more than seven different polymorphs. The most abundant forms are: tetragonal anatase and rutile. Anatase is commonly observed at film deposition temperatures of 350-700°C, while higher temperature promotes the growth of rutile phase. The orthorhombic brookite is another commonly known polymorph of titania material. However, it is unstable phase which can only be obtained by complex preparation but never as a single phase [22]. Thus, the description of the properties of TiO_2 material will be concentrated only on rutile and anatase phases.

A summary of the most important physical properties of these phases is depicted in Table I-1. Their unit cells are presented in Figure I-1. As it can be seen from this Figure, the rutile structure is the simplest one. It crystallizes in a tetragonal structure with space group P42/mnm (N° 136) [23]. The unit cell contains two titanium situated at (0,0,0) and (½a,½a,½c) and four oxygen atoms, where the latest form a distorted octahedron around every Ti atom. However, anatase phase crystallizes in the tetragonal structure with space group I41/amd (N° 141) [23], where the unit cell contains four titanium located at (0,0,0), (½a,½a,½c), (0,½a,¼c) and (-½a, 0,-¼c) and eight oxygen atoms. The basic building units of both rutile and anatase structures are constituted from TiO_6 octahedra. However, they differ from each other by the arrangement and the distortion of the octahedra. In order to improve the structural properties of titanium dioxide thin films, several investigations were interested by the study of different growth parameters:

Effect of the deposition time: A group of research reported the effect of the deposition time on the structural properties of TiO_2 thin films elaborated by chemical bath deposition (CBD) [11]. The obtained XRD patterns are depicted in Figure I-2.

Table I-1. Physical properties of rutile and anatase phases of TiO$_2$

Phase	Rutile	Anatase
Crystal structure	Tetragonal	
Space group	P4$_2$/mnm	I4$_1$/amd
a (A$^\circ$)	4.5937 [24]	3.7843 [25]
c (A$^\circ$)	2.9587 [24]	9.5146 [25]
Unit cell volume (nm^3)	0.0624	0.1363
Density (Kg cm^{-3})	4.25	3.89
Electrical properties at RT	n-type semiconductor	n-type semiconductor
Indirect band gap (eV)	3.02-3.24	3.23-3.59
Refractive index	2.79-2.90	2.49 -2.54
Solubility in HF	Insoluble	Soluble
Solubility in H$_2$O	Insoluble	Insoluble
Bulk modulus (GPa)	206	183

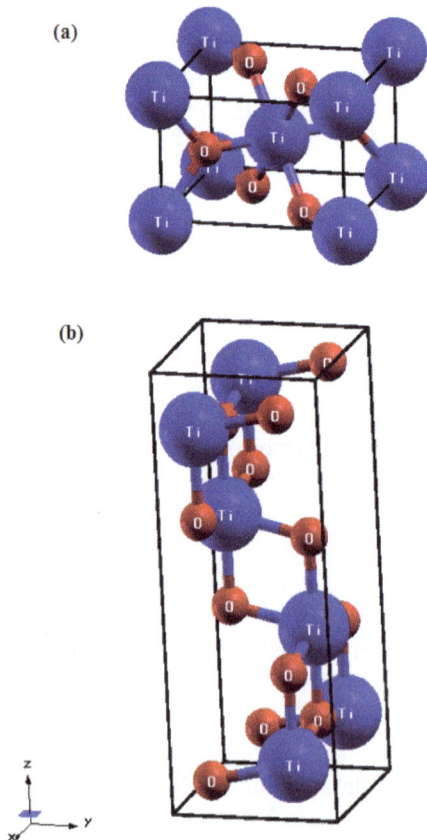

(a)

(b)

Figure I-1: The unit cell of (a) rutile and (b) anatase TiO$_2$ crystal structures.

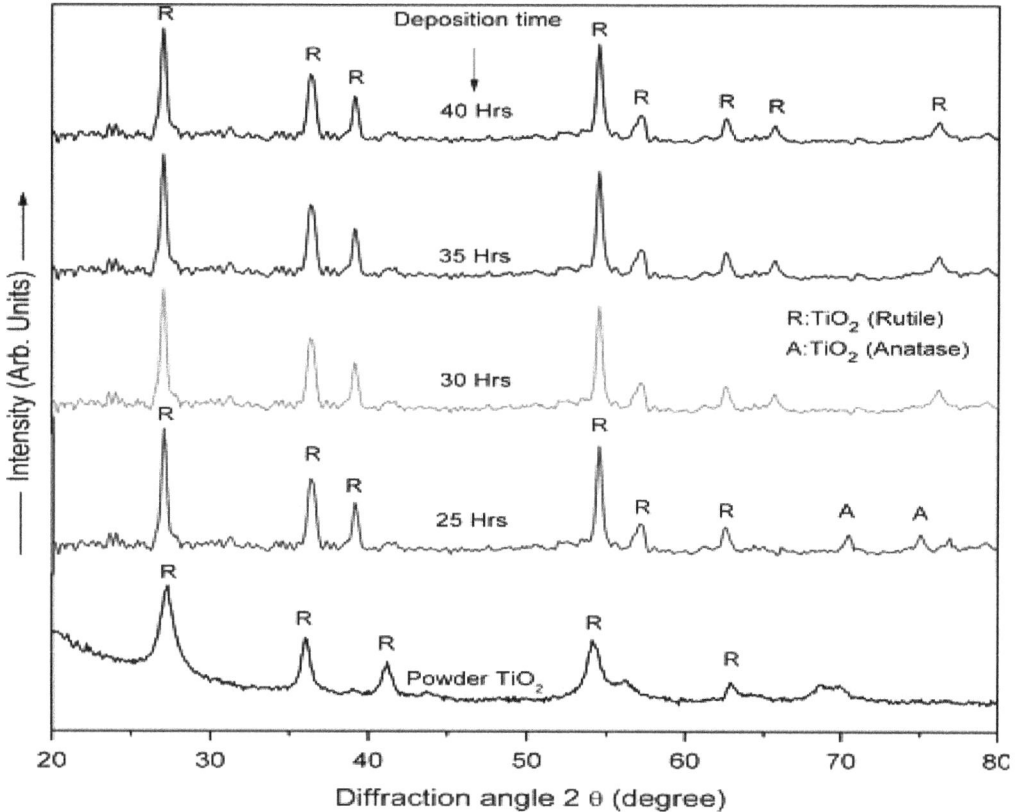

Figure I-2. XRD patterns of TiO_2 thin films prepared on glass substrates for different deposition times [11].

As it can be seen from this Figure, that both tetragonal rutile and anatase phases are identified in the case of films deposited for 25 h. However, for TiO_2 thin films grown using longer deposition time, only tetragonal rutile phase persists. This means that a simple increase of deposition time can lead to a transition from biphasic anatase-rutile to pure rutile phase.

Effect of the substrate temperature

Several groups have studied the effect of substrate temperature on the structural properties of titanium dioxide thin films. Among them, Moses Ezhil Raj *et al.* [19] have prepared TiO_2 thin films using spray pyrolysis technique. Figure I-3 displays the XRD patterns of TiO_2 films deposited at different substrate temperatures (T_s).

It is observed that the films deposited at Ts $\leq$ 450°C crystallized into the anatase phase where the (101) is the preferred orientation. However, for further increase of the substrate temperature, the rutile phase arises directly without any heat treatment. Nevertheless, this result is not consistent with that published by Abou-Helal *et al.* [20]. In fact, they found that even for an increase of substrate temperature up to 550°C, the chemically sprayed films crystallized into the anatase phase.

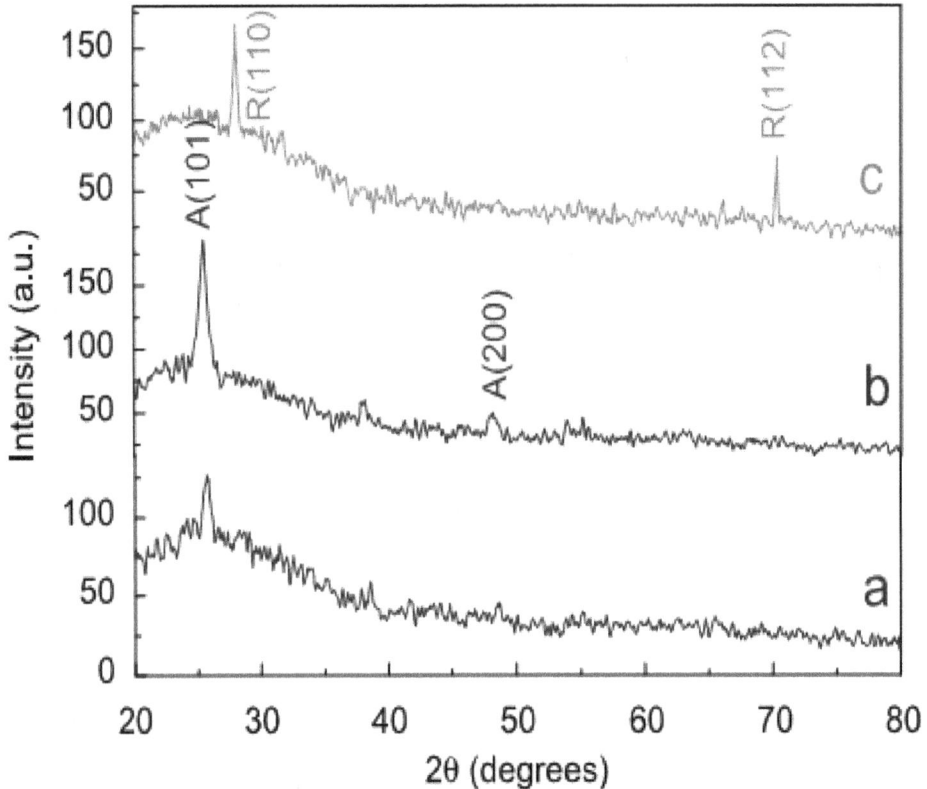

Figure I-3. XRD pattern of TiO$_2$ thin films deposited at different substrate temperatures: (a) 400, (b) 450 and (c) 500°C [19].

Effect of the film thickness

The study of the influence of film thickness was also investigated. In this context, P. Shinde *et al.* [21] have fabricated TiO$_2$ thin films on soda lime glass using spray pyrolysis technique for different film thicknesses ranging from 25 to 785 nm. According to XRD studies (Figure I-4), the formation of the anatase phase is revealed for thickness higher than 123 nm with a crystallite size on the order of 40.5 ± 2.5 nm. Furthermore, we can remark that the crystalline quality of the films improves with further increase of film thickness. This last finding is in agreement with that published by other report [26].

Effect of the annealing process

The heat treatment is an important parameter which can be useful in order to improve the crystalline quality of the films. Numerous reports were concentrated on the study of the effect of annealing process on the structural properties of titanium dioxide thin films. Thus, Elfanaoui *et al.* [12] reported that TiO$_2$ thin films elaborated using CBD technique crystallize into the anatase phase after annealing in air for 1 h, at annealing temperature < 700°C. However, rutile phase appears when the annealing temperature reached 700°C. This result is consistent with other investigations [27,28,29].

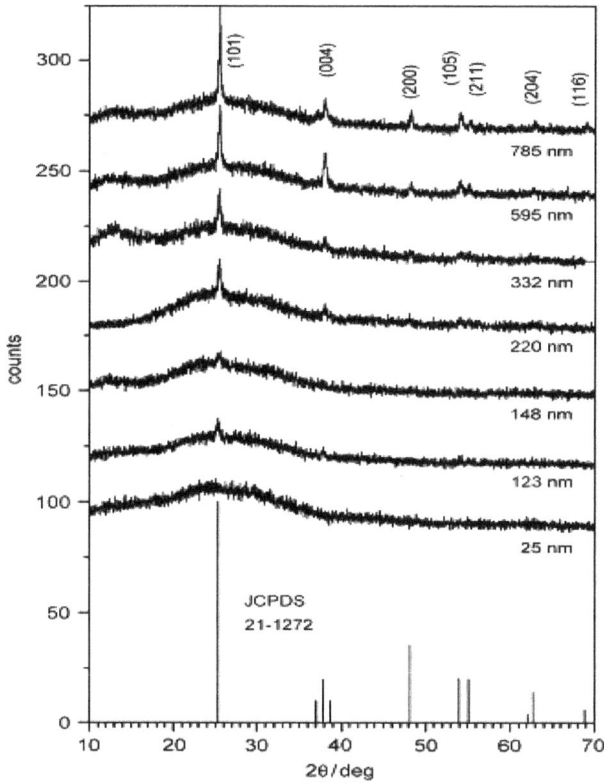

Figure I-4.
X-ray diffraction patterns of TiO_2 thin films deposited on soda lime glass, with different thicknesses [21].

Ben Naceur *et al.* [5] prepared TiO_2 thin films on Si substrates by sol gel technique. The films were annealed in air for 1 h. As shown in Figure I-5, the transition from anatase to rutile phase occurs at 1000 °C. The crystallite size increases from 12.58 to 35.17 nm when the annealing temperature increases from 400 to 1000°C.

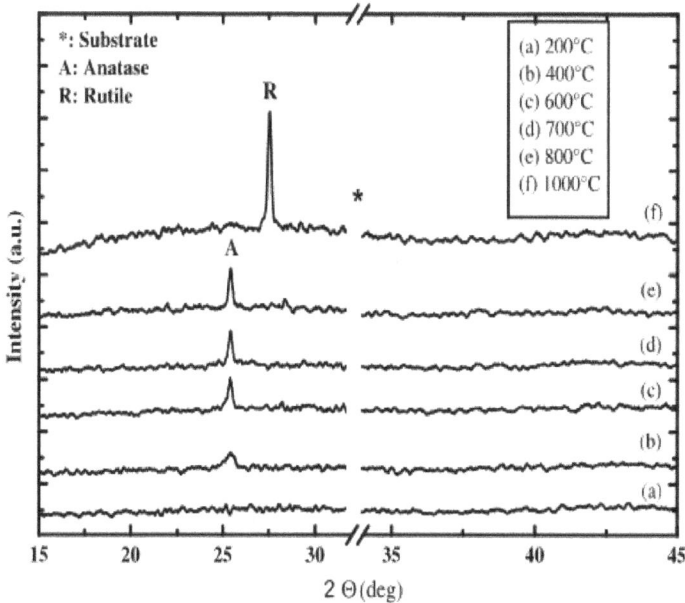

Figure I-5.
X-ray diffraction patterns of TiO_2 thin films as a function of annealing temp, A = anatase, R = rutile and * = substrate [5].

I.4. Morphological properties of TiO₂ thin films

Effect of deposition time

It is certain that the deposition time may affect the surface topography of TiO$_2$ thin films, as shown in Figure I-6. In fact, Mayabadi *et al.* [11] published that there is a significant difference in structure of TiO$_2$ films deposited by CBD technique as the deposition time increases. Indeed, films grown at low deposition time are formed by grains with an irregular shape and size. This behavior may be explained by the premature termination of the growth process at low deposition time. However, we remark the formation of semi-spherical particles with increasing deposition time.

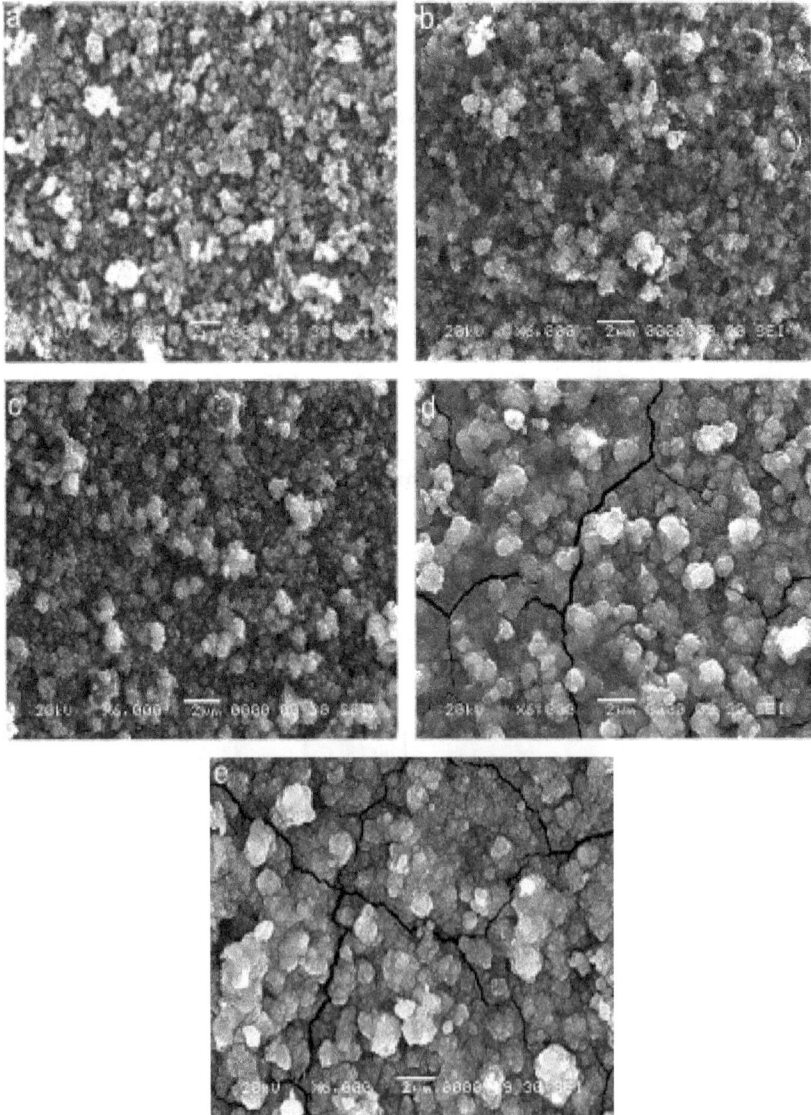

Figure I-6. SEM images of TiO$_2$ films for deposition times of (a) 25, (b) 30, (c) 35, (d) 40 h and (e) 45 h [11].

Effect of substrate temperature

Moses Ezhil Raj *et al.* [19] reported that for low substrate temperature, the film surfaces are very smooth, as displayed in Figure I-7. However, as the substrate temperature increases, the grains begin to grow in size reflecting an improvement of the crystalline quality of sprayed TiO_2 thin films.

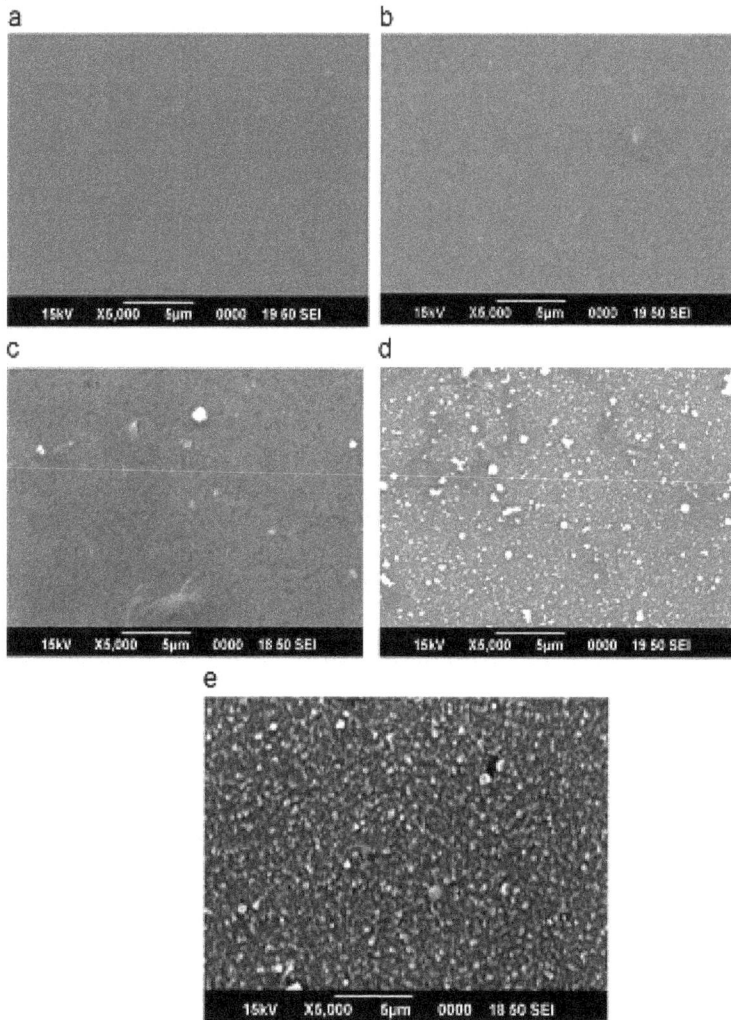

Figure I-7. SEM micrographs of TiO_2 films deposited at different substrate temperatures: (a) 300, (b) 350, (c) 400, (d) 450 and (e) 500°C [19].

Effect of film thickness

Shinde *et al.* [21] showed that film thickness may affect the surface topography of sprayed TiO_2 thin films. Indeed, Figure I-8 depicts the SEM images of TiO_2 thin films deposited on FTO at various film thicknesses. It is clear that the surface morphology changed from granular to platelet shape with increasing film thickness.

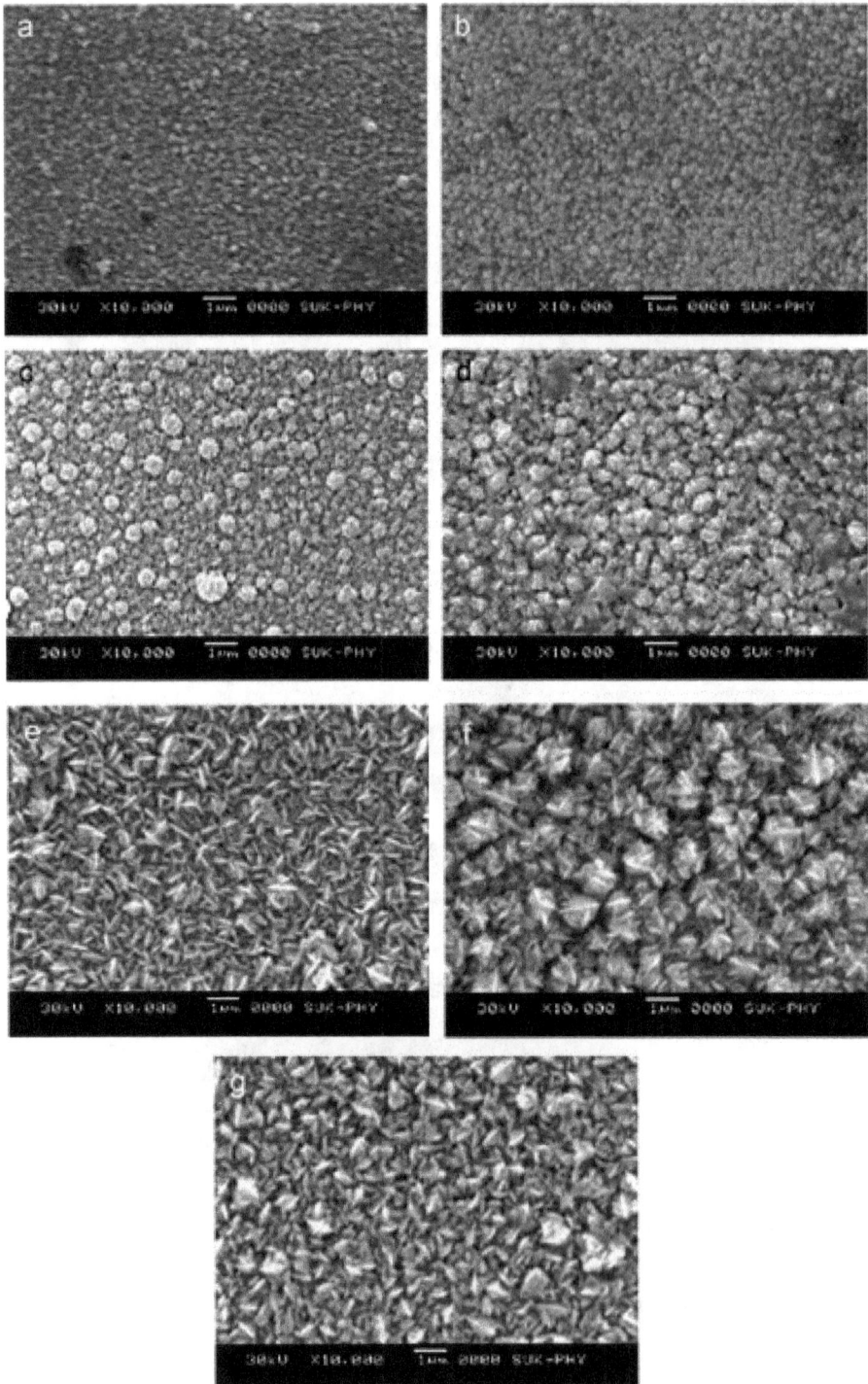

Figure I-8. SEM images of sprayed TiO$_2$ films on FTO with different thicknesses, (a) 25 nm, (b) 123 nm, (c) 148 nm, (d) 220 nm, (e) 332 nm, (f) 595 nm and (g) 785 nm [21].

Effect of annealing process

Ben Naceur *et al.* [5] demonstrated that heat treatment can significantly improve the crystalline quality of TiO_2 thin films elaborated by sol gel process on Si substrates. In fact, at low annealing temperature, the film surface is very smooth. Then, with increasing annealing temperature, a granular morphology starts to appear as seen in Figure I-9.

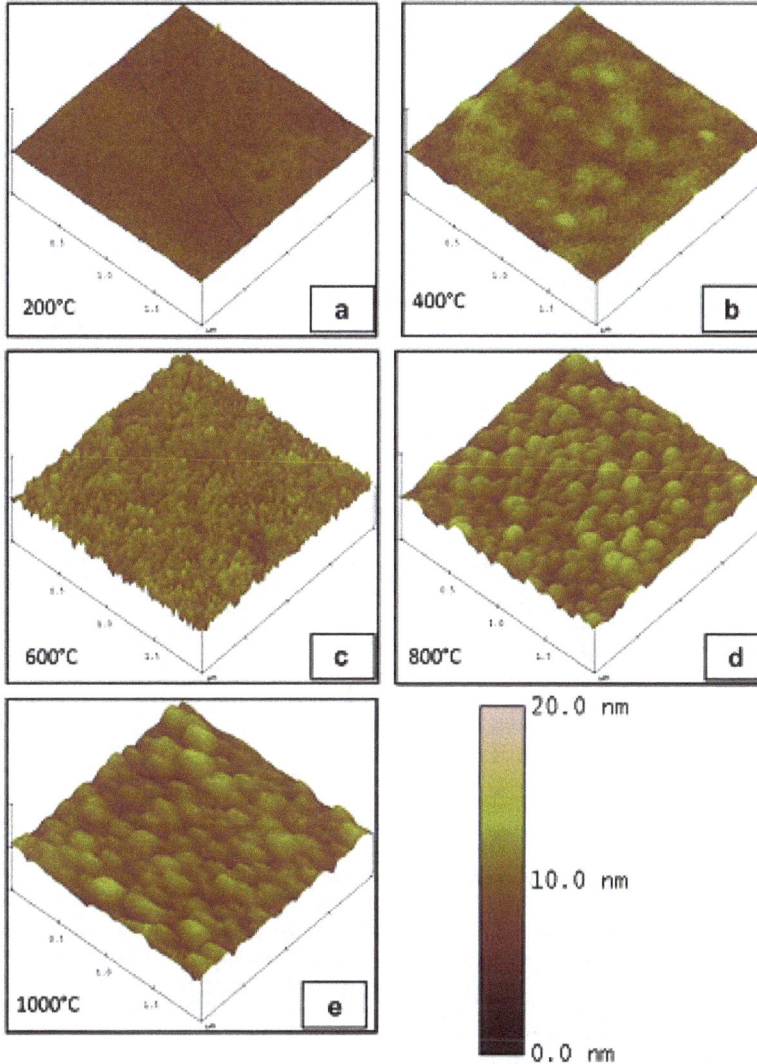

Figure I-9. 3D AFM images of TiO_2 thin films obtained after annealing temperatures of (a) 200, (b) 400, (c) 600, (d) 800 and (e) 1000 °C [5].

I.5. Optical properties of TiO_2 thin films

Effect of the deposition time

Mayabadi *et al.* [11] studied the effect of deposition time on the optical properties of titanium dioxide thin films elaborated by CBD technique. Indeed, the extrapolation of the linear part of $(\alpha h\nu)2$ versus $(h\nu)$ plots presented in Figure I-10 yields the direct band gap Eg.

The obtained values are listed in the table below. It is clear that the band gap value decreases from 3.29 to 3.07 eV with deposition time increasing. This result may be explained by the high crystallinity of the elaborated films.

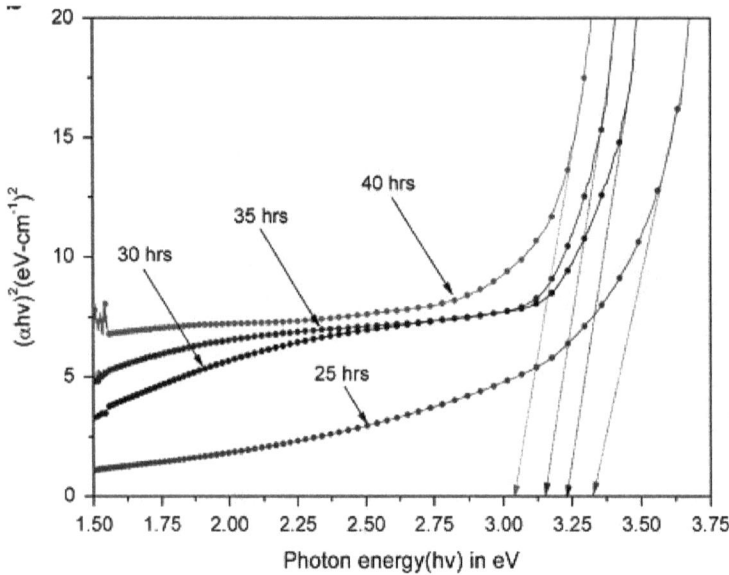

Figure I-10. $(\alpha h v)^2$ versus hv plots for TiO$_2$ thin films on glass prepared using different deposition times [11].

Table I-2. Optical properties for CBD-deposited TiO$_2$ thin films [11].

Deposition time (h)	Phase	Thickness (μm)	E$_g$ (eV)
25	Rutile-anatase	0.5	3.29
30	Rutile	0.7	3.21
35	Rutile	1.1	3.14
40	Rutile	1.4	3.07

Effect of the substrate temperature

Abou-Helal [20] reported that the substrate temperature had an effect on transmission spectra of sprayed TiO$_2$ thin films. Thus, as shown in Figure I-11, the optical transmission of the elaborated films is high around 75% in the visible range, which allows the use of this material as an optical window in photovoltaic devices. Moreover, there is a slight enhancement of transmission with substrate temperature increment. This behavior may be due to the improvement of crystalline quality of TiO$_2$ thin films.

Effect of the film thickness

A numerous investigations studied the effect of film thickness on the optical properties of titanium dioxide thin films. Among them, Shinde *et al.* [21] published that the transmission of sprayed films decreased with film thickness increment due to scattering. Moreover, the transmission is maxim for the thinnest layer and it is around 90 %. All spectra show a strong decrease at 400 nm due to the absorption edge of TiO$_2$ material.

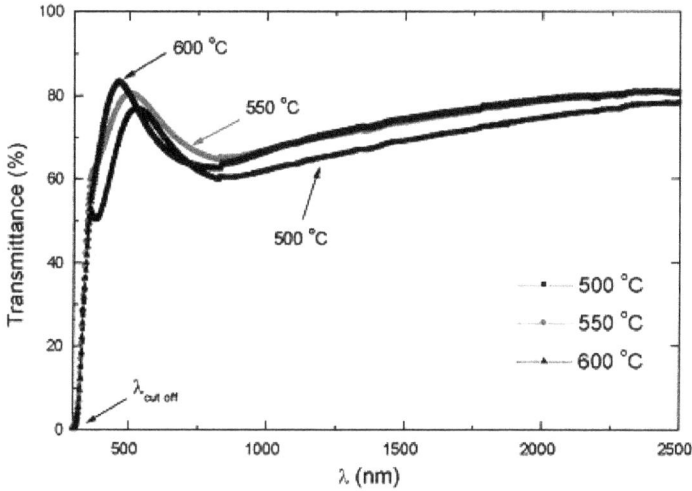

Figure I-11. Transmission spectra of sprayed TiO_2 films elaborated on glass substrates at various substrate temperatures [20].

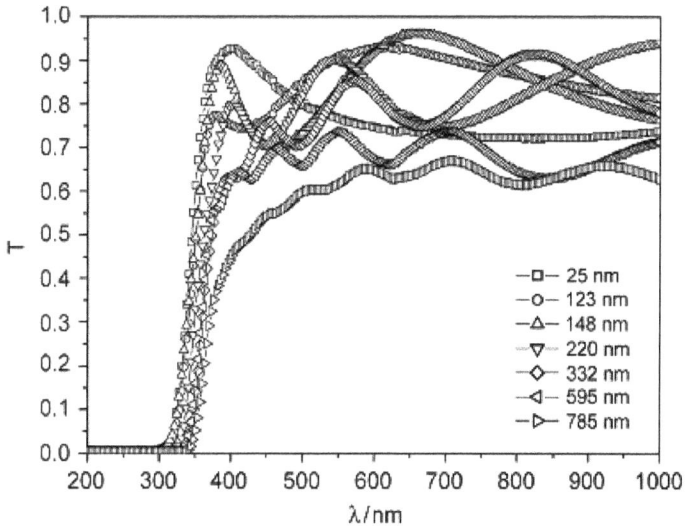

Figure I-12. Plot of transmittance versus wavelength for sprayed TiO_2 thin films elaborated with different thicknesses ranging from 25 to 785 nm [21].

Effect of the annealing process

Elfanaoui *et al.* [12] investigated the effect of annealing temperature on the optical properties of TiO_2 thin films deposited on glass substrates using CBD technique. It is clear from Figure I-13 that the optical transmission is around 50% for films annealed at $500°C$, then it reaches 75% when the annealing temperature increases to $600°C$. This increase may be due to the improvement of film crystallinity with annealing temperature increment. However, for films annealed at $700°C$, the optical transmission decreases and reaches 50%.

Figure I-14 depicts $(\alpha h\nu)^{1/2}$ versus $h\nu$ plots of TiO_2 thin films annealed at various annealing temperatures. The estimated indirect band gap decreases from 3.22 to 2.88 eV when annealing temperature increases from 500 to $700°C$.

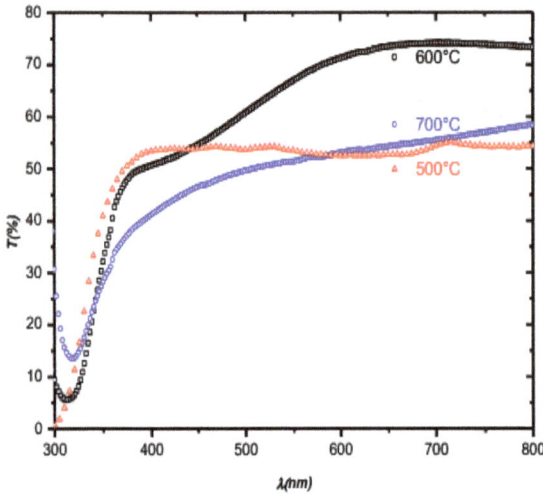

Figure I-13. Optical transmission spectra of TiO₂ thin films deposited by CBD technique on glass substrates and annealed at 500, 600 and 700°C [12].

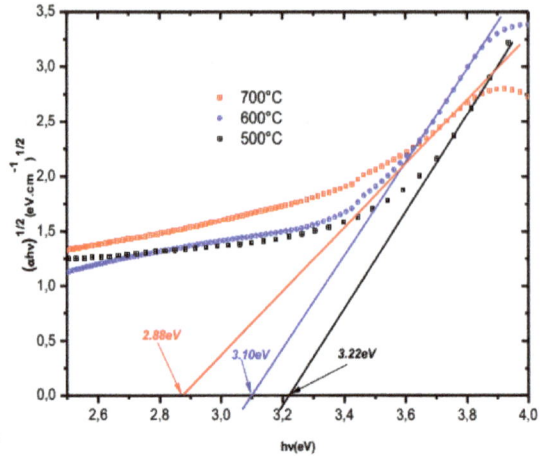

Figure I-14. The $(\alpha h\nu)^{1/2}$ versus hν plots of TiO₂ thin films annealed at 500, 600, and 700°C [12].

Moreover, Vishwas *et al.* [14] published the effect of heat treatment on photoluminescence spectra of TiO₂ thin films deposited on quartz substrates using electron beam evaporation method. The films were excited with an excitation wavelength of about 325 nm. As depicted in the Figure I.15 bellow, three prominent emission peaks can be identified. These peaks are located at 410, 542 and 664 nm and corresponded to UV, green and red emissions, respectively. When the annealing temperature is higher than 400°C, a red shift of the emission band maxima can be detected and an enhancement of PL intensity can be observed. This behavior may be due to crystalline quality improvement.

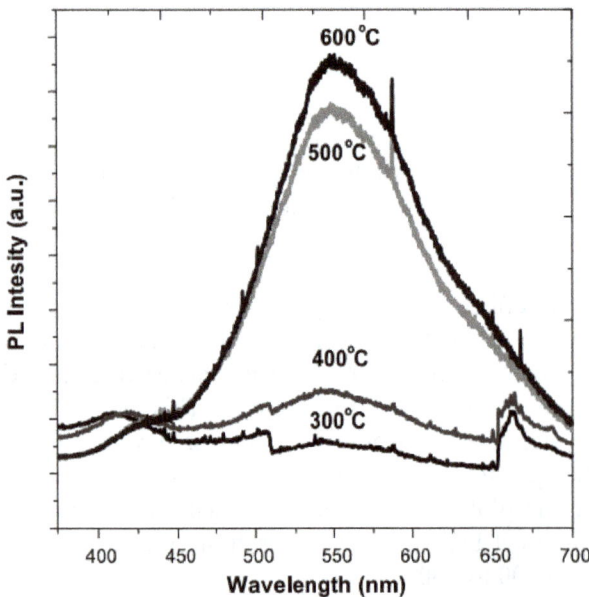

Figure I-15. Photoluminescence spectra of TiO₂ films annealed at different annealing temperatures for 4 h [14].

I.6. Experimental setup

A thin film is a material whose thickness has been greatly reduced, so that it can reach in some cases the order of a nanometer. It can therefore be considered as being two-dimensional. Generally, this reduction leads to a significant disruption in the physical properties of these thin films. In fact, the effects related to the surface boundary can be dominant for the material elaborated as thin film while they are negligible in the solid state.The deposition of thin films is always realized on another material, called "substrate". Both electronic semiconductor devices and photovoltaic cells are the main applications benefiting from thin film elaboration for their insulating or conductive properties.

Description of the elaboration technique "Chemical Spray Pyrolysis"

TiO_2 thin films were deposited using chemical spray pyrolysis technique. This method consists on the pulverization of a transparent solution containing the reagents on a heated substrate in order to form the desired material, as shown in Figure I-16.

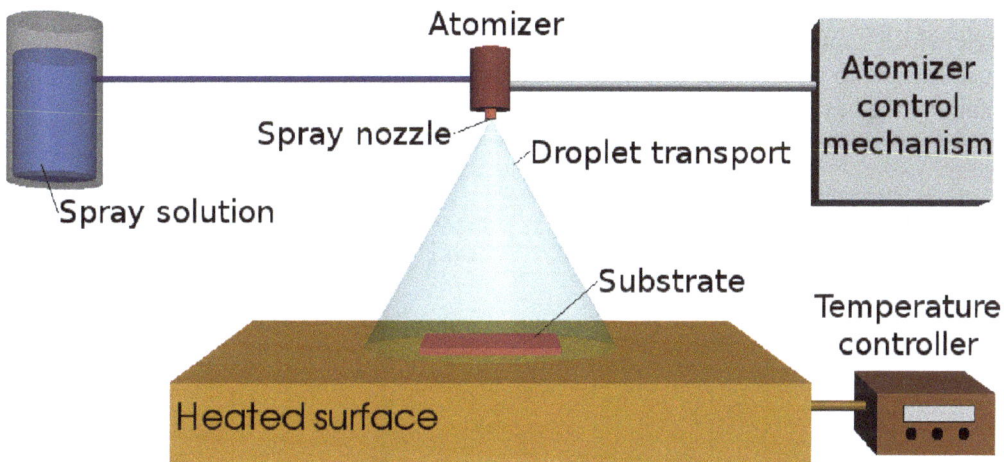

Figure I-16. Schematic diagram of chemical spray pyrolysis technique.

The equipment of this technique is composed of:

The heated plate: The used heated plate is an iron cuboid block heated using resistors placed in parallel and attached to a thermocouple. This thermocouple is fixed at the center of the front side of the iron plate and it is used to measure the substrate temperature. A temperature controller connected to the thermocouple serves to monitor the temperature of the hot plate. In order to evaluate the homogeneity of the temperature over the surface of the heated plate, the variation of the temperature controller Tr versus the substrate temperature Ts has been plotted in Figure I-17. It is noted that the variation is quasi-linear which signifies an uniformity of the temperature over the surface of the heated plate.

The pump: The flow rate of the sprayed solution is controlled by a suction pump (Masterflex cosole drive). The selected carrier gas is air or nitrogen. The choice of flow rate is the result of a compromise between the speed of growth of thin films which must be the greatest possible and the good crystallinity of these layers.

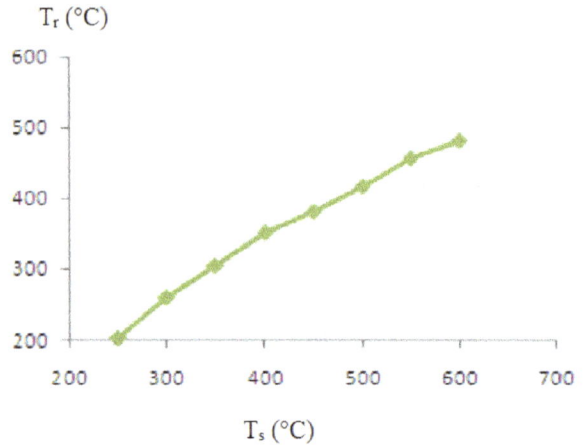

Figure I-17.
Calibration curve of the temperature controller T$_r$ versus the substrate temperature T$_s$.

The nozzle: The sprayed solution is pulverized onto the heated substrates through a nozzle which contains two pipes for liquid and carrier gas, as presented in Figure I-18. The nozzle is fixed to an atomizer allowing an uniaxial scanning over the whole isothermal zone containing the heated substrates.

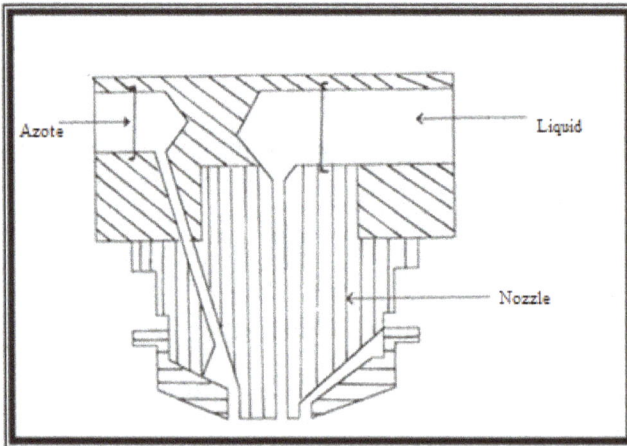

Figure I-18. The nozzle.

Deposition process and TiO$_2$ thin film preparation

The idea is to suck a transparent solution using a pump. Then, an aerosol will be generated from the conversion of the pumped solution to fine droplets, at the nozzle tip with the help of compressed air. These fine droplets will be sent towards the substrate surface. It is important to mention that the sprayed solution should be transparent in order to prevent the clogging of the nozzle. During the movement from the nozzle towards the heated substrate, the fine droplets (aerosols) undergo various physical and chemical changes as displayed in Figure I-19. In fact, after leaving the nozzle, the solvent is completely evaporated during the flight of the aerosols toward the substrate. Before reaching the heated substrate, a vaporization of the precursors containing chemical species takes place. Then, this vapor diffuses to the heated substrate and a chemical reaction takes place leading to the formation of the desired material.

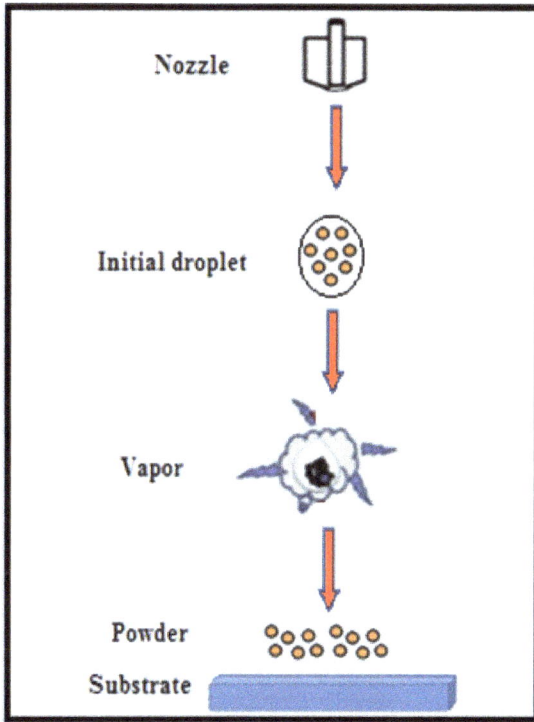

Figure I-19.
Spray pyrolysis droplets modifying as they are transported from the atomizing nozzle to the substrate.

Compared to various synthetic routes used to produce titanium dioxide thin films, the spray pyrolysis technique is relatively inexpensive, simple, and suitable method for large-scale manufacturing. Furthermore, it allows an easy control of the experimental parameters such as sprayed solution flow rate, substrate temperature, substrate nature, sprayed volume, solvent, solution pH, etc. Therefore, all the experimental parameters used to depose TiO_2 thin films will be provided in the next paragraph.

TiO_2 thin film preparation

Sprayed titanium dioxide thin films are deposited using a solution containing ethanol (C_2H_6O), acetylacetonate (AcAc, $C_5H_8O_2$) and titanium tetraisopropoxide (TTIP, $C_{12}H_{28}O_4Ti$) as depicted in Table I-3.

Table I-3. Precursors for preparation of TiO_2 thin films [30].

Precursors	C_2H_6O	$C_5H_8O_2$	$C_{12}H_{28}O_4Ti$
Volume (mL)	54	3.6	2.4

The obtained solution is transparent with yellowish color. The TTIP is used as Ti source while ethanol and AcAc are used as solvent and stabilizer, respectively [31]. In order to remove excess organics and to improve the crystalline quality, a post thermal treatment was performed at 500°C for 2 hours in ambient atmosphere for all the deposited films. The deposition of titanium dioxide thin films results at high temperature according to the following reaction mechanism [32]:

$$Ti(OC_3H_7)_4 \rightarrow TiO_2 + 2\ C_3H_6 + 2\ C_3H_7OH$$

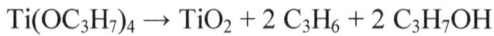

During these experiments, three steps were performed in order to optimize the elaboration process, as follows:

- **First step:** TiO$_2$ thin films were deposited on glass substrates at different solution flow rates (FR = 2, 7, 10 and 15 ml/min) where the other parameters are kept constant [30] (Ts = 350°C and Vsol = 60 ml).

- **Second step:** Using the optimized flow rate (FR = 10 ml/min), TiO$_2$ thin films were grown on glass substrates at different sprayed solution volumes (Vsol = 30, 60, 90 and 120 ml). The substrate temperature was kept constant (Ts = 350°C).

- **Third step:** In this experiment, the films were deposited using the optimized parameters (FR = 10 ml/min and Vsol = 60 ml) where the substrate temperature varied from 250 to 400°C by step of 50°C.

The optimization of the elaboration process was followed by a heat treatment under ambient atmosphere for two hours at different annealing temperatures (500, 700, 900, 1000 and 1100°C). Doping with samarium and manganese at different doping ratios was also investigated.

I.7. Thin film characterization techniques

Several relevant characterization techniques were used throughout the work of the present chapter in order to investigate the physical properties of titanium dioxide thin films. Indeed, the film crystallinity is studied by X-ray diffraction (XRD), using XPERT-PRO Diffractometer. Both atomic force microscope (AFM) and scanning electron microscope (SEM) were used for surface topography and roughness measurements. The optical measurements were carried out using Perkin–Elmer Lambda 950 spectrophotometer. The photoluminescence measurements were performed, at room temperature, using a Perkin-Elmer LS 55 Fluorescence spectrometer. A short description of these techniques is given below.

X-ray diffraction "XRD"

X-ray diffraction technique is used as a tool to crystallographic analysis. It is the most suitable, versatile and non-destructive method for structural analysis which can provide detailed crystallographic information such as crystallite size, lattice parameters, planes orientation and microstrain. It consists on the bombardment of a crystal with a monochromatic beam of X-ray (wavelengths in the A° region) in order to produce the corresponding diffraction pattern. This incident beam will be scattered in all directions by the atoms (Figure I-20). Owing to their regular arrangement, the scattering waves will produce constructive interferences, in certain directions, when conditions satisfy Bragg's Law:

$$n\lambda = 2d \sin\theta \quad \text{(I-1)}$$

Where n is an integer, λ is the X-ray wavelength, d is the inter-atomic spacing and θ is the angle between the incident x-ray beam and the crystal lattice planes.

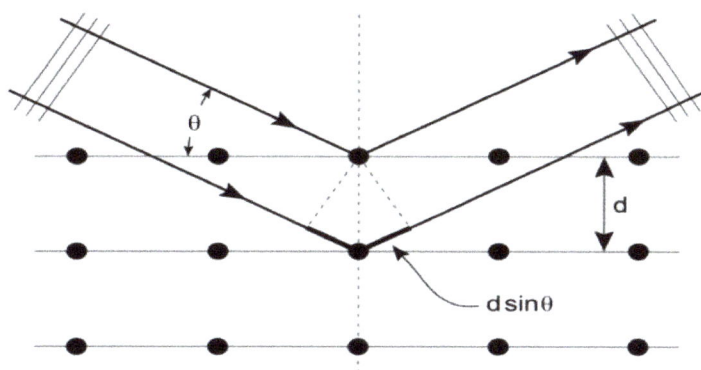

Figure I-20.
Scheme of X-ray
diffraction.

During this work, all XRD measurements were performed using an XPERT-PRO Diffractometer system for 2Θ values over 20–60 degrees. The wavelength, the accelerating voltage and the current were 1.5418 Å, 40 kV and 30 mA, respectively. The measurements were conducted at room temperature.

Atomic force microscopy "AFM"

Atomic force microscopy (AFM) is one of the most powerful tools for topographical imaging. The principle of the AFM microscope is depicted in Figure I-21. It consists on scanning the specimen surface using a mechanical probe which is a pyramidal tip mounted at the end of a cantilever. Generally, both tip and cantilever are made of Si or Si3N4. Bringing the tip into proximity of the sample surface will give rise to attractive or repulsive forces in the order of 10-9-10-5 N between the tip and the sample causing the deflection of the cantilever. This deflection is most often measured by focusing a laser beam on the back side of the cantilever and then the reflected light is detected using a position sensitive detector, as displayed in Figure I-21.

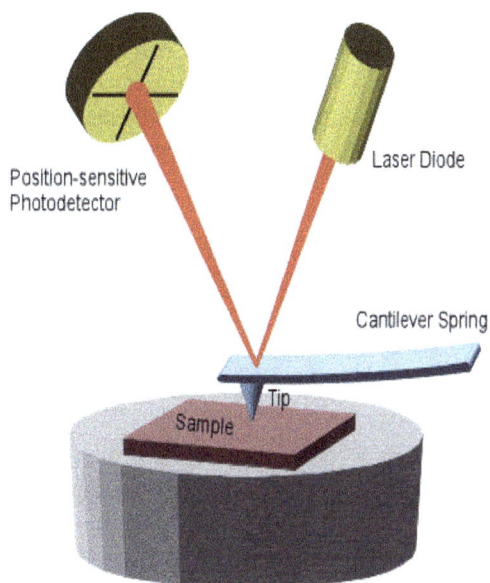

Figure I-21.
Principle of the
Atomic Force Microscopy.

Three surface topography modes performed by AFM can be used:

- Contact mode, or static mode, where the tip is in contact with the sample. It does not vibrate but it is moving in X-Y plane over the surface to scan the defined area line by line. Then, by putting all these line scans together, the whole image is created on the computer. This allows a high resolution but it is more likely to damage both the tip and the sample surface.

- Tapping mode, which is called also vibration mode. In this mode, the tip oscillates and touches the tops of the surface. Damaging both tip and surface sample, in this mode, is lower compared to the contact mode.

- Non-contact mode, where the tip oscillates without having any contact with the surface sample.

All the AFM images presented in this work were recorded using a NT-MDT Solver Pro scanning probe microscope equipped with standard silicon cantilevers (Nanosensors) having a nominal force constants of 50 N/m. Image processing and analysis of AFM data were performed using the free software Gwyddion (Czech Metrology Institute).

Scanning electron microscopy "SEM"

Basing on the principle of electron-matter interactions, scanning electron microscope (SEM) provides a highly magnified image of sample surface by scanning a beam of high energy electrons of 1-40 keV across the analyzed surface. It consists on an electron gun on the top of the column generating an energetic electron beam focused onto the surface sample by magnetic lenses. The interaction between the valence electrons of sample material and the incident beam generates secondary electrons, namely back scattered electrons. These electrons are accelerated towards a secondary electron detector. Because of the low energy of secondary electrons, only those emitted from the very top of surface are able to come out and be detected. Thus, the scattered signal is amplified and the corresponding specimen surface information is obtained. During this work, all SEM images were recorded using a ZEISS Supra 40 FEG-SEM microscope working at 5 kV.

Spectrophotometry

Transmission and reflection measurements were collected using a Perkin Elmer Lambda 950 spectrophotometer (Figure I-22) over a wavelength range of 250 to 2500 nm. Both WI lamp (visible range) and a Deuterium lamp (ultraviolet range) were used as light source. The transmission spectra background was taken against substrate background.

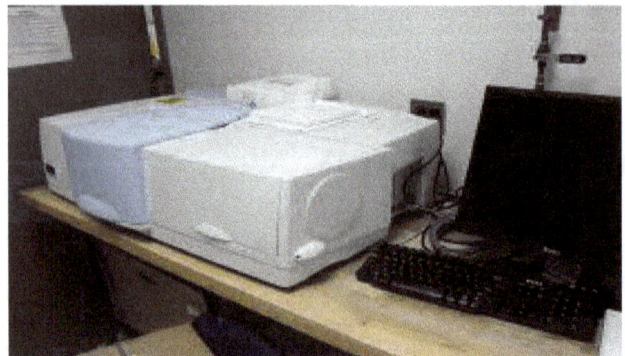

Figure I-22.
Spectrophotometer
Perkin Elmer lambda 950
The spectrophotometry is a very

interesting tool to determine the optical properties of a material because it is quick, accurate, non-destructive and does not require any sample preparation.

The principle of spectrophotometer is illustrated in Figure I-23. It is based on double beam system which divides the monochromatic light into two beams. Both of them are guided to the reference and the sample, respectively, using a rotational mirror. Then, these two beams are received by a detector and converted to computer data.

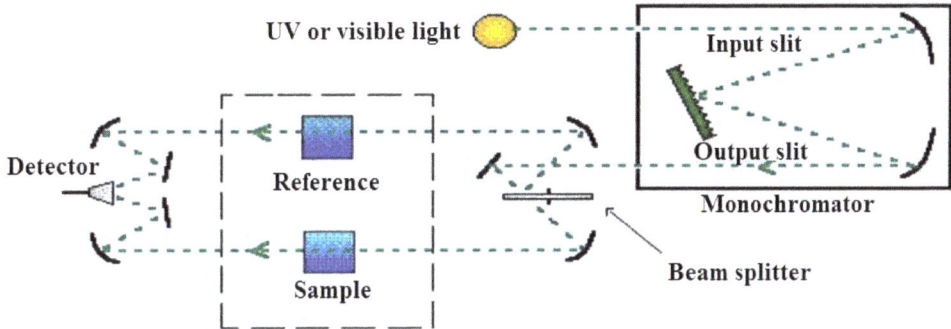

Figure I-23. Constituent elements of a spectrophotometer.

Spectrofluorimetry

Photoluminescence (abbreviated as PL) is originated from an absorption/emission process between different electronic energy levels in the semiconductor. It occurs when an electron returns from an excited state to the electronic ground state and loses its excess energy as a photon. It is an interesting tool which may determine optically active impurities in semiconductors. It is formally divided into two categories depending on the nature of the excited state:

Fluorescence: When a photon is, spontaneously, emitted from the singlet excited state (all electrons are spin-paired) to the singlet ground state (Figure I-24).

Phosphorescence: When the emission occurs after few seconds between a triplet excited state (One set of electron spins is unpaired) and a singlet ground state (Figure I-24).

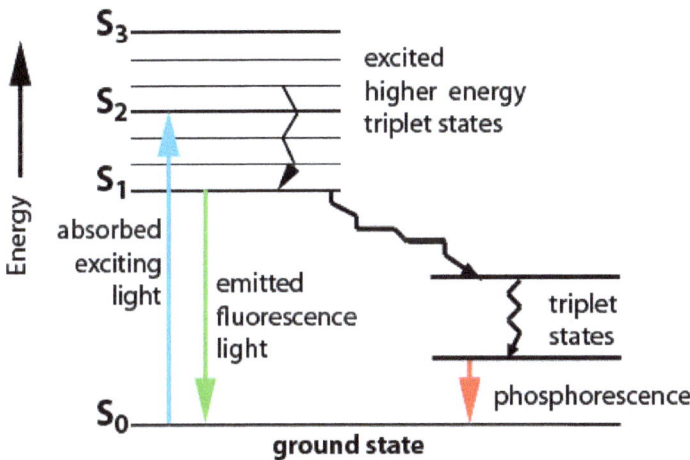

Figure I-24. Mechanism of fluorescence and phosphorescence.

In order to

investigate the PL properties of chemically sprayed TiO_2 thin films, a Perkin Elmer LS55 spectrofluorimeter equipped with an Xenon lamp has been used (Figure I-25). The spectrofluorimetry is a simple, versatile, non-contact and non-destructive method. The basic elements of a spectroflorimeter are: an optical excitation source, an emission monochromator and a detector. A typical PL setup is illustrated in Figure I-26. In fact, the incident beam is focused on the sample through an excitation monochromator. Then, the emitted light will be collected by an emission monochromator and guided to the detector.

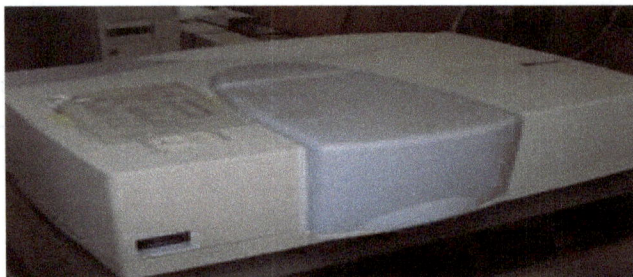

Figure I-25. Spectrofluorimeter « Perkin Elmer LS 55 ».

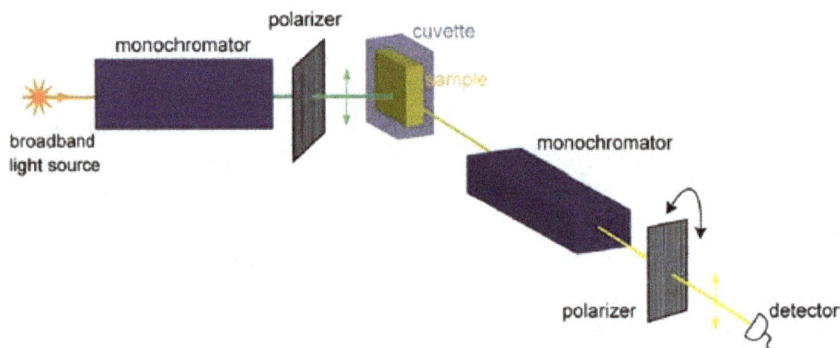

Figure I-26. Constituent elements of a spectrofluorimeter.

Conclusion

In summary, Titanium dioxide (TiO_2) semiconductor attracts a great interest in material science due to its wide range of technological applications as well as its interesting physical properties. It presents a rich phase diagram including tetragonal anatase, rutile and orthorhombic brookite which are its most abundant forms in the nature. The next section deals with a theoretical study of anatase titanium dioxide compound. A particular attention will be paid to the calculation of the band structure, the total and partial density of states, the optical, thermal and transport properties of TiO_2, using the WIEN2k package.

References

[1] A. Ramadossa, S.J. Kim, J. Alloys Comp. 561, 262 (2013).
[2] S. Ding, B. Gao, D. Shan, Y. Sun, S. Cosnier, Biosens. Bioelectron. 39, 342 (2013).
[3] Z. Lu, X. Jiang, B. Zhou, X. Wu, L. Lu, Appl. Surf. Sci. 257, 10715 (2011).
[4] J.Ben Naceur, R.Mechiakh, F.Bousbih, R.Chtourou, Appl. Surf. Sci. 257, 10699 (2011).

[5] J.Ben Naceur, M.Gaidi, F.Bousbih, R.Mechiakh, Curr. App. Phys. 12, 422 (2012).

[6] L. Hou, P. Liu, Y. Li, C. Wu, Thin Solid Films 517, 4926 (2009).

[7] D. S. Han, D. K. Choi, J. W. Park, Thin Solid Films 552, 155 (2014).

[8] I. Oja Acik, A. Katerski, A. Mere, J. Aarik, A. Aidla, T. Dedova, M. Krunks, Thin Solid Films 517, 2443 (2009).

[9] A. Bahramian, D. Vashaee, Sol. Energy Mater. Sol. Cells 143, 284 (2015).

[10] J. H. Lee, C. H. Park, J. P. Jung, J. H. Kim, J. Power Sources 298, 14 (2015).

[11] A. H. Mayabadi, V. S. Waman, M. M. Kamble, S. S. Ghosh, B. B. Gabhale, S. R. Rondiya, A. V. Rokade, S. S. Khadtare, V. G. Sathe, H. M. Pathan, S. W. Gosavi, S. R. Jadkar, J. Phys. Chem. Solids 75, 182 (2014).

[12] A. Elfanaoui, A. Ihlal, A. Taleb, L. Boulkaddat, E. Elhamri, M. Meddah, K. Bouabid, X. Portier, M. J. condensed matter 13 (2011).

[13] B. Choudhury, A. Choudhury, Curr. Appl .Phys. 13, 1025 (2013).

[14] M. Vishwas, K. Narasimha Rao, R. P. S. Chakradhar, Spectrochim. Acta Part A: Molecular and Biomolecular Spectroscopy 99, 33 (2012).

[15] L. Lin, H. Wang, H. Luo, P. Xu, J. Photochem. Photobiol. A. Chem. 307, 88 (2015).

[16] A.Ishii, Y.Nakamura, I.Oikawa, A.Kamegawa, Appl. Surf. Sci.347, 528 (2015).

[17] M. Mazur, J. Morgiel, D. Wojcieszak, D. Kaczmarek, M. Kalisz, Surf. Coat. Technol. 270, 57 (2015).

[18] V. Bukauskas, S. Kaciulis, A. Mezzi, A. Mironas, G. Niaura, M. Rudzikas, I. Simkiene, A. Satkus, Thin Solid Films 585, 5 (2012).

[19] A. Moses Ezhil Raj, V. Agnes, V. Bena Jothy, C. Sanjeeviraja, Mater. Sci. Semicond. Process. 13, 389 (2010).

[20] M. O. Abou-Helal, W. T. Seeber, Appl. Surf. Sci. 195, 53 (2002).

[21] P. S. Shinde, S. B. Sadale, P. S. Patil, P. N. Bhosale, A. Bruger, M. Neumann-Spallart, C. H. Bhosale, Sol. Energy Mater. Sol. Cells 92, 283 (2008).

[22] Y. Hu, H.L. Tsai, C.L. Huang, J. Eur. Ceram. Soc. 23, 691 (2003).

[23] J. Muscat, V. Swamy, N. M. Harrison, Phys. Rev. B 65, 224112 (2002).

[24] S.C. Abrahams and J.L. Bernstein, J. Chem. Phys. 55 (1971), 3206.

[25] MHorn, CFSchwerdtfeger, EPMeagher, Zeitschrift für Kristallographie, 1972,136, 273

[26] I. A. Alhomoudi, G. Newaz, Thin Solid Films 517, 4372 (2009).

[27] S. Sankar, K. G. Gopchandran, Indian J. Pure Appl. Phys. 46, 791 (2008).

[28] A. Nakaruk, D. Ragazzon, C.C. Sorrell, Thin Solid Films 518, 3735 (2010).

[29] M.D. Wiggins, M.C. Neison, C. R. Aita, J. Vac. Sci. Technol. A14 (3), 772 (1996).

[30] A.Mani, C.Huisman, A.Goossens, J.Schoonman, J. Phys. Chem. B 112, 10086 (2008).

[31] I. Oja, A. Mere, M. Krunks, C-H. Solterbeck, M. Es-Souni, Solid State Phenomena Vols. 99-100 (2004) pp 259-264.

[32] C. P. Fictorie, J. F. Evans, W. L. Gladfelter, J. Vac. Sci. Technol. A 12, 1108 (1994).

II. Ab-initio studies of fundamental properties of anatase titanium dioxide compound

This section deals with theoretical calculation of the structural, electronic, optical, thermal and transport properties of anatase TiO_2 compound with the help of density functional theory and Boltzmann theory. The fully optimized structure was obtained by minimizing the total energy. Both band structure and density of states were discussed. Some optical constants were investigated, too. The variations of the volume, bulk modulus, Debye temperature, heat capacities, entropy, Grüneisen parameter and thermal expansion coefficient as a function of the pressure and temperature were all obtained and analyzed in detail. Boltzmann theory calculations have been used to evaluate important transport properties such as Seebeck coefficient, electrical conductivity, electronic thermal conductivity and power factor as a function of the chemical potential.

2.1. The density functional theory

It is well known that the solid is described by a many-particle wave function since it typically contains around 1023 atoms. Thus, this huge number of degrees of freedom makes the resolution of the corresponding Schrödinger equation very difficult neither analytically nor numerically. In the 1960s, Hohenberg, Kohn and Sham made the treatment of solid systems possible thanks to a new method of calculation of macroscopic properties of many-electron systems, namely density functional theory (DFT) [1,2]. Nowadays, DFT has becoming among the most versatile and widely used method in condensed-matter physics. It treats not only bulk compounds but also complex systems like films, proteins, molecules and nanoparticles basing on their crystal structure. The main idea of DFT is that the use of only the electron density is sufficient to describe an N-body system. Thus, it leads to reduction of its 3N degrees of freedom to uniquely three spatial coordinates, which simplifies enormously the problem of a many-body system. This approximation is based on the two Hohenberg-Kohm theorems [2]: the first one stipulates that the ground state properties of N-body system are described only using the electron density which can be determined by a variational principle, as stated by the second theorem.

Figure II-1 gives us an idea about the evolution of the number of articles published using DFT compared to the other methods, over the last decades. As it can be seen from this Figure , DFT was released in 1964 then it became popular at the end of the eighties, with the emergence of personal computer as well as modern supercomputers. In the early 21st century, the DFT overtook all the other methodologies and became the most popular and successful method for calculations in solid-state physics. Besides, the improvement of computer capabilities and the development of several codes, such as: WIEN2k, VASP and CASTEP, allow the study of different properties of matter with higher accuracy.

The WIEN2k package

The WIEN2k package was developed by Peter Blaha and Karlheinz Schwarz from the Institute of Materials Chemistry at the Technical University of Vienna (Austria) and it was distributed for the first time in 1990. It consists on a program written in FORTRAN 90 which requires a UNIX operating system. It is based on the full-potential (linearized) augmented plane-wave ((L)APW) + local orbitals (lo) method in order to solve the Kohn-Sham equations. The main purpose for using this package is to perform electronic, structural and optical calculations of solids basing on the density functional theory (DFT).

Computational details

Anatase titanium dioxide TiO_2 compound crystallizes in a tetragonal structure with space group I41/amd (N° 141) with lattice constants a = 3.785 A° and c = 9.514 A°. The muffin-tin (MT) spheres radii adopted for Ti and O atoms were chosen to be 1.91 and 1.73

a.u., respectively. The Full-Potential Linearized Augmented Plane Wave (FP-LAPW) method within the framework of density functional theory (DFT) as implemented in WIEN2k package [4], was used to solve the Khon-Sham equations. Generalized gradient approximation based on Perdew–Burke–Ernzerhofand (PBE-GGA) approximation was used to calculate the exchange-correlation potential for structural properties. Both electronic and optic calculations were conducted using Tran and Blaha's modified Becke-Johnson (MBJ) approximation. The calculations were carried out by using 6000 k points in the first Brillouin zone, corresponding to 475 k points in the reduced Brillouin zone. The self-consistent calculations converged when the total energy is less than 0.1 mRy. The thermal response of the material has been evaluated using the quasi-harmonic Debye approach as embodied in the GIBBS code [5]. The analysis of electronic bands and the semi-classical Boltzmann theory as implemented in the BoltzTrap package [6] were employed to perform some transport coefficients. The constant relaxation time (τ) approximation was performed during the calculations. Within this method, the Seebeck coefficient S does not depend on τ. However the electrical conductivity σ, the electronic thermal conductivity Kel and the power factor σS^2 need to be evaluated with respect to relaxation time.

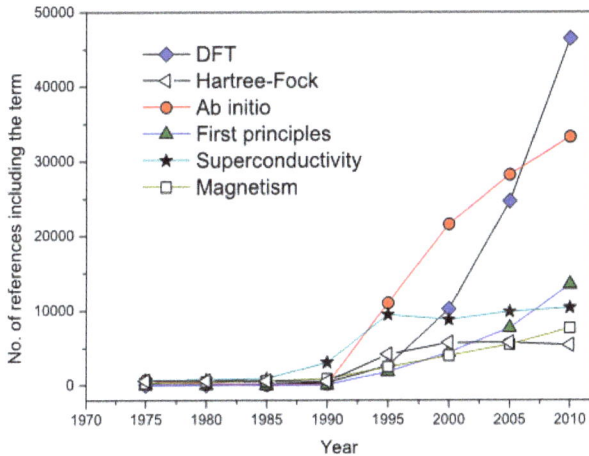

Figure II-1: Total number of published articles including the terms "density functional theory", "Hartree-Fock", "ab initio", "first principles", "superconductivity" and "magnetism" over 40 years, accumulating every 5 years [3].

2.2. Structural properties of anatase TiO$_2$ compound

Structural properties of anatase TiO$_2$ have been carried out by energy minimization process with respect to the other geometrical parameters using the (PBE-GGA) approach. The minimization is done in two steps:

- Volume optimization: Calculate the total energy versus the unit cell volume using the experimental ratio of c/a in order to estimate the lattice parameters which correspond to the lowest energy. The unit cell energy versus the volume is fitted by the Birch–Murnaghan's equation of state [7].
- c/a Optimization: Calculate the total energy as a function of c/a ratio in order to obtain the optimized structure corresponding to the lowest energy.

The calculated total energy versus volume of unit cell of anatase TiO$_2$ crystal is plotted in Figure II-2.

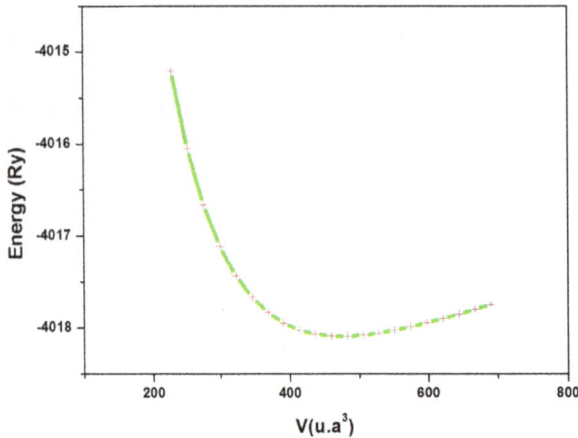

Figure II-2.
Calculated total energy versus unit cell volume for anatase TiO$_2$.

A summary of the obtained results is presented in Table II-1, together with available experimental and theoretical information. It is clearly seen that the lattice constants found in this work are in close agreement with the reported data [8,9,10].

Table II-1. Calculated equilibrium lattice constants, bulk modulus (B0) and pressure derivative of bulk modulus (B_0') of anatase Ti TiO$_2$ compound.

	a (A°)	c (A°)	(c/a)	B0 (GPa)	B_0'
This work	3.814	9.583	2.512	198.210	3.411
PBE-GGA [8]	3.798	9.852	2.593	188	3.7
GGA [9]	3.823	9.612	2.514	-	-
Exp. [10]	3.785	9.514	2.513	-	-

2.3. Electronic properties of anatase TiO$_2$ compound

In solid state physics, the electronic band structure of a solid is a property which distinguishes semiconductors from metals and insulators. Figure II-3 presents the band structure of anatase TiO$_2$ compound plotted along the high-symmetry lines of the first Brillouin zone calculated with MBJ approximation. Fermi level is taken as the origin of the energy scale. The top of the valence band is located near the X point and the bottom of the conduction band at the Γ point, which reveals an indirect band gap of about 3.05 eV. This result is in line with other DFT calculations [11]. However, this obtained value is less than the experimental one [12]. Indeed, it is empirically well known that theoretical values of the gap energy are underestimated compared to the experimental ones due to the discontinuity in the exchange-correlation potential [13], which is not taken into account within the DFT framework.

For further electronic analyzing, it is important to investigate total and partial density of states (DOS) in order to identify the contribution of various orbitals in the conduction. The calculated MBJ total and partial DOS for anatase TiO$_2$ are plotted in Figure II-4. In lower part of the valence band ranged from -18 to -17 eV, the O-s states show an important contribution with a small admixture of Ti-s and Ti-d states. However, between -4.5 to 0 eV, the higher part of the valence band is at most originated from both O-p and Ti-d states. In the energy range of 4 to 8 eV, the conduction band is mainly consists of Ti-d orbitals.

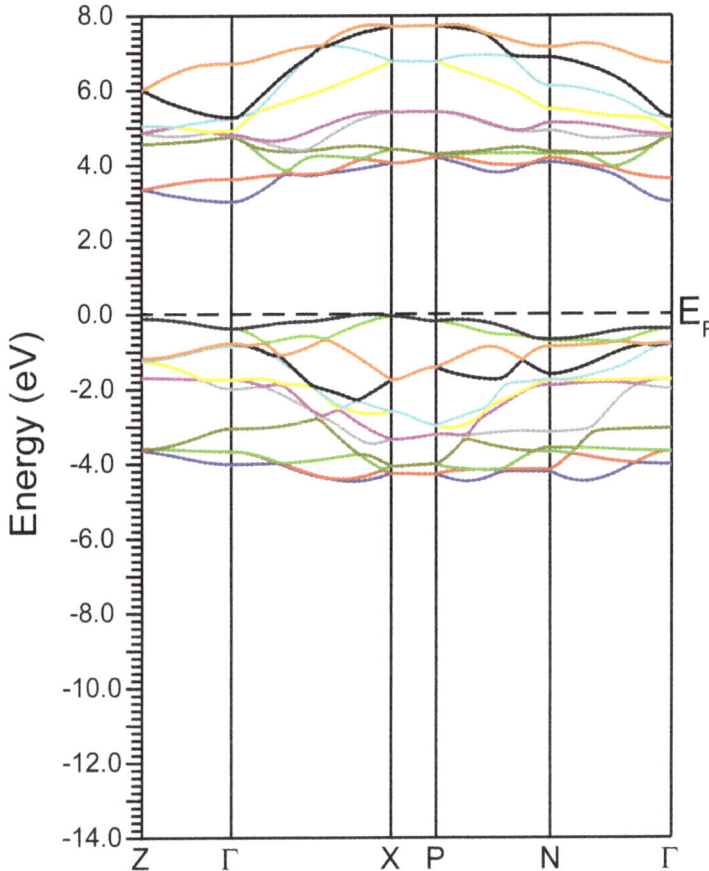

Figure II-3.
MBJ calculated band structure of anatase TiO_2.

2.4. Optical properties of anatase TiO_2 compound

The optical behavior of a matter can be described using its complex dielectric constant ε ($\varepsilon = \varepsilon_1 + i\varepsilon_2$) which presents the response of a medium to an external electromagnetic field. Its real and imaginary parts ($\varepsilon1$ and $\varepsilon2$) are related to the electronic polarizability and the electronic absorption of the material, respectively. Their dependence on photon energy is displayed in Figure II-5a. As it can be seen from this Figure, there is a slight anisotropic behavior at low energy which may be explained by the tetragonal structure of anatase TiO_2 compound. However, a high isotropic behavior was obtained at higher energy regions. Furthermore, a threshold of optical absorption situated at $\sim 3.4\ eV$ occurs in the dispersion spectra of $\varepsilon2$ which corresponds to fundamental absorption of anatase TiO_2. This result is consistent with experimental data [12].

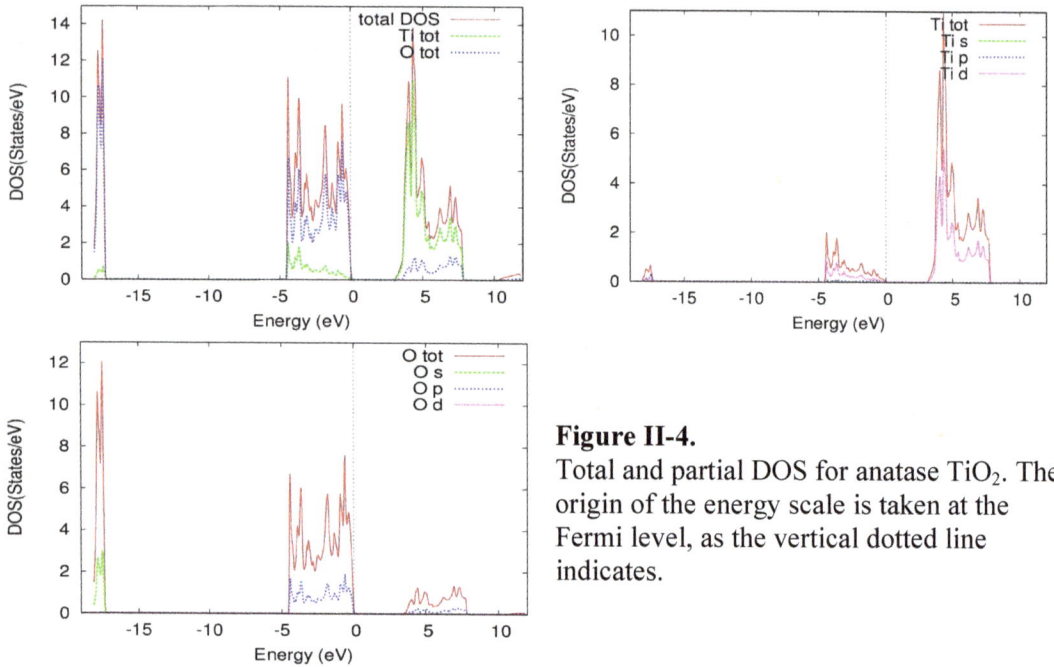

Figure II-4.
Total and partial DOS for anatase TiO$_2$. The origin of the energy scale is taken at the Fermi level, as the vertical dotted line indicates.

For further optical analysis, the complex refractive index was investigated. It is defined as $N = n + ik$, where n and k are the refractive index and the extinction coefficient, respectively. n describes how light propagates through a medium. It is defined as follows:

$$n = \frac{c}{V_\varphi}$$

$(II-1)$

where c is the light celerity and Vφ is the phase velocity of light in the medium.

k is related to light absorption. Both n and k can be calculated in terms of real and imaginary parts of dielectric constant values as follows [14,15]:

$$n = \frac{(\varepsilon_1 + (\varepsilon_1^2 + \varepsilon_2^2)^{1/2})^{1/2}}{\sqrt{2}}$$

$(II-2)$

$$k = \frac{(-\varepsilon_1 + (\varepsilon_1^2 + \varepsilon_2^2)^{1/2})^{1/2}}{\sqrt{2}}$$

$(II-3)$

The obtained refractive index and extinction coefficient are shown in Figure II-5b. Both of them exhibit an anisotropic behavior. Furthermore, we note that n and k closely follow $\varepsilon 1$ and $\varepsilon 2$, respectively. Moreover, n reaches a maximum value of about 3.25 at around 3 eV then it falls down for photon energy higher than 9 eV. This means that anatase TiO$_2$ compound cannot behave as transparent material at high photon energy. Furthermore, we note that in this region (energy > 9 eV), n is less than 1, so, c is smaller than Vφ, according to Eq. II-1, which is in disagreement with relativity. This behavior can be explained by the fact that the signal propagates in a dispersive medium as a wave packet at the group velocity Vg ($V_g = \frac{d\omega}{dk}$) rather than as a monochromatic wave at phase velocity Vφ ($V_\varphi = \frac{\omega}{k} = \frac{c}{n}$). Vg is related to V$\varphi$ using this relation:

$$V_g = V_\varphi \left(1 - \frac{k}{n}\frac{dn}{dk} \right)$$ (II-4)

Thus, according to this equation, Vg is usually less than Vφ. Similar behavior was reported by Sai *et al.* [11].

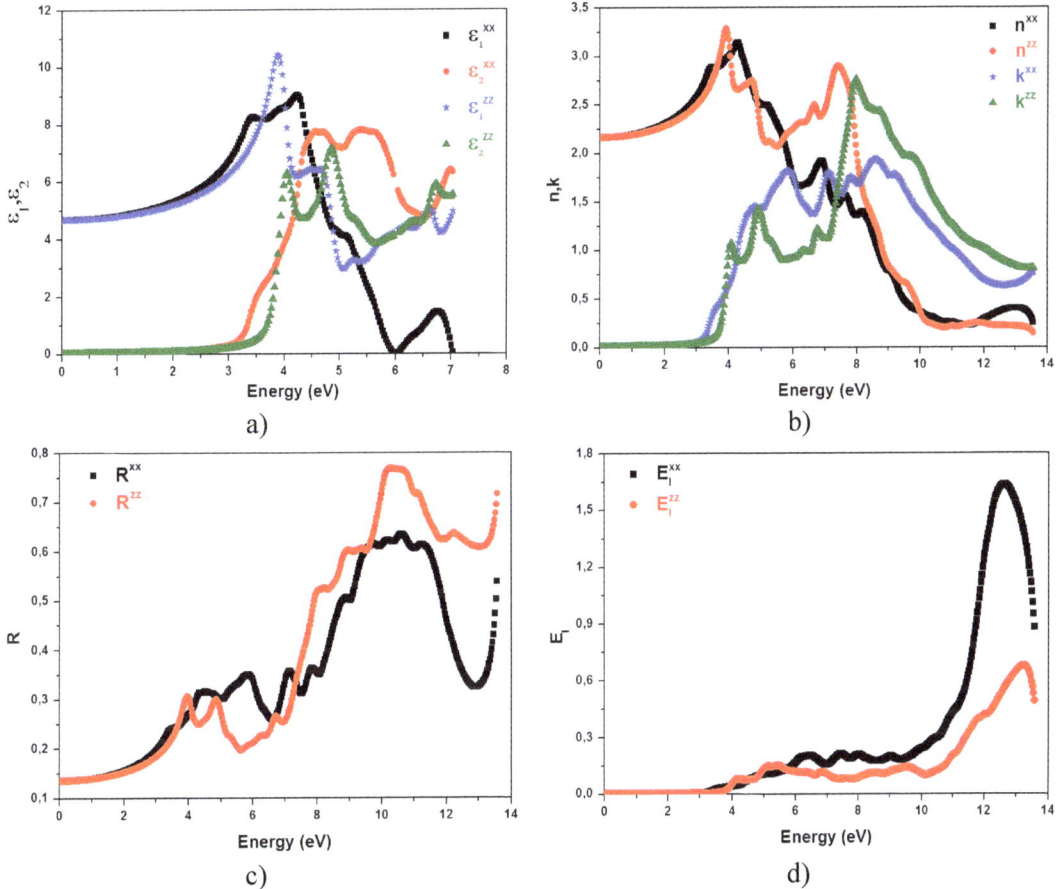

a)   b)

c)   d)

Figure II-5: Calculated dispersion spectra of (a) real and imaginary part of dielectric function, (b) refractive index and extinction coefficient, (c) reflectivity and (d) energy loss function of TiO_2 compound.

To further investigate the optical properties of anatase TiO_2 compound, the reflectivity R and the electron energy loss function El are obtained using the following expressions [16]:

$$R = \left| \frac{\sqrt{\varepsilon}-1}{\sqrt{\varepsilon}+1} \right|^2$$ (II-5)

$$E_l = \frac{\varepsilon_2}{\varepsilon_1^2+\varepsilon_2^2}$$ (II-6)

The dependence of R and El on photon energy is depicted in Figure II-5c and d, respectively. We note that at zero frequency, R(0) is about 0.14. However, it reaches a maximum value around 0.77 at 10 eV. Furthermore, the electron energy loss function El exhibits a higher value of about 1.6 at 12 eV.

2.5. Thermal properties of anatase TiO₂ compound

The study of the thermal properties of compounds is an important topic since it provides an insight about the effect of several constraints such as high temperature and high pressure on their behaviors. Therefore, we have investigated the temperature and pressure effects on thermal properties of anatase TiO₂, through the quasi-harmonic Debye model as implemented in the Gibbs program [5]. In this approach, the non-equilibrium Gibbs function $G^*(V; P, T)$ is defined as follow:

$$G^*(V; P, T) = E(V) + PV + A_{vib}(\Theta(V); T) \tag{II-7}$$

where E(V) presents the total energy per unit cell, PV corresponds to the constant hydrostatic pressure condition and $\Theta(V)$ is the Debye temperature. The vibrational Helmholtz free energy A_{vib} can be expressed using the Debye model of the phonon density of states [17]:

$$A_{vib}(\Theta; T) = nk_B T \left[\frac{9}{8} \frac{\Theta}{T} + 3\ln\left(1 - e^{-\frac{\Theta}{T}}\right) - D\left(\frac{\Theta}{T}\right) \right] \tag{II-8}$$

n being the number of atoms per formula unit and D(y) represents the Debye integral, it takes the form of :

$$D(y) = \frac{3}{y^3} \int_0^y \frac{x^3}{e^x - 1} dx \qquad \text{And} \qquad x = \frac{\hbar\omega}{k_B T} \tag{II-9}$$

The Debye temperature Θ, in the case of an isotropic solid, is given by [17]:

$$\Theta = \frac{\hbar}{k_B} \left[6\pi V^{1/2} n \right]^{1/3} f(\sigma) \sqrt{\frac{B_s}{M}} \tag{II-10}$$

where M is the molecular mass per unit cell, σ is the Poisson ratio and BS presents the adiabatic bulk modulus approximated by the static compressibility as [5]:

$$B_s \cong B(V) = V \left\{ \frac{d^2 E(V)}{dV^2} \right\} \tag{II-11}$$

E being the total energy for each crystal volume, obtained using the FP-LAPW method at T = 0 K (static conditions).

G*(V; P, T), the non-equilibrium Gibbs function, can be minimized with respect to volume as follow:

$$\left(\frac{\partial G^*(V; P, T)}{\partial V} \right)_{P, T} = 0 \tag{II-12}$$

By solving this equation, the isothermal bulk modulus BT, the heat capacities CV, CP, the entropy S and the thermal expansion coefficient α are obtained, respectively, using the following relations [18]:

$$B_T(P, T) = V \left(\frac{\partial^2 G^*(V; P, T)}{\partial V^2} \right)_{P, T} \tag{II-13}$$

$$C_V = 3nk_B \left[4D\left(\frac{\Theta}{T}\right) - \frac{3\Theta/T}{e^{3\Theta/T} - 1} \right] \tag{II-14}$$

$$C_p = C_V(1 + \alpha \gamma T) \tag{II-15}$$

$$S = nk\left[4D\left(\frac{\theta}{T}\right) - 3\ln\left(1 - e^{-\theta/T}\right)\right]$$ (II-16)

$$\alpha = \frac{\gamma C_V}{B_T V}$$ (II-17)

Where γ is the Grüneisen parameter defined as:

$$\gamma = -\frac{d\ln\theta(V)}{d\ln V}$$ (II-18)

Figure II-6 depicts the equilibrium volume as a function of pressure for various temperatures. As it can be seen, the volume increases as the temperature increases too, but with very moderate rate. This increase can be explained by the fact that the crystal becomes more dilated with augmenting temperature. However, at a given temperature the volume decreases as the pressure increases. This behavior can be explained by the fact that anatase TiO_2 compound looks to be more compressible with increasing pressure.

Figure II-7 presents the bulk modulus, which signifies the average strength of coupling between the neighboring atoms, versus temperature at various pressures. From this Figure , we can remark that B decreases slightly with increasing temperature. However, a linear shift has been observed with changing pressure. As expected, anatase TiO_2 compound becomes more compressible with increasing pressure. This behavior is consistent with the trend of volume discussed previously.

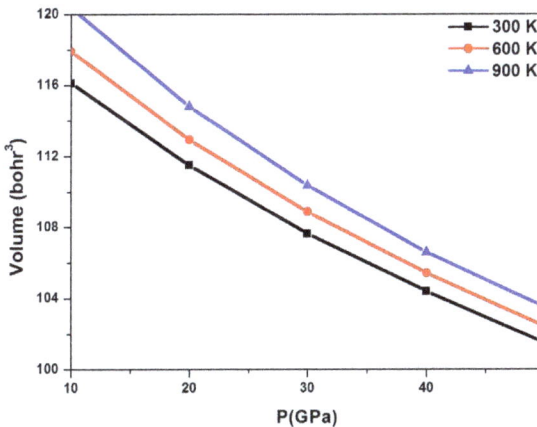

Figure II-6. Calculated P-V-T relationship of anatase TiO_2.

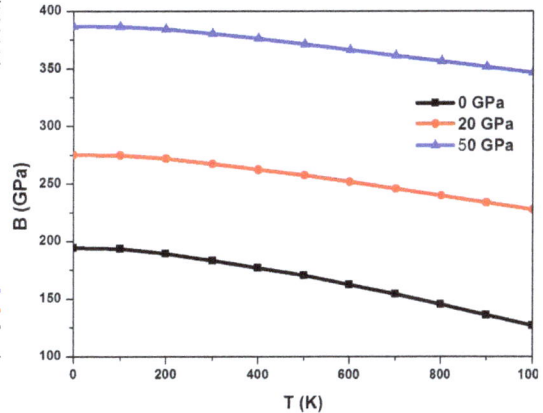

Figure II-7. Bulk modulus *vs* temperature at various pressures for anatase TiO_2.

The Debye temperature as a function of pressure and temperature is represented in Figure II-8. As it can be seen in this Figure, a remarkable increasing of the Debye temperature with the applied pressure is noticed, in contrast to the nearly linear reduction resulted from the temperature impact. This result can be due to the equivalence of decreasing temperature to increasing pressure at the level of the crystal volume. Moreover, we note that anatase TiO_2 possesses a high Debye temperature which allows us to classify this material among the hard compounds.

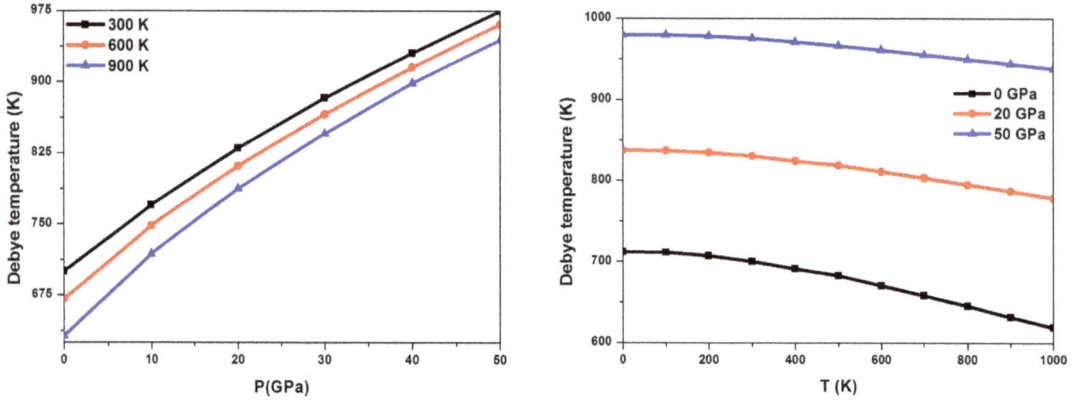

Figure II-8. Debye temperature at different pressure and temperature values.

The heat capacity of a substance is considered to be an important parameter since it provides essential idea about atoms vibration inside the investigated material. Figure II-9 presents the calculated specific heats at constant volume (Cv) and constant pressure (Cp) of anatase TiO$_2$ versus temperature for various pressures. We can remark that both of them have the same behavior and follow a T3 law at sufficiently low temperatures, which is due to the used anharmonic approximation of the Debye model, and then both of them vanish when the temperature tends to zero.

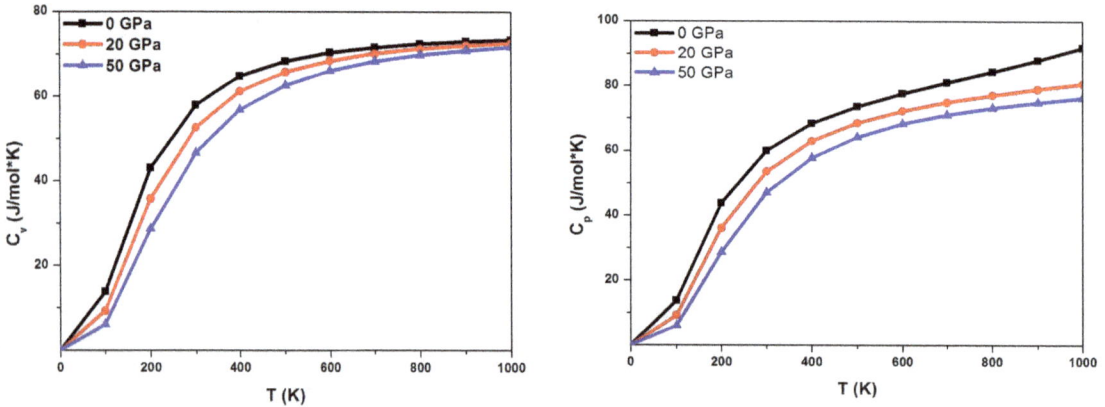

Figure II-9. Variation of the specific heats Cv and Cp with temperature and pressure.

This appearance is compared with experimental data reported by Levchenkoin *et al.* [19] in the range of 0 < T < 350 K in the case of specific heat at constant pressure (Cp). At higher temperatures, Cv tends to the Dulong-Petit limit (Cv (T) ≈ 3R for mono-atomic solids) which is a common property for all solids [20,21]. As it can be seen from Figure II-9, the heat capacities decrease with pressure and increase with temperature, but the effect of pressure is less significant than that of temperature.The entropy S, which can be defined as the measure of disorder of a system, versus temperature for various pressures is displayed in Figure II-10. As it can be seen, the entropy increases monotonously with increasing temperature. However, it decreases with pressure increment. This behavior may be explained by the fact that the increase of temperature leads to an increase of vibrational contribution and then an increase of the entropy.

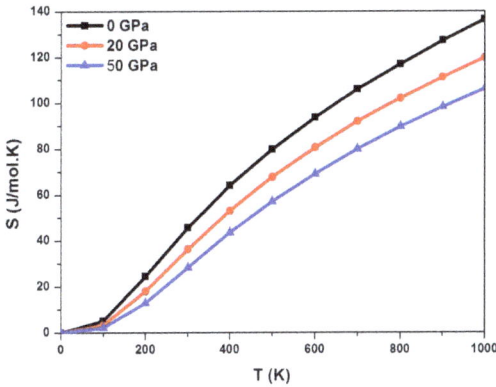

Figure II-10. The entropy versus temperature at different pressures for anatase TiO₂.

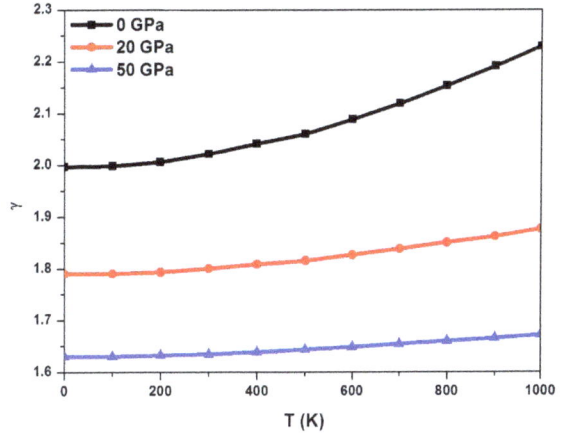

Figure II-11. Grüneisen parameter versus temperature at various pressures for anatase TiO₂.

The Grüneisen parameter γ describes the effect of volume change on the dynamics of a crystal. Its variation versus temperature at various pressures is presented in Figure II-11. We can remark that γ values remain practically constant at lower temperature, then, it increases linearly with increasing temperature. However, it decreases with increasing pressure at a given temperature.

The investigation of the thermal expansion coefficient is an important topic since it is essential to predict the thermodynamic equation of state. Figure II-12 depicts the temperature and pressure dependences of the volume thermal expansion coefficient (α) of anatase TiO₂. We found that the thermal expansion coefficient values increase quickly with temperature at a given pressure, and then, for a temperature higher than 400 K, the increase tends to be linear. However, at a certain temperature, the thermal expansion coefficient decreases gradually with augmenting pressure.

2.6. Transport properties of anatase TiO₂ compound

The transport coefficients of anatase TiO₂ such as the Seebeck coefficient S and the electronic conductivity σ were investigated basing on the band structure as well as the resolution of Boltzmann transport equation using the condition of constant relaxation time approximation. The main expressions are given by [22]:

$$\sigma_{(\alpha\beta)}(T,\mu) = \frac{1}{\Omega} \int d\varepsilon \; \sigma_{\alpha\beta}(\varepsilon) \left(-\frac{\partial f_\mu(T,\varepsilon)}{\partial \varepsilon} \right) \tag{II-19}$$

The Seebeck coefficient is related to the electrical conductivity with the following relation:

$$S_{\alpha\beta} = \sum_\gamma (\sigma^{-1})_{\alpha\gamma} \; v_{\beta\gamma} \tag{II-20}$$

With: $v_{\beta\gamma}(T,\mu) = \frac{1}{eT\Omega} \int d\varepsilon \; \sigma_{\beta\gamma}(\varepsilon)(\varepsilon - \mu) \left(-\frac{\partial f_\mu(T,\varepsilon)}{\partial \varepsilon} \right) \tag{II-21}$

where T is the absolute temperature, µ is the chemical potential, e is the charge of electron, ε is the band energy, fµ is the Fermi distribution function and τ is the relaxation time.

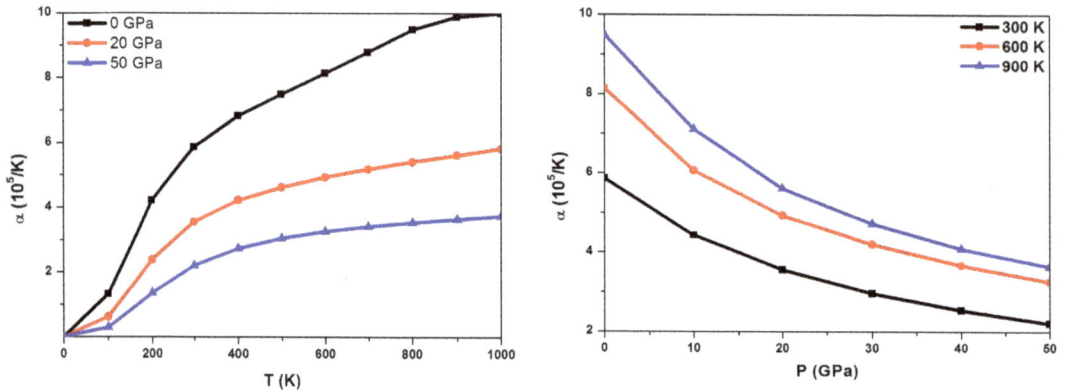

Figure II-12. Variation of the volume thermal expansion coefficient (α) of anatase TiO$_2$.

The electronic thermal conductivity can be defined as:

$$K_{\alpha\beta}(T,\mu) = \frac{1}{e^2 T} \int (\varepsilon - \mu)^2 \sigma_{\alpha\beta}(\varepsilon) \left(-\frac{\partial f_\mu(T,\varepsilon)}{\partial \varepsilon}\right) d\varepsilon$$

(II-22)

Figure II-13 displays the variation of the Seebeck coefficient S versus chemical potential, at different temperatures. As seen in this Figure, the Seebeck coefficient ranges from -300 µV/K to 300 µV/K, which reflects a high efficiency thermoelectric material [23]. A relatively similar result was obtained experimentally, at room temperature, by Brahimi *et al.* [24]. Moreover, a decrease of the absolute seebeck coefficient with temperature augmenting is noticed and higher value is obtained at room temperature (300 K). This behavior can be explained by the fact that as the temperature increases, the number of carrier concentration increases too which leads to a decrease of the seebeck coefficient [25]. It is important to signal that along the chemical potential axis, the curve below the Fermi level depicts the valence band maximum (VBM) and above presents conduction band minimum (CBM). We note that the Seebeck coefficient has two opposite peaks around the Fermi level. In fact, its sign define the contribution of charge carrier, so, for n-type doping system we have a negative Seebeck coefficient, whereas, a p-type doping is noted for a positive Seebeck coefficient of anatase TiO$_2$ compound.

Electrical conductivity estimates the flow of free charge carriers present in a compound. Therefore, a good thermoelectric compound should have high electrical conductivity which leads to reduction of Joule heating effect [26]. The calculated electrical conductivity with respect to relaxation time of anatase TiO$_2$ at various temperatures is presented in Figure II-14. The result reveals that (σ/τ) decreases slightly with increasing temperature from 300 K to 900 K. We note that (σ/τ) average is greater for p-type doping as compared to the n-type doping.

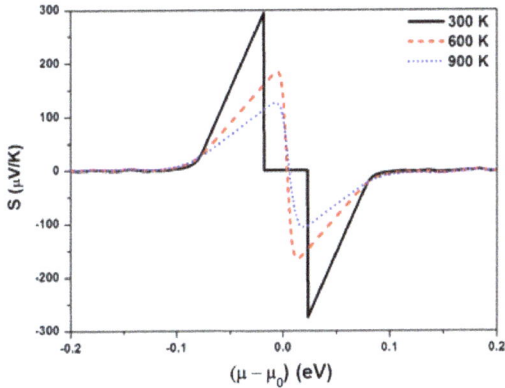

Figure II-13. Calculated Seebeck coefficient S of anatase TiO_2 as a function of the chemical potential at various temperatures.

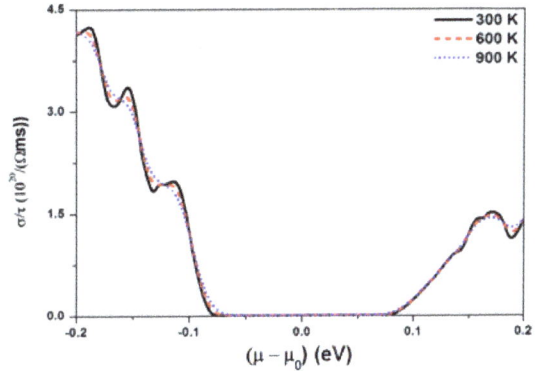

Figure II-14. Calculated electrical conductivity with respect to scattering time (σ/τ) of anatase TiO_2 as a function of the chemical potential at various temperatures.

The thermal conductivity gives us an idea about the ability of a material to conduct heat caused by electrons and lattice vibrations. Thus, the total thermal conductivity (K = Kel + Kl) consists of two parts: the electronic and lattice terms caused by electrons and lattice vibrations, respectively. We note that only electronic parts Kel could be calculated using the BoltzTrap code. Figure II-15 depicts the calculated electronic thermal conductivity as a function of chemical potential for anatase TiO_2 at various temperatures. It is clear from this Figure that for both n and p-type anatase TiO_2, the electronic thermal conductivity is higher at 900 K than 300 K. This behavior may be explained by the fact that an increase of temperature leads to an increment in the number of charge carriers. Further, we can remark that electronic thermal conductivity is higher for p-type material comparing to n-type.

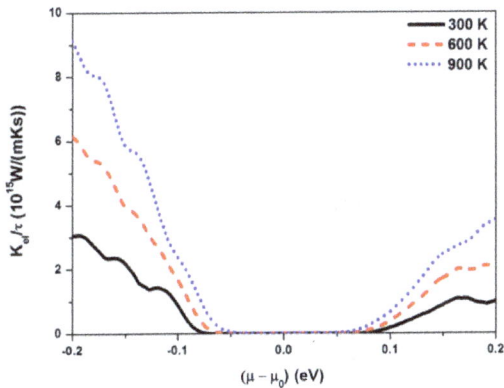

Figure II-15.
Calculated electronic thermal conductivity as a function of chemical potential for anatase TiO_2 at various temperatures.

Power factor presents a measure of the ability of compounds to generate electric power. The calculated power factor with respect to scattering time $S^2\sigma/\tau$ of anatase TiO_2 compound versus chemical potential is shown in Figure II-16. It is noted that the power factor increases as the temperature increases too. In fact, at higher temperature (900 K), the power factor average presents greater values which can be due to the fact that the value of carrier concentration increases with augmenting temperature. Moreover, we note the presence of numerous peaks in the whole range of chemical potential; suggesting that transport performance of anatase TiO$_2$ material can be improved using an appropriate doping [27]. However, as depicted in Figure II-16, the power factor average is higher in the hole doping region than in the electron one. Consequently, we can conclude that p-type doped region is more efficient than n-type doped region in anatase TiO$_2$ compound.

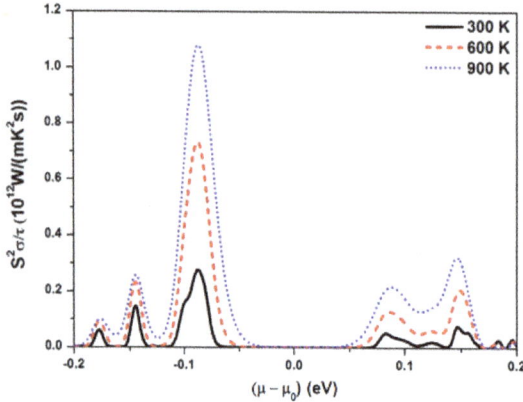

Figure II-16.
Calculated power factor with respect to scattering time (S2σ/τ) of anatase TiO$_2$as a function of the chemical potential at various temperatures.

2.7. Conclusions

In summary, we have investigated the structural, electronic, optical, thermal and transport properties of anatase TiO$_2$using the FP-LAPW approach. The obtained lattice parameters are in good agreement with experimental data. The MBJ calculated band structure revealed an indirect band gap of about 3.05 eV. Some optical parametric quantities such as dielectric constant, refractive index, extinction coefficient, reflectivity and electron energy loss function, were also investigated. By applying the Debye quasi-harmonic approximation, the effect of temperature and pressure on the thermal properties of anatase TiO$_2$ has been studied and discussed. Some transport coefficients of anatase TiO$_2$ compound were performed using BoltzTraP code as embodied in the WIEN2k package. The present results depict that the calculated Seebeck coefficient for anatase TiO$_2$ compound is higher at room temperature. The electrical conductivity is found to decrease slightly with augmenting temperature. However, electronic thermal conductivity gives higher values with increasing temperature. The calculated power factor average was appreciable. We note that p-doping would greatly enhance the thermoelectric properties of the investigated material compared to n-doping. This study may be useful guidelines to design some thermoelectric devices based on anatase TiO$_2$ compound.

References

[1] P. Hohenberg, W. Kohn, Phys. Rev. B 136, 864 (1964).
[2] W. Kohn, L. S. Sham, Phys. Rev. A 140, 1133 (1965).
[3] A.P.Rodriguez, "Electronic, magnetic and structural properties of systems close to a metal-insulator transition: an ab initio study", University of Santiago de Compostela, Spain,2013
[4] P.Blaha, K.Schwarz, G.K.H.Madsen, D.Kvasnicka, J.Luitz, WIEN2K An Augmented Plane Wave + Local Orbitals Program for Calculating Crystal Properties, Karlheinz Schwarz, Techn. Universitat, Austria, Vienna, (2001).

[5] M.A. Blanco, E. Francisco, V. Luana, Comput. Phys. Commun. 158, 57 (2004).
[6] G.K.H. Madsen, D.J. Singh, Comput. Phys. Commun. 175, 67 (2006).
[7] F.D. Murnaghan, Proc. Natl. Acad. Sci. USA 30, 244 (1944).
[8] T. Mahmood, C. Cao, R. Ahmed, M. Ahmed, M. A. Saeed, A. A. Zafar, T. Husain, M. A. Kamran, Sains Malaysiana 42, 231 (2013).
[9] Q. Chen, H. H. Cao, J. Mol. Struct. THEOCHEM 723, 135 (2005).
[10] W. Zeng, T. Liu, Z. Wang, Sensors Actuat. B 166–167, 141–149 (2012).
[11] G. Sai, L. B. Gui, Chin. Phys. B 21, 057104 (2012).
[12] W.Naffouti, T.Ben Nasr, O.Briot, N.Kamoun-Turki, J. Electron. Mater. 44, 3661 2015
[13] J.P. Perdew, M. Levy, Phys. Rev. Lett. 51, 1884 (1983).
[14] M.Fox, "Optical Properties of Solids", Oxford University Press, New York (2001).
[15] M. Dressel, G. Gruner, "Electrodynamics of Solids: Optical Properties of Electrons in Matter", Cambridge University Press, UK (2002).
[16] S.Saha, TP.Sinha, "Electronic structure, chemical bonding, and optical properties of paraelectric BaTiO3", Phys. Rev. B, 62, pp.8828-8834 (2000).
[17] M.A. Blanco, A. Martín Pendás, E. Francisco, J.M. Recio, R. Franco, J. Mol. Struct. (Theochem.) 368, 245 (1996).
[18] E. Francisco, M.A. Blanco, G. Sanjurjo, Phys. Rev. B 63, 094107-1 (2001).
[19] A. A. Levchenko, A. I. Kolesnikov, N. L. Ross, J. Boerio-Goates, B. F. Woodfield, G. Li, A. Navrotsky, J. Phys. Chem. A 111, 12584 (2007).
[20] L. Y. Lu, Y. Cheng, X. R. Chen, J. Zhu, Physica B 370, 236 (2005).
[21] J. Chang, X. R. Chen, W. Zhang, J. Zhu, Chin. Phys. B 17, 1377 (2008).
[22] L.J. Zhang, D.J. Singh, Phys. Rev. B 81, 245119 (2010).
[23] K. Ahn, E. Cho, J.-S. Rhyee, S. I. Kim, S. Hwang, H.-S. Kim, S. M. Lee, K. H. Lee, J. Mater. Chem. 22, 5730 (2012).
[24] R. Brahimi, Y. Bessekhouad, M. Trari, Physica B 407, 3897 (2012).
[25] W. Khan, A.H. Reshak, Comput. Mater. Sci. 89, 52 (2014).
[26] H.A.R.Aliabad, M.Ghazanfari, I.Ahmad, M.A.Saeed, Comput.Mater. Sci. 65, 509,2012
[27] H.Y.Lv, H.J.Liu, L.Pan, YW.Wen, X.J.Tan, J.Shi, Appl. Phys. Lett. 96, 142101 (2010).

III. Structural and morphological properties of titanium dioxide thin films

It is important to have an idea about the structure and the surface morphology of chemically sprayed titanium dioxide thin films in order to optimize the elaboration process. Indeed, this section reports the effect of the growth parameters and mechanism conditions such as the sprayed solution flow rate, sprayed solution volume, the substrate temperature and such doping, on the structural and morphological properties of TiO_2 thin films.

3.1. Effect of spayed solution flow

Figure III-1 shows the effect of solution flow rate on the X-ray diffraction patterns of titanium dioxide films. At a flow rate of 2 ml/min, no diffraction peaks in the entire recorded range are observed indicating the amorphous nature of the film. Raising the solution flow rate (> 2ml/min), the anatase phase (JCPDS N° 89-4921), with preferred orientation along (101) plane, is identified. This result is in good agreement with previous reported workers [1,2]. At a flow rate of 10 ml/min, four distinctive anatase TiO$_2$ peaks can be assigned to (101), (004), (200) and (105) crystal planes. In fact, the preferred peak intensity increases with increasing the flow rate until 10 ml/min and decreases above revealing degradation of the film crystallinity.

In fact, it is well known that the flow rate is proportional to the drop size of the sprayed solution [3]. Therefore, if this last increases, then the drop size becomes larger. However, in reality, there is an optimal drop size, which should be small enough to avoid excessive wetting, but large enough to prevent the solvent evaporation before arriving to substrate surface [3]. Indeed, drying dominates for sprayed solution flow rate lower than 10 ml/min, and wetting dominates for higher flow rates (> 10 ml/min).

The crystallite size values calculated using Scherrer's formula [4] for the dominant peak are listed in Table III-1. An average size value of 28.60 nm is determined at 10 ml/min and above the grain size decreases to reach a value of 12.09 nm.

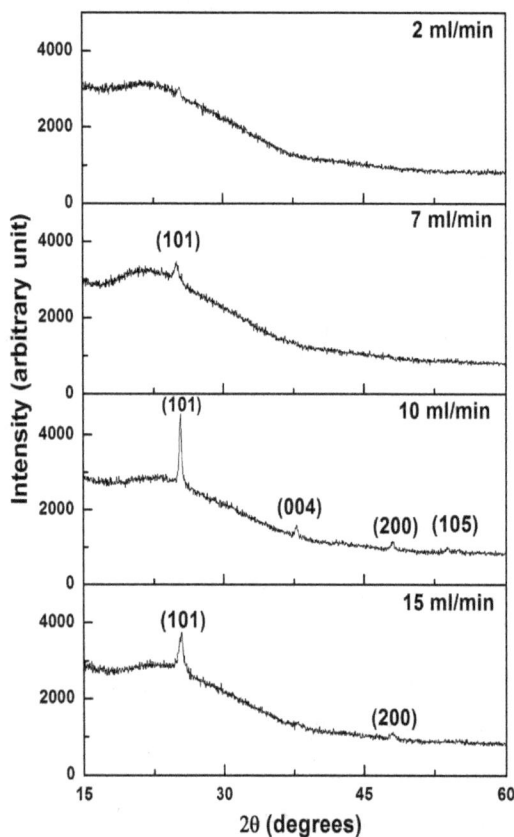

Figure III-1.
X-ray spectra of TiO$_2$ thin films obtained at different solution flow rates.

Table III-1. Variation of the crystallite size (D) of TiO$_2$ thin films as a function of solution flow rates.

Flow rate (ml/min)	2	7	10	15
D (nm)	-	17.01	28.60	12.09

Using grain size values, the dislocation density (δdis), defined as the imperfection in crystal, has been calculated using the Williamson and Smallman's formula [5]:

$$\delta_{dis} = \frac{1}{D^2} \qquad (1)$$

where "D" is the grain size. The micro strain of thin films is also calculated with the help of the following relation [6]:

$$\epsilon = \frac{\beta \cos \theta}{4} \qquad (2)$$

where ϵ is the micro strain of thin films, β is Full Width at Half Maximum (FWHM) and Θ is the peak position.

The variations of the micro strain (ϵ) and dislocation density (δdis) versus solution flow rate are presented in Figure III-2. We note that the film deposited at 10 ml/min possesses larger crystallite size, less strain and minimum dislocation density. The probable reason may be the enhancement of crystallinity in this case; consequently, there is a simultaneous movement of interstitial atoms from grain boundary to crystallites which causes a decrease of dislocation density as well as strain in TiO$_2$ thin films [7].

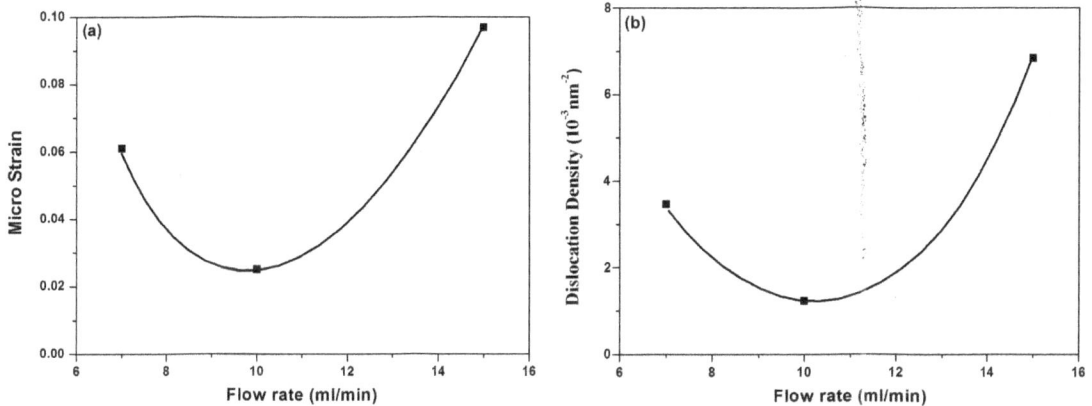

Figure III-2. Variation of micro strain and dislocation density with solution flow rates for TiO$_2$ thin films, the line drawn is a guide for the eye.

Figure III-3 shows AFM image topography of TiO$_2$ films prepared at different solution flow rates. It provides direct visual information on both polycrystalline grain size and surface roughness. Based on these micrographs, the surface roughness (RMS) values of anatase TiO$_2$ films are in the range of 2.45 5.30 nm. We remark that the RMS value increases as the solution flow rate increases up to 10 ml/min, and then it decreases to reach 3.49 nm at flow rate of 15 ml/min as shown in Table III-2. The reason may be that the thickest film is obtained at a solution flow rate of 10 ml/min, as it will be described later (Section IV). Moreover, the film surface, in this case, is more homogenous and well covered by uniformly distributed grains. It can be seen a considerable decrease in grain size from 10 to 15 ml/min. This may be explained by the fact that, at 15 ml/min, films do not adhere to the substrate.

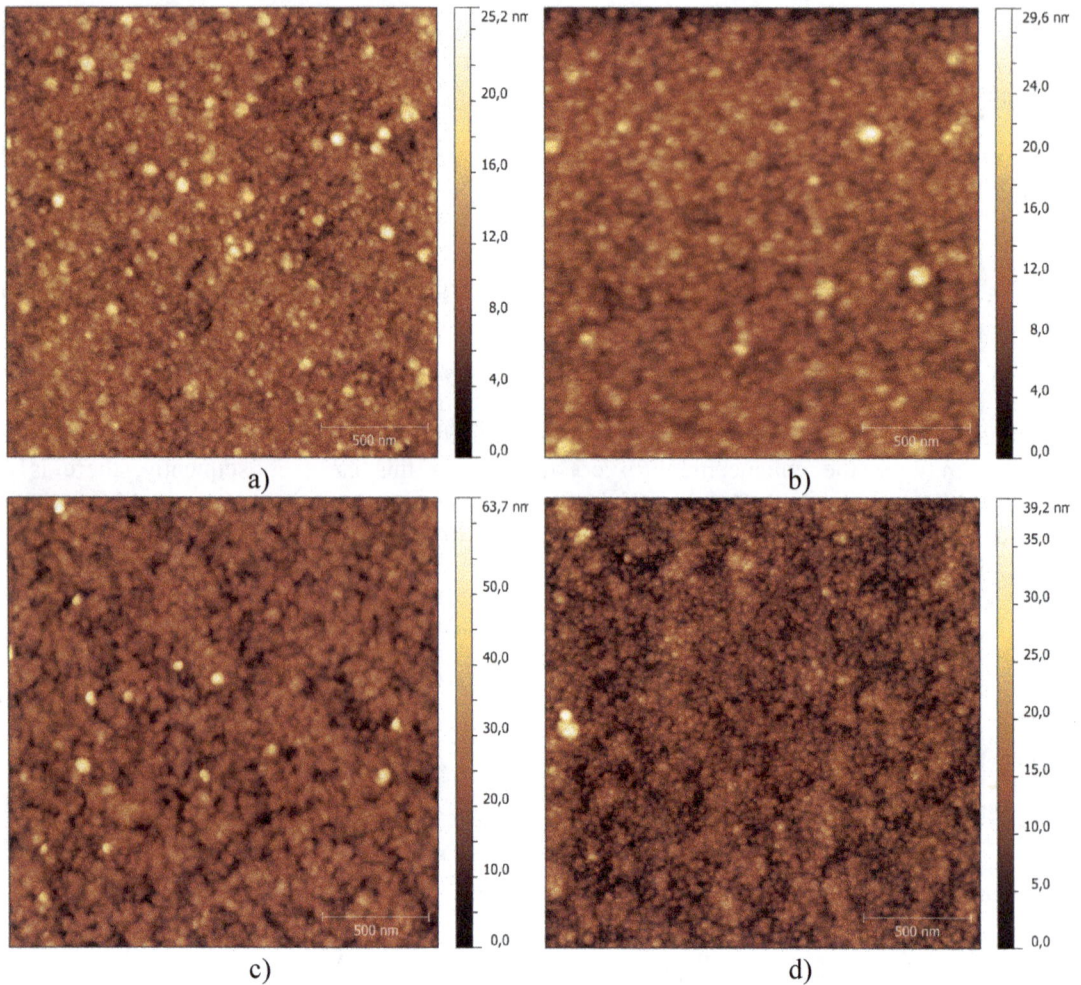

Figure III-3. AFM images of TiO$_2$/glass thin films obtained at solution flow rates of: (a) 2 ml/min, (b) 7 ml/min, (c) 10 ml/min and (d) 15 ml/min.

Table III-2. Variation of the surface roughness of TiO$_2$/glass thin films as a function of solution flow rates.

Solution flow rate (ml/min)	2	7	10	15
RMS (nm)	2.45	2.54	5.30	3.49

3.2. Effect of sprayed solution volume

At a sprayed solution flow rate equals to 10 ml/min, according to our previous research [8], a variation of sprayed solution volume is investigated. In fact, Figure III-4 depicts the XRD patterns of TiO$_2$ thin films deposited with various solution volumes. One can remark that there is only a small peak, observed at the diffraction angle of 25.30° for the TiO$_2$ film prepared in the case of sprayed solution volume equals to 30 ml.

This poor crystallinity with amorphous background is due to an incomplete growth of the crystallites. For sprayed solution volume of about 60 ml, a thicker film is obtained and a crystallinity improvement is revealed. In fact, the preferential peak becomes narrower and its intensity becomes higher. Four diffraction peaks are observed, in this case, at $2\Theta \sim$ 25.30°, 37.75°, 47.96° and 53.87° and are indexed as (101), (004), (200) and (105) crystal planes. These peaks are assigned to the tetragonal anatase phase of TiO_2 (JCPDS card [89-4921]) as reported by other works [9]. Indeed, by increasing the sprayed volume, an increment of film thickness is obtained which may lead to higher crystallization probability [10]. In this context, Trofimov *et al.* [10] have investigated a model for crystallization kinetics of thin film versus thickness by extension the Kolmogorov–Johnson–Mehl–Avrami (KJMA) model. Thus, it is found that the thinner film crystallizes slower with an inhomogeneous structure compared to the thicker one. For sprayed solution volume greater than 60 ml, a noticeable degradation of the film crystallinity is revealed by the broadening of (101) peak and a dominance of the amorphous background. This behavior may be related to an increase in the number of defects with the increase in film thickness.

This result is in accordance with other works [11]. In fact, Mortezaali *et al.* [11] reported that the density of defects increases in the case of thicker films by increasing the amount of material sprayed on the heated substrate surface. These defects can be assigned to oxygen vacancies which may be reduced by heat treatment. However, it is important to signal that annealing process affects more the vicinity of the surface than the volume [12]. Thus, the density of defects stills relatively high in thicker layers even after air annealing.

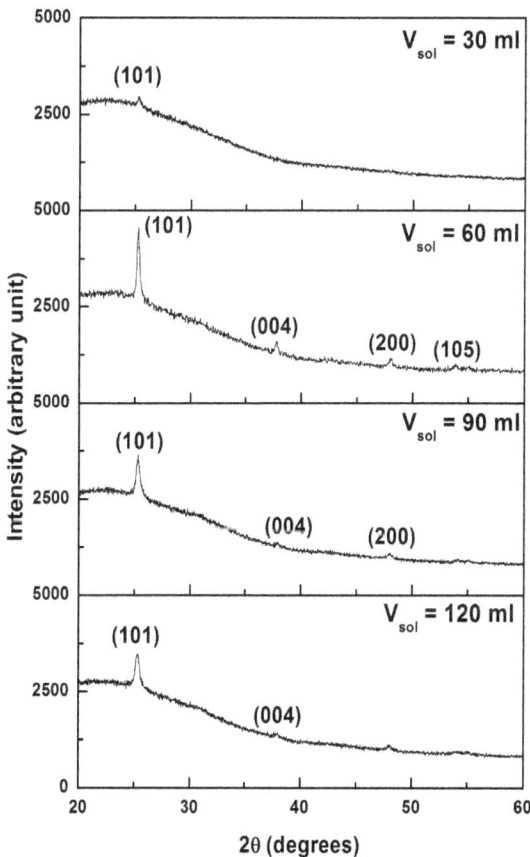

Figure III-4.
X-ray spectra of anatase TiO_2 thin films deposited on glass substrate for different solution volumes.

A full-pattern X-ray quantitative Rietveld refinement is utilized to calculate the lattice parameters with the help of Material Analysis Using Diffraction (MAUD) software program. The fitting procedure was controlled by the Marquardt least square method to reduce the difference between calculated and experimental diffraction spectra [13]. In fact, Figure III-5 presents the experimental and calculated XRD patterns for TiO_2 thin film deposited with sprayed solution volume equals to 60 ml. The lattice parameters (a,c) of TiO_2 crystallizing into anatase phase, with (101) preferential orientation situated at 25.31°, are found to be equal to 3.781 A° and 9.507 A°, respectively. This last finding is in good agreement with the JCPDS card n° 89-4921.

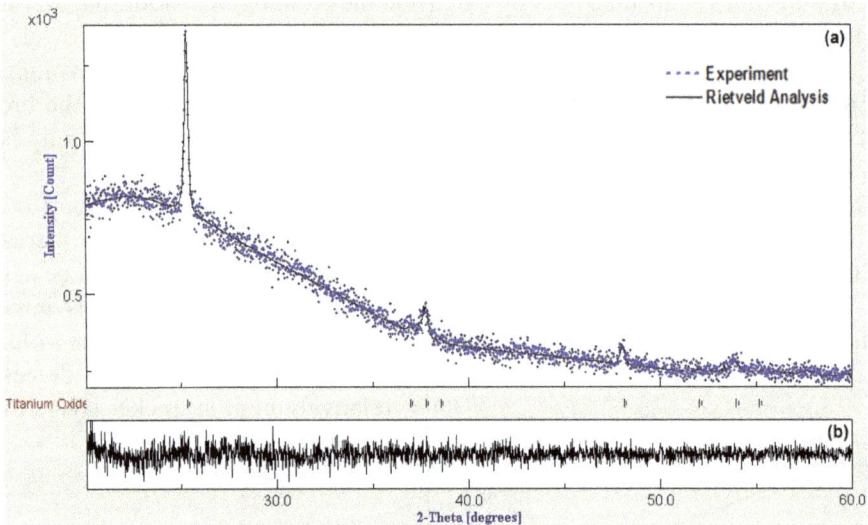

Figure III-5. (a) Experimental and calculated curves of the X-ray diffraction of the TiO₂/glass thin film grown at sprayed solution volume equals to 60 ml, (b) Difference between the experimental and the calculated curves.

Figure III-6 shows the variation of crystallite size, calculated from Scherrer's formula [4], of the TiO_2 sample versus sprayed solution volume, the sample elaborated at 30 ml is not discussed owing to its poor crystallinity. The crystallite size is found to be of the nanoscale order. It varies from 15.2 nm to 28.6 nm, where the maximum value is obtained in the case of film deposited with solution volume equals to 60 ml.

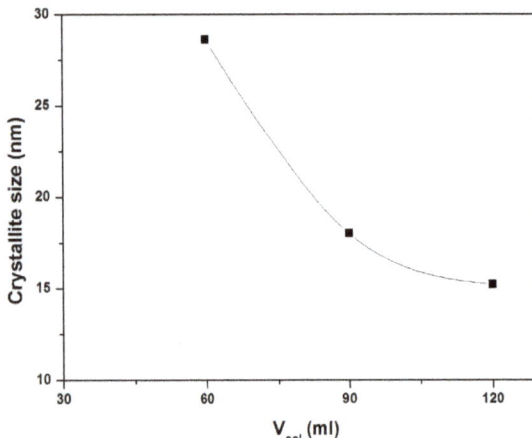

Figure III-6.
Crystallite size of TiO₂/glass thin films for different solution volumes, the line drawn is a guide for the eye.

Figure III-7 displays the dislocation density (δdis) and the micro strain (ϵ) versus sprayed solution volume. It is clear that both of them increase with increasing of film thickness, this means that defects are formed during the growth of the layers. It is observed that the film grown at sprayed solution volume equals to 60 ml is found to have maximum value of crystallite size and minimum value of strain and dislocation density.

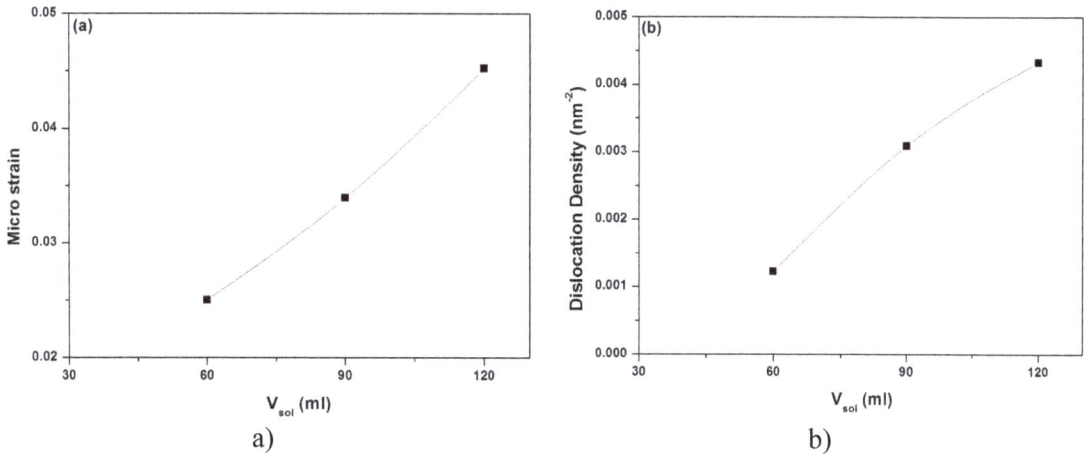

a)   b)

Figure III-7. Plots of (a) micro strain and (b) dislocation density of TiO_2/glass thin films versus sprayed solution volumes, the line drawn is a guide for the eye.

AFM micrographs are used to study the evolution and the development of morphological state of the growing TiO_2 thin films as a function of sprayed solution volume. Figure III-8 shows two and three dimensional AFM images of anatase TiO_2 thin film grown at sprayed solution volume of about 60 ml.

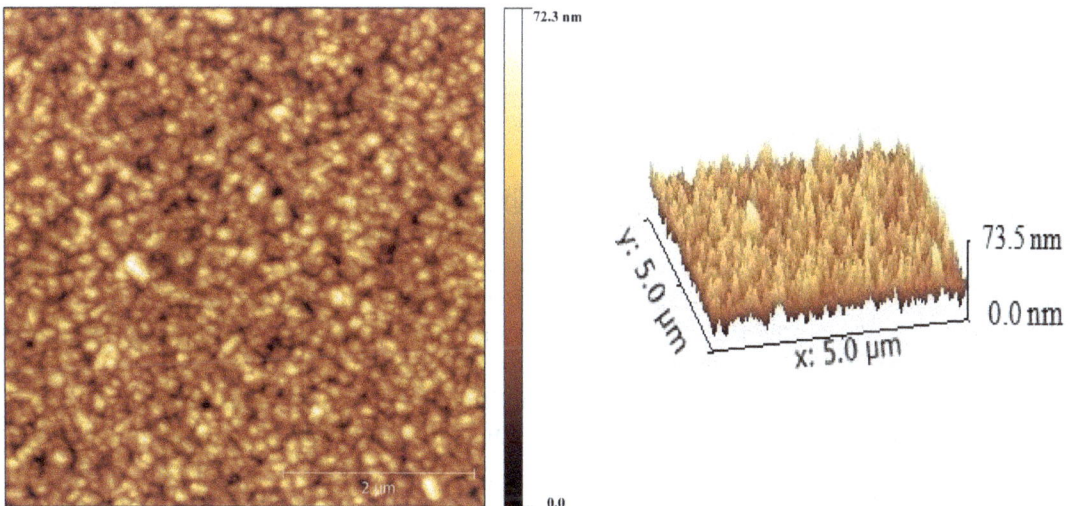

Figure III-8. AFM plane and 3D views of TiO_2/glass thin film deposited at 60 ml.

It gives us a panoramic view of the grain size and the aggregation state of the grown film over a surface of 5µm×5µm. As it can be seen from this Figure , the film exhibits a uniform surface morphology and practically no voids are present. It is composed of small particles with circular shape, a uniform distribution and a relatively good connectivity over the glass surface. Their size ranges from 14 nm to 27 nm, which is in agreement with that calculated using Scherrer's formula (Figure III-6). Based on the AFM images, the variation of roughness is presented in Figure III-9. It ranges from 5.64 nm to 6.38 nm. The smoothest surface is obtained for anatase TiO$_2$ thin films grown at 60 ml.sprayed solution

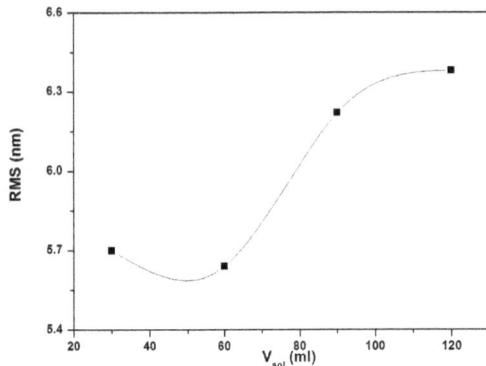

Figure III-9.
Root-mean-square (RMS) roughness of TiO$_2$/glass thin layers deposited using various solution volumes; the line drawn is a guide for the eye.

3.3. Effect of substrate temperature

Figure III-10 shows the XRD patterns of titanium dioxide thin films deposited on glass substrate at different substrate temperatures. It is observed that all the layers show a pronounced diffraction peak around 25.30°, indicating the anatase phase with (101) preferred orientation. Moreover, an increase of (101) peak intensity with substrate temperature increment to 300°C is revealed, which indicates the improvement of crystal quality with substrate temperature increment. This behavior can be explained by the fact that with the increase of substrate temperature, the surface species mobility is enhanced, thus they form more stoichiometric crystalline phase by migration through the lattice and deposition at the appropriate lattice sites. Consequently, an improvement of crystalline quality is obtained. However, a further increase in the temperature up to 300°C leads to deterioration of the crystal quality. In fact, the high substrate temperature induces defects in the TiO$_2$ thin films which lead to the poor crystalline quality.

From Figure III-11, one can notice that the average grain size calculated through the Scherrer formula varies from 9.4 to 32.5 nm where the maximum value is obtained for film grown at 300°C. In fact, the increase of the grain size with substrate temperature is related to the crystalline quality improvement, which may be explained by the fact that high substrate temperature enhances the surface mobility by providing sufficient energy required in the crystallization process.

The effect of the substrate temperature on the surface morphologies of TiO$_2$/glass thin films is depicted in Figure III-12. It is observed that all the grown layers exhibit dense and compact surfaces which are practically covered with equal sized grains uniformly distributed. Furthermore, we note that the grain size is proportionate to the substrate temperature. In fact, it practically increases with substrate temperature increment which indicates an improvement of crystalline quality.

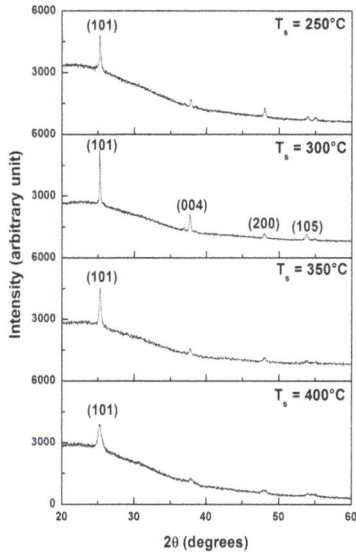

Figure III-10. X-ray spectra of TiO_2 thin films on glass substrate at different substrate temperatures.

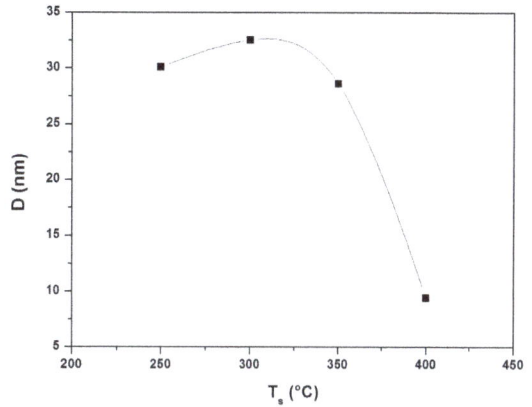

Figure III-11. Crystallite size of TiO_2 thin films deposited at various substrate temperatures

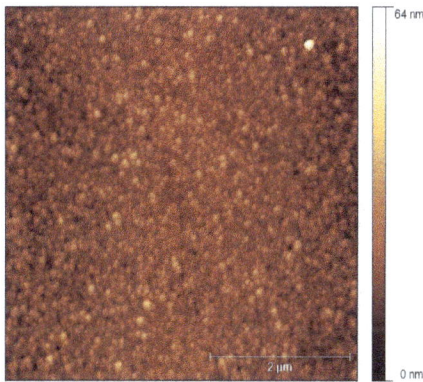

250 °C

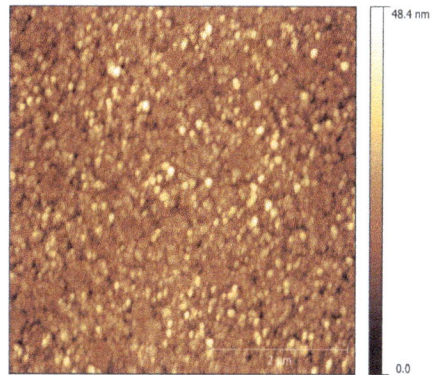

300 °C

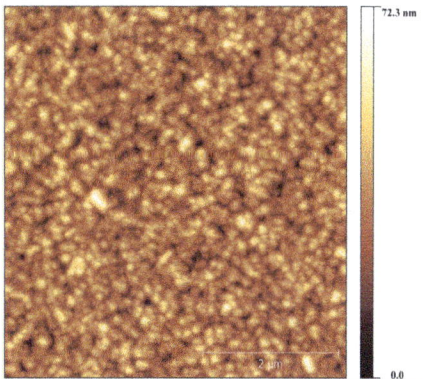

350 °C

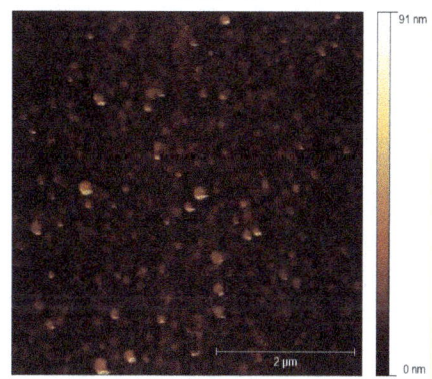

400 °C

Figure III-12. AFM micrographs of TiO_2/glass thin film deposited at various temperatures.

A significant enhancement of RMS roughness with substrate temperature increment can be noticed (Figure III-13). Thus, the film deposited at substrate temperature of about 300°C exhibits the smallest RMS roughness (4.68 nm).

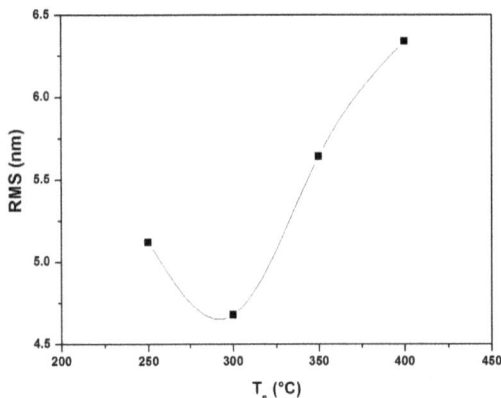

Figure III-13.
Root-mean-square (RMS) roughness of TiO₂ thin layers deposited using various substrate temperatures, the line drawn is a guide for the eye.

Effect of air annealing

Heat treatment usually plays significant role in enhancing the physical properties of thin films. Titanium dioxide thin films are chemically sprayed on quartz substrate and annealed in air for 2 hours at different temperatures (Ta) in the range [500-1100]°C. Figure III-14 presents the XRD spectra of TiO₂ thin films annealed in air at various annealing temperatures. Indeed, for heated temperature ≤ 900°C, three peaks are detected at 25.25°, 37.72° and 47.98° corresponding to (101), (004) and (200) planes, respectively. These peaks are assigned to anatase phase according to JCPDS card n°89-4921.However, at higher annealing temperature (Ta ≥ 1000°C), a phase transition is observed from the anatase to the rutile phase. This result is similar to that reported by Ben Naceur at al. [14].

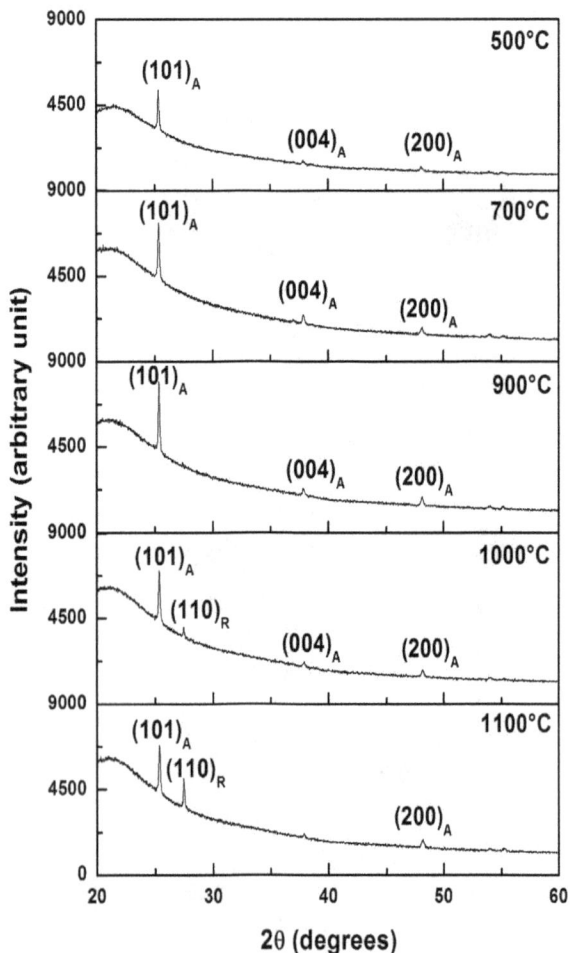

Figure III-14.
XRD patterns of TiO₂/quartz thin films annealed in air at various temperatures.

Figure III-15 depicts the variation of crystallite size (D) of TiO_2 thin films versus annealing temperatures. It is found that D increases from 37.1 to 59.98 nm when the annealing temperature increases from 500 to 900°C, implying an improvement of film crystallinity. For annealing temperature greater than 900°C, a decrease of crystallite size is noticed due to the transition from anatase to rutile phase.

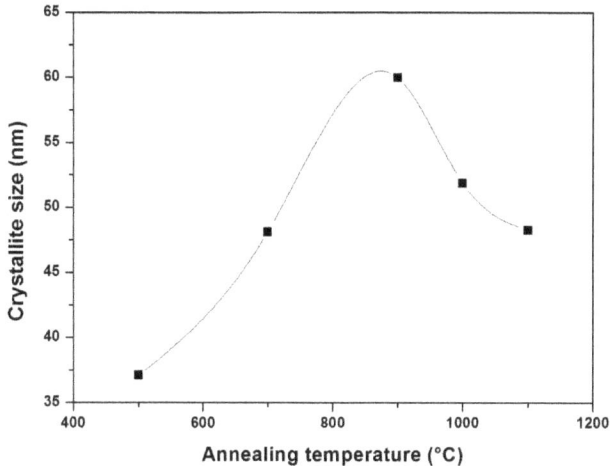

Figure III-15.
Crystallite size of TiO_2/quartz thin films annealed in air at different temperatures, the line drawn is a guide for the eye.

In order to investigate the effect of air annealing on the morphological properties of TiO_2/quartz thin films, their AFM plane views are shown in Figure III-16 as a function of annealing temperature. It can be seen from this Figure that both films annealed at 500 and 700°C exhibit a smooth surface. However, when the annealing temperature is higher than 700°C, a granular morphology consisting of spherical shaped grains homogenously distributed without visible pores, relatively, starts to appear. Thus, an improvement of crystallinity with an increase of crystallite size is revealed. This finding is already obtained by XRD analysis (Figure III-15). Similar behavior is reported by Ben Naceur et al. [14].

For further morphological analysis, the root mean square (RMS) roughness of the annealed films is displayed in Table III-3. It is found to increase slowly from 3.0 to 4.9 nm when annealing temperature increases from 500 to 700°C, and then it increases drastically and reaches 18.6 nm for 1100°C. This increment may be caused by the phase transition as well as the growth of the grain size.

Effect of Samarium doping

Recently, numerous studies have been carried out on titanium dioxide rare earth (RE) doping (e.g., Sm, Ce, Eu and Er). Indeed, RE doping has been proven to be an efficient route to improve the luminescence properties of TiO_2 films due to the effective emission in the visible region [15]. Therefore, RE doped TiO_2 thin films can be a promising candidate for several photoelectric devices and optical communications such as active planar waveguides, flat plane displays, high energy radiation detectors, optical data storage, etc. [16-18]. Furthermore, thanks to its 4f electron configurations, RE elements are promising for environmental detoxification, thus, it can significantly enhance the photocatalytic activity of TiO_2 based composites [19-21]. Among various rare earth used for doping, samarium ions have been chosen for their high efficiency in improving the physical properties of TiO_2 [1].

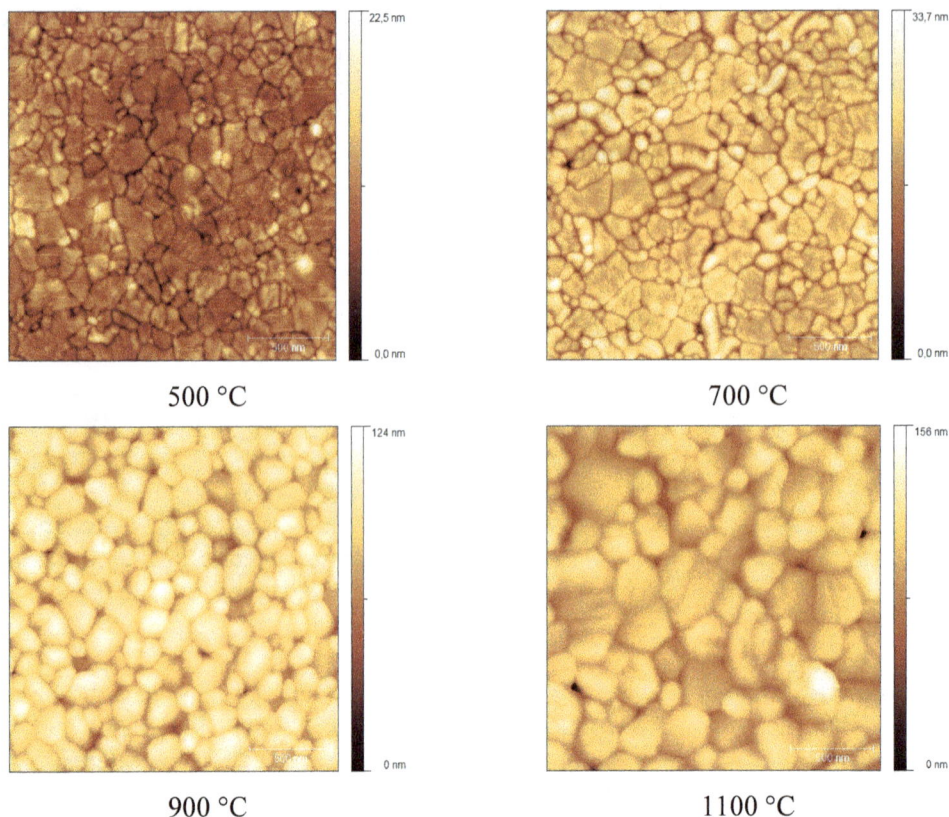

Figure III-16. AFM plane views of TiO$_2$ thin film annealed in air at various temperatures.

Table III-3. Variation of the surface roughness of TiO$_2$/quartz thin films as a function of annealing temperature.

Annealing temperature (°C)	500	700	900	1100
RMS (nm)	3.00	4.9	12.7	18.6

Therefore, titanium dioxide thin films are doped with samarium ion at different concentration ratios (y = ([Sm^{3+}]/[Ti^{4+}]) in the range of 0-1 at.% by a step of 0.25 at.%) and chemically sprayed on glass substrates.

Figure III-17 depicts the XRD spectra of TiO$_2$:Sm thin films deposited on glass substrates for various samarium doping concentrations. All the samples show the (101) diffraction peak of the tetragonal anatase phase (JCPDS n° 89-4921). Its intensity is significantly high suggesting a (101) preferred crystalline orientation for the films, particularly those with a nominal Sm doping ratio smaller or equal to 0.25 at.%. For these films in particular, the (004), (200) and (105) diffraction peaks are also detected. This preferred orientation corresponds to an improved crystal quality. In the same time, increasing the nominal Sm doping ratio at a value higher than 0.25 at.% seems to induce a degradation of this preferred orientation and may be a degradation of the overall crystalline quality of the sprayed films, suggesting that the introduction of large amount of Sm atoms promotes severe crystallographic disorder. Full-pattern X-ray quantitative Rietveld refinement was performed using the MAUD software program [13]. The fit quality of the achieved refinements is illustrated in Figure III-18 for undoped and 0.25 at.% samarium doped TiO_2 thin film.

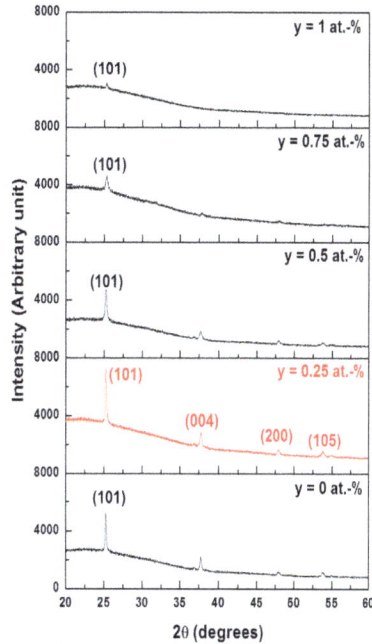

Figure III-17.
XRD patterns of the sprayed TiO_2 thin films for different samarium doping ratios.

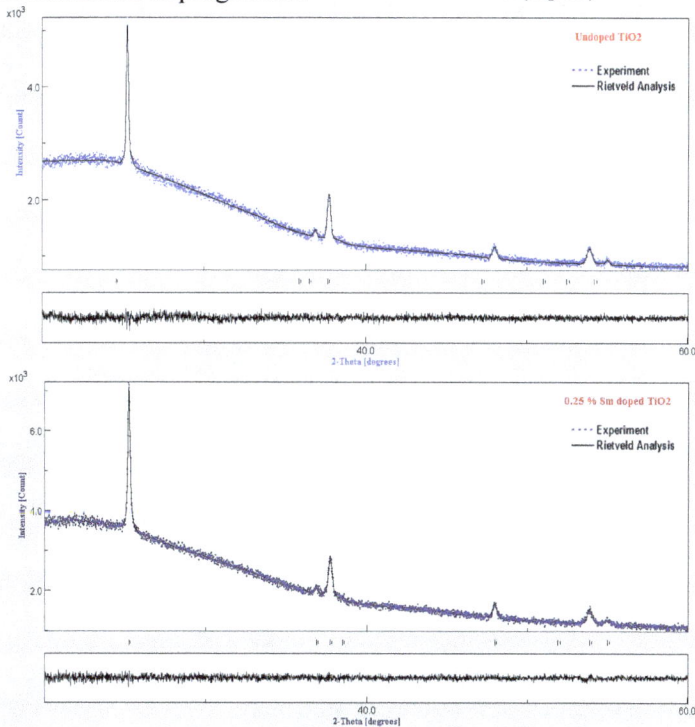

Figure III-18. Rietveld refinement for XRD patterns of undoped and 0.25 at.% Sm doped TiO_2 thin film.

The inferred lattice parameters a and c from these refinements are summarized in Table III-4 for all the produced films. They are found to be very close to those tabulated for bulk anatase phase, namely 3.791 and 9.526 Å, respectively. An attentive lecture of the obtained values makes evidence a slight increase of these parameters when y increases from 0 to 1 at.% in agreement with the progressive incorporation of large Sm^{3+} (Table III-4).

Table III-4. Cell parameters, average crystal size and microdeformation inferred from Rietveld refinements for all the produced films.

y (at.%)	Crystal size (nm)	Microdeformation (%)	Lattice parameters	
			a (A°)	c (A°)
0	36,84	0,10	3.791	9.526
0.25	39,47	0,14	3.792	9.534
0.5	34,80	0,17	3.791	9.538
0.75	24,96	0,18	3.796	9.539
1	22,06	0,22	3.874	9.652

Interestingly, the inferred average crystal size and microstrain induce lattice deformation of the sprayed films changes when their nominal atomic Sm content increases as illustrated in Figure III-19. The former decreases from 39 to 22 nm (with an accuracy of about 1 nm) while the latter increases from about 0.1 to 0.2 % when y increases from 0 to 1 at.%. These deformations are assumed to be due to local defects related to the replacement of Ti^{4+} cations by larger Sm^{3+} ones in the TiO$_2$ tetragonal lattice.

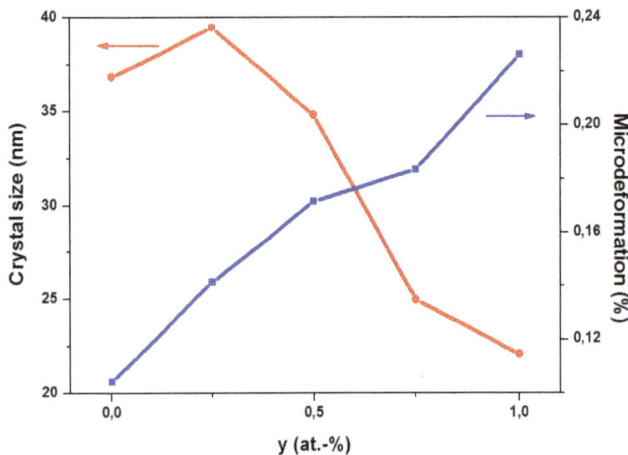

Figure III-19. Variation of the average crystal size and microdeformation of the sprayed TiO$_2$:Sm thin films as a function of their nominal atomic Sm doping ratio.

AFM micrographs allow us to study the evolution of the morphology of the deposited TiO$_2$ thin films versus their Sm content. Figure III-20 depicts AFM plane views of v (y = 0 at.%) and TiO$_2$:Sm (y = 0.25 at.%) sprayed films. It allows us to have a panoramic view about the grain size and the density of the deposited layer over an area of 5μm×5μm. We note that the films are uniform, dense and practically covered with spherical grains that are regularly distributed and relatively connected over the glass surface. Their size varies from 19 to 41 nm, which deals clearly with that calculated previously from XRD analysis (Figure III-19). Note the surface quality of the films is quite equivalent. All exhibit almost the same morphology.

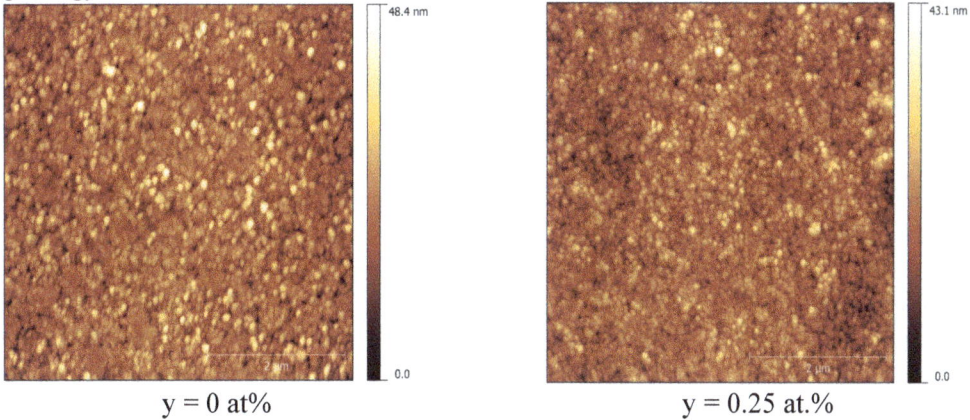

| y = 0 at% | y = 0.25 at.% |

Figure III-20. AFM plane view of Sm doped TiO_2 (y = 0 & o.25at.%)

Their roughness is also evaluated by calculating from their recorded AFM micrographs, the Root Mean Square roughness (RMS) parameter. The obtained values are summarized in Table III-5. They range between 4.3 and 5.5 nm, the smallest value being measured on the poorest Sm doped sample, namely y = 0.25 at.%. Focusing on this film, the comparison of its SEM topography to that of undoped one (Figure III-21) reveals that both have a uniform morphology and a good surface coverage, but the former clearly exhibits less voids on its top layer than the latter.

Table III-5. RMS roughness of the sprayed TiO_2:Sm thin films as a function of their nominal atomic Sm doping ratio (y).

y (at.%)	0	0.25	0.5	0.75	1
RMS (nm)	5.30	4.30	4.42	4.75	5.51

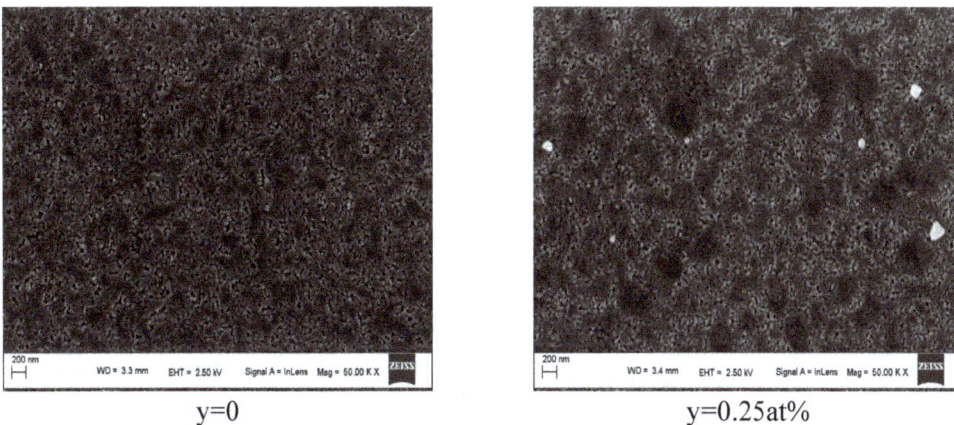

| y=0 | y=0.25at% |

Figure III-21. SEM topography of undoped and 0.25 at.% Sm doped TiO_2 thin films.

Effect of Manganese doping

Recently, several investigations paid attention to titanium dioxide transition metals doping (e.g., Fe [22], Co [23], Mn [22,25], and Ni [26]) since it can improve the photocatalytic efficiency as well as absorption in the visible region. Several reports suggest that Mn doping is one of the most suitable ions for doping TiO_2 owing to its ability to enhance the photocatalytic efficiency [24].

Thus, Mn-doped TiO_2 thin films are deposited on glass substrates by spray pyrolysis technique at different manganese doping concentrations (y = ([Mn^{2+}]/[Ti^{4+}] = 0, 2, 4, 6, 10%).

The influence of Mn doping on the titanium dioxide structure is monitored through the XRD measurements. Figure III-22 displays the XRD spectra of TiO_2 thin films for different manganese doping ratio. Pure TiO_2 spectra exhibits four peaks located at 25.27, 37.7, 47.98 and 53.8°, attributed to anatase (101), (004), (200) and (105) planes, respectively. Moreover, we note that Mn doping ratio of about 2 at.% is found to increase the preferred orientation intensity (101), implying an improvement of film crystallinity. However, a noticeable degradation of film crystallinity is revealed for an amount of Mn ions greater than 2 at.%. It is worth noting that the (101) diffraction peak totally disappears as y reaches 10 at.%.

Table III-6 presents the relative intensity of XRD peaks I(hkl)/I(101) of TiO_2:Mn sprayed thin films. Indeed, it is clear that the minimum ratios are obtained for 2 at.% Mn doping ratio and all of them are much less than 1. Thus, we can confirm that films doped at 2 at.% Mn are preferentially oriented along the direction (101).

For further crystallographic investigation, Figure III-23 displays the (101) peak position versus Mn doping ratio. Indeed, with increasing the amount of Mn element, a shift of (101) peak position towards higher angle is observed, implying a decreasing of the interplanar spacing according to Bragg's law. This tendency may be due to the contraction of the crystalline lattice.

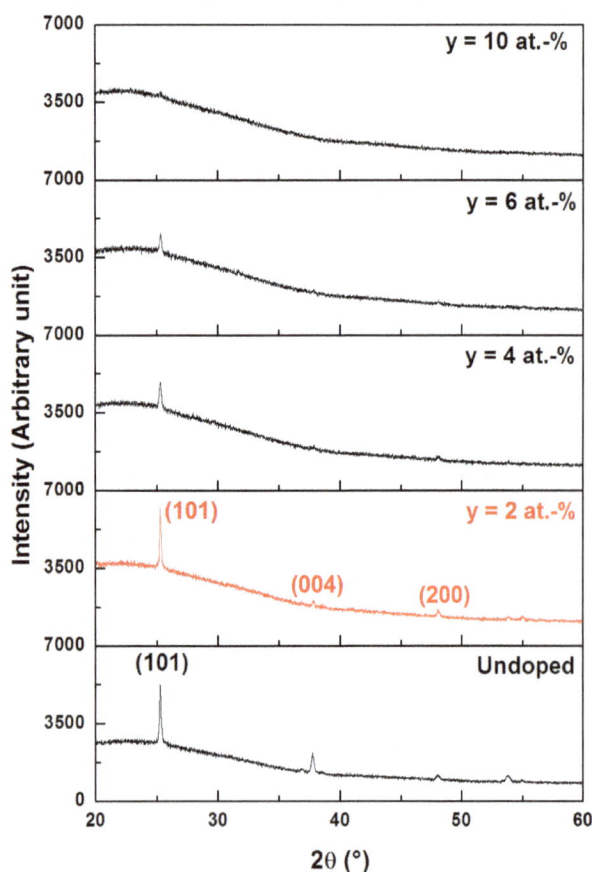

Figure III-22
XRD patterns of TiO_2 thin films for different manganese doping ratio.

Table III-6. Relative intensity of XRD peaks I(hkl)/I(101) of TiO_2:Mn sprayed thin films.

y (at.%)	I(004)/I(101)	I(200)/I(101)	I(105)/I(101)
0	0.31	0.10	0.20
2	0.08	0.09	0.04
4	0.072	0.14	-
6	-	-	-

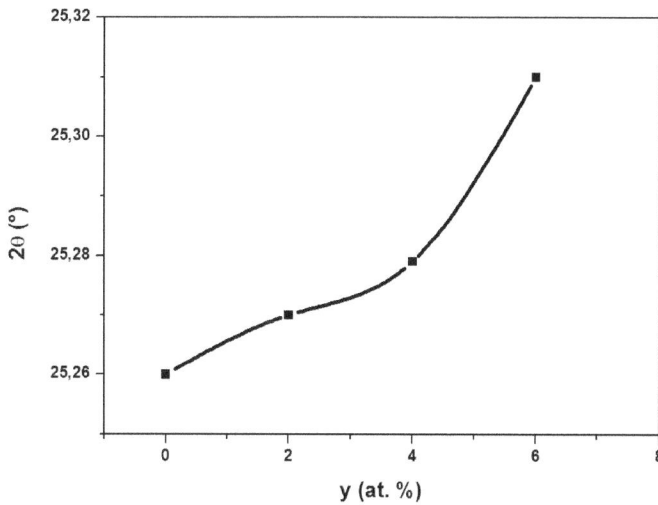

Figure III-23. Position of the (101) peak versus Mn doping ratio.

Table III-7. Variation of the crystallite size (D) of TiO_2:Mn thin films for different manganese doping ratio.

y (at.%)	0	2	4	6	10
D (nm)	36.84	66.57	46.84	38.70	-

The variation of the crystallite size (D) of TiO_2 thin films for different manganese doping ratio is summarized in Table III-7. We note that a largest D of about 66.57 nm is obtained at 2 at.% Mn doping concentration.

AFM plane views of undoped and 2 at.% manganese doped TiO_2 thin film are presented in Figure III-24. All the films exhibit uniform and compact surfaces which are practically covered with equal sized spherical grains. However, it is worth mentioning that there is a slight difference between the surface morphologies of undoped and Mn doped TiO_2 thin films. Indeed, for sample doped at 2 at.% Mn, the surface is more homogenous and the grains are more connected over the glass substrate. Therefore, an improvement of the surface state is revealed. Basing on the AFM study, a slight enhancement of RMS roughness may be noticed for films doped at 2 at.% Mn, according to Table III-8.

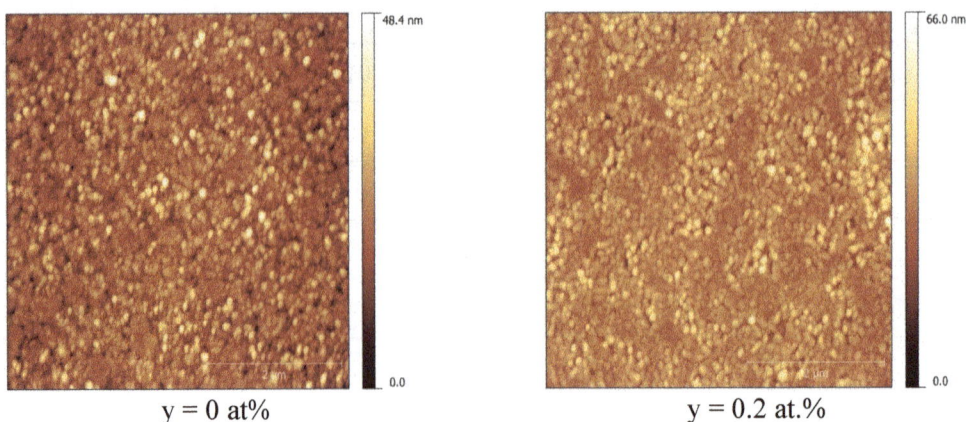

y = 0 at%　　　　　　　　　　　y = 0.2 at.%

Figure III-24. AFM plane views of undoped and 2 at.% manganese doped TiO$_2$ thin film.

Table III-8. Root-mean-square (RMS) roughness of TiO$_2$:Mn thin layers as a function of Mn doping ratio.

y (at.%)	0	2	4	6	10
RMS (nm)	5.45	5.30	6.02	7.40	6.90

In order to achieve more direct insight into the morphological features of the TiO$_2$:Mn films, scanning electron microscopy (SEM) is employed to investigate the evolution of the morphological state of the deposited TiO$_2$ thin layers as a function of Mn doping ratio. Figure III- 25 depicts the SEM images of undoped and 2 at.% Mn doped TiO$_2$ thin films. The SEM micrographs clearly show that manganese doping induces changes on the surface morphology of TiO$_2$ thin films. In fact, pure films are smooth in surface and there are some voids spread non-uniformly on the top of the layer. However, the presence of the Mn dopant practically decreases the voids density and the film surface becomes more uniform.

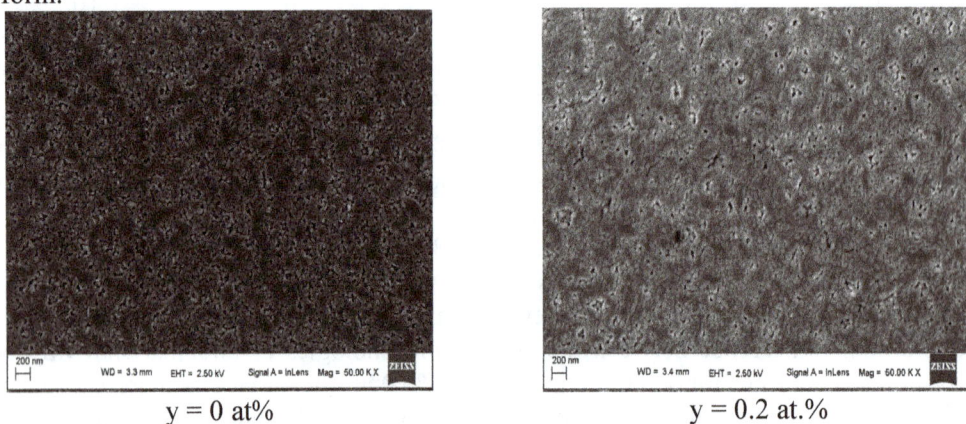

y = 0 at%　　　　　　　　　　　y = 0.2 at.%

Figure III-25. SEM topography of undoped and 2 at.% Mn doped TiO$_2$ thin films.

Conclusion

To summarize, the effect of various experimental parameters and annealing treatment on the structural and morphological properties of chemically sprayed titanium dioxide thin films has been investigated. It is found that a sprayed solution flow rate of about 10 ml/min presented a good growth of TiO_2 thin films. Moreover, TiO_2 thin film deposited by spraying 60 ml exhibited the best physical properties. Besides, substrate temperature of about 300°C was found to improve the structural properties of the grown layers. The optimization of the deposition parameters has been followed by an annealing treatment in air at various temperatures. It appeared that TiO_2 thin films are very sensitive to heat treatment conditions. In fact, starting from the annealing at 500°C, the thin films crystallized in the anatase phase. At higher heat temperatures (=1000°C), it started to change into the rutile structure. Furthermore, in order to enhance the physical properties of the growth films, both samarium and manganese doping have been investigated. Thus, better crystalline quality was obtained for Sm and Mn doping ratios of about 0.25 at.% and 2 at.%, respectively. Finally, the variation of experimental conditions not only affects the structural and morphological properties of TiO_2 thin films, but also it has a considerable influence on their optical properties which will be discussed later in the next section.

References

[1] I. Oja Acik, V. Kiisk, M. Krunks, I. Sildos, A. Junolainen, M. Danilson, A. Mere, V. Mikli, Appl. Surf. Sci. 261, 735 (2012).
[2] D. Y. Lee, J. T. Kim, J. H. Park, Y. H. Kim, I. K. Lee, M. H. Lee, B. Y. Kim, Curr. Appl. Phys. 13, 1301 (2013).
[3] J. H. Woo, H. Yoon, J. H. Cha, D. Y. Jung, S. S. Yoon, J. Aerosol Sci 54, 1 (2012).
[4] M. J. Buerger, X-ray crystallography (Wiley, Inc., New York, 1960), p. 23.
[5] D. P. Padiyan, A. Marikani, K. R. Murali, Mat. Chem. and Phys. 78, 51 (2002).
[6] K. Girija, S. Thirumalairajan, S.M. Mohan, J. Chandrasekaran, Chalco. Lett. 6, 351 (2009).
[7] A. Akkari, M. Reghima, C. Guasch, N. Kamoun-Turki, Mater. Sci. 47, 1365 (2012).
[8] W. Naffouti, T. Ben Nasr, A. Mehdi, N. Kamoun-Turki, J. Electron. Mater. 43, 4033 (2014).
[9] A. P. Huang, Z. F. Di, P. K. Chu, Surf. Coat. Technol. 201, 4897 (2007).
[10] V.I. Trofimov, I.V. Trofimov, J. Kim II, Thin Solid Films 495, 398 (2006).
[11] A. Mortezaali, O. Taheri, Z.S. Hosseini, Microelectron. Eng. 151, 19 (2016).
[12] K. Iijima, , M. Goto, S. Enomoto, H. Kunugita, K. Ema, M. Tsukamoto, N. Ichikawa, H. Sakama, J. Lumin. 128, 911 (2008).
[13] S. Louidi, F. Z. Bentayeb, W. Tebib, J. J. Suñol, A. M. Mercier, J. M. Grenèche, J. Non-Cryst. Solids 356, 1052 (2010).
[14] J. Ben Naceur, M. Gaidi, F. Bousbih, R. Mechiakh, R. Chtourou, Current App. Phys. 12, 422 (2012).
[15] E.L. Prociow, J. Domaradzki, A. Podhorodecki, A. Borkowska, D. Kaczmarek, J. Misiewicz, Thin Solid Films 515, 6344 (2007).
[16] C. Mignotte, Appl. Surf. Sci. 226, 355 (2004).
[17] R. P. Merino, A. C. Gallardo, M. G. Rocha, I. H. Calderon, V. Castano, R. Rodriguez, Thin Solids Films 401, 118 (2001).

[18] C. Jia, E. Xie, A. Peng, R. Jiang, F. Ye, H. Lin, T. Xu, Thin Solid Films 496, 555 (2006).

[19] A. Charanpahari, S. S. Umare, R. Sasikala, Appl. Surf. Sci. 282, 408 (2013).

[20] S. Shogh, R. Mohammadpour, A. I . Zad, N. Taghavinia, Mater. Lett. 159, 273 (2015).

[21] J. Xie, D. Jiang, M. Chen, D. Li, J. Zhu, X. Lu, C. Yan, Colloids Surf., A : Physicochem. Eng. Aspects 372, 107 (2010).

[22] L. Lin, H. Wang, H. Luo, P. Xu, J. Photochem. Photobiol., A Chemistry 307, 88 (2015).

[23] N. Ahmadi, A. Nemati, M. S. Hashjin, Mater. Sci. Semicond. Process. 26, 41 (2014).

[24] C. Y. W. Lin, D. Channeai, P. Koshy, A. Nakaruk, C. C. Sorrell, Physica E 44 (2012) 1969-1972.

[25] B. Choudhury, A. Choudhury, *Curr. Appl. Phys.* 13, 1025 (2013).

[26] A. Ranjitha, N. Muthukumarasamy, M. Thambidurai, Dhayalan Velauthapillai, R. Balasundaraprabhu, S. Agilan, Solar Energy 106,159 (2014).

IV. Optical properties of titanium dioxide thin films

This section describes the effect of growth and mechanism conditions in particular flow rate, sprayed volume, substrate temperature, heat treatment and both samarium and manganese doping on the optical properties of chemically sprayed titanium dioxide thin films. Photoluminescence spectra are also investigated for further enhancement of the optical performance of deposited films.

4.1. Optical properties of titanium dioxide thin films

4.1.1. Effect of sprayed solution flow rate

In order to investigate the effect of sprayed solution flow rate on the optical properties of TiO$_2$ films, transmission and reflection measurements are performed. Reflection R(λ) and transmission T(λ) spectra are displayed on Figure IV-1. It reveals that all the films have a high transmission in the visible region over 60% and a broad cut-off towards short-wavelengths which corresponds to the absorption edge of TiO$_2$ material. Anatase TiO$_2$ thin film exhibits interference fringes in the transmission spectrum, which indicates a smooth film surface with no much scattering loss. Moreover, as expected, we can remark that the transmission of TiO$_2$ thin films decreases slightly with increasing film thickness. This behavior may be attributed to the increase of both grain size and film thickness leading to an increment of scattering in thicker films. In fact, as the thickness is increased as more and more light is absorbed by the deposited films.

There are contradictory reports regarding the nature of the band gap, in case of anatase TiO$_2$ thin films. Zhang *et al.* [1] reported a direct band gap in the range of 3.5-3.7 eV, whereas *Lee et al.* [2] presented an indirect band gap of 3.25 eV. However, the investigation of the band structure of anatase TiO$_2$ crystal using WIEN2k package (Figure II-3, Section II) revealed an indirect band gap (Eg.th) in the order of 3.05 eV.

Moreover, energy band gap and optical transition type of the film can be determined by the following relation [3,4]:

$$\alpha h\nu = A(h\nu - E_g)^m \qquad\qquad \text{(IV-1)}$$

where A is an energy-independent constant, Eg is the optical band gap and m is a constant that determines type of optical transitions. For direct allowed transition, m=1/2 and for direct forbidden transition, m=3/2; for indirect allowed transition, m=2, and for indirect forbidden transition, m=3. To determine the type of the optical transition, Ln(αhv) versus Ln(hv-Eg.th) is plotted in Figure IV-2.

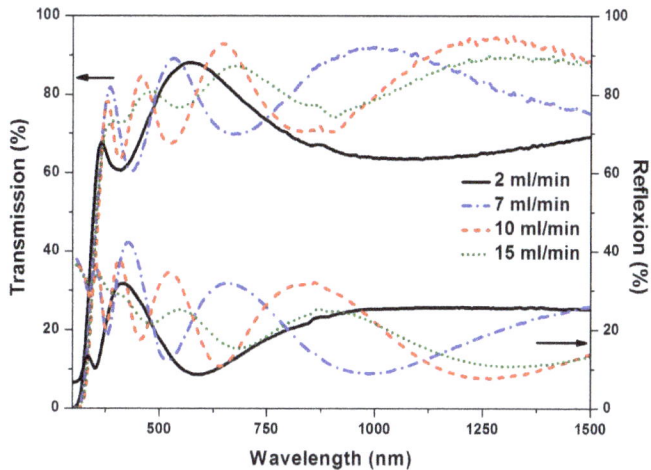

Figure IV-1. Transmission and reflection spectra of TiO_2 thin films obtained at different flow rates.

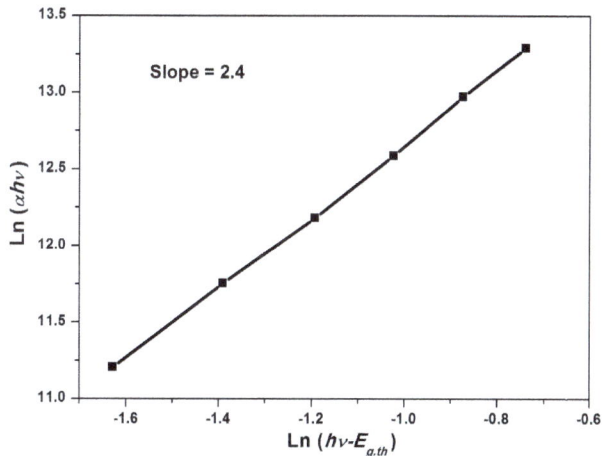

Figure IV-2. Plot of Ln(αhv) versus Ln(hv-E$_{g.th}$).

The m value can be directly determined from slope of the straight line, it is found to be about 2. This suggests that indirect-allowed transitions are dominant in the fundamental absorption edge of the anatase TiO_2. Based on the optical transmission and reflection measurements, (αhv)(1/2) is plotted as a function of photon energy (hv) at a solution flow rate of 10 ml/min (Figure IV-3). An indirect band gap energy of about 3.46 eV is obtained by extrapolating the linear portion of the plot to the crossing with hv axis.

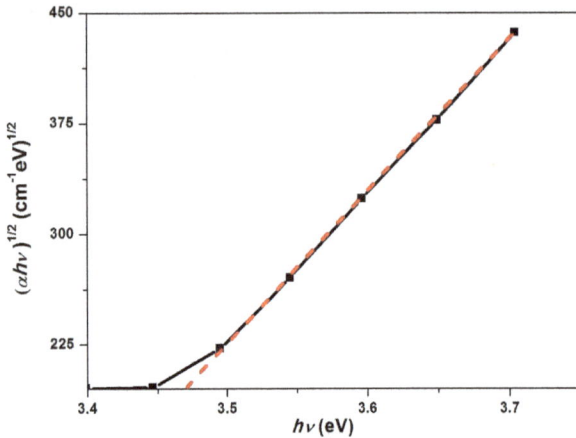

Figure IV-3.
Plot of (αhυ)(1/2) versus (hυ) for TiO₂ obtained at 10 ml/min.

A theoretical simulation of transmission T(λ) spectra using the envelope method [5] allows us to calculate the thickness and some optical constants of TiO₂ thin films. Figure IV-4 shows experimental and theoretical transmission curves obtained for films deposited at 10 ml/min.

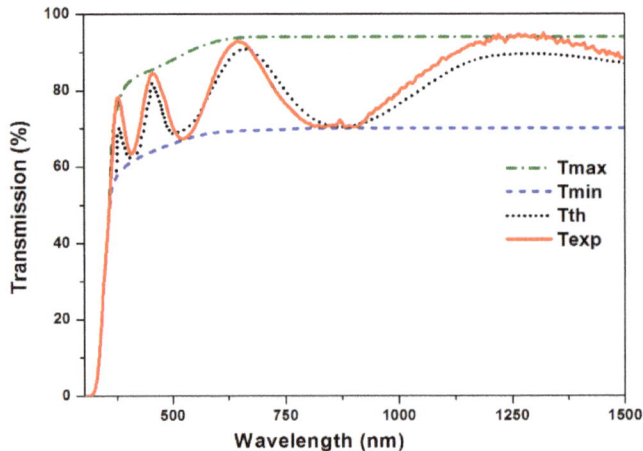

Figure IV-4. Experimentally (Texp) and theoretically (Tth) transmission of TiO₂ prepared at a flow rate of 10 ml/min.

The interference phenomena between the wave fronts generated at two interfaces (air and substrate) defines the sinusoidal behavior of the transmission curves. The thickness eth of anatase titanium dioxide layer calculated from two maxima or minima is given by [5]:

$$e_{th} = \frac{M\lambda_1\lambda_2}{2(\lambda_2 n(\lambda_1) - \lambda_1 n(\lambda_2))} \qquad (IV-2)$$

where M is the number of oscillations between two extrema, $\lambda 1$, $n(\lambda 1)$, $\lambda 2$ and $n(\lambda 2)$ are the corresponding wavelengths and refractive indices.

Optical constants such as refractive index (n) and extinction coefficient (k) are the parameters which characterize how a material responds to an electromagnetic field excitation at a given frequency. The expression of refractive index is calculated using this equation [5]:

$$n = \left[N + (N^2 - n_0^2 n_1^2)^{1/2} \right]^{1/2} \qquad \text{(IV-3)}$$

With:

$$N = 2n_0 r_1 \left[\frac{(T_{max} - T_{min})}{T_{max} T_{min}} \right] + \left(\frac{n_1^2 + n_0^2}{2} \right) \qquad \text{(IV-4)}$$

Where n is the refractive index of anatase titanium dioxide thin film, $n0$ is air refractive index, and $n1$ is the refractive index of glass substrate.

Knowing n and eth values, the extinction coefficient k in the transparency region is given by [5]:

$$k = \frac{\lambda}{4\pi} \alpha \qquad \text{(IV-5)}$$

Where α is the absorption coefficient calculated using the following formula:

$$\alpha = -\frac{1}{e_{th}} ln \left(\frac{c_1 \left[1 - \left(\frac{T_{max}}{T_{min}} \right)^{1/2} \right]}{c_2 \left[1 + \left(\frac{T_{max}}{T_{min}} \right)^{1/2} \right]} \right) \qquad \text{(IV-6)}$$

where: $C_1 = (n + r_0)(n_1 + n)$ and $C_2 = (n - n_0)(n_1 - n)$

The fundamental excitation spectrum of the film is described by means of a frequency dependence of the complex electronic dielectric constant. The dielectric constant is defined as:

$$\varepsilon = \varepsilon_r + i\varepsilon_i \qquad \text{(IV-7)}$$

The real part εr and imaginary part εi are related to n and k by the following expression [6]:

$$\varepsilon_r = n^2 - k^2 \qquad \text{(IV-8)}$$

$$\varepsilon_i = 2nk \qquad \text{(IV-9)}$$

The thickness values of anatase TiO_2 films are calculated from Eq. IV-2, IV-3 and summarized in Table IV-1.

Table IV-1. Calculated values of film thickness (eth), for different solution flow rates, of TiO_2 thin films based on the envelope method

Solution flow rate (ml/min)	2	7	10	15
eth (nm)	100 ± 4	240 ± 9	300 ± 12	280 ± 11

It varies from 100 nm to 300 nm for different solution flow rates where the thickest layer is obtained at a flow rate of 10 ml/min. The obtained value, in this case, is in good agreement with scanning electron microscopy (SEM) cross section image (Figure IV-5).

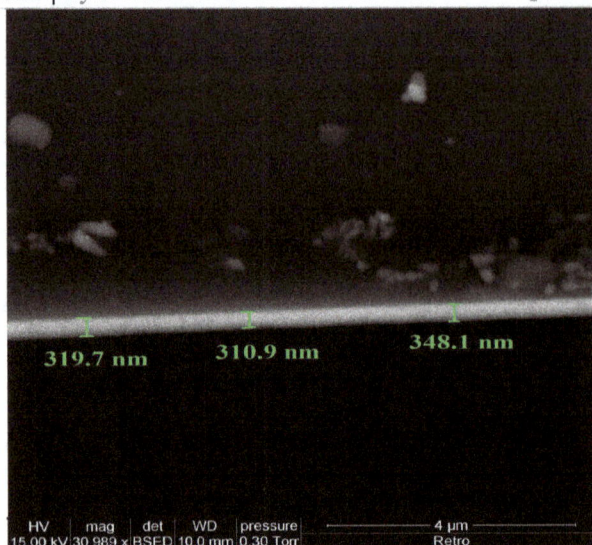

319.7 nm   **310.9 nm**   **348.1 nm**

| HV | mag | det | WD | pressure |
| 15.00 kV | 30 989 x | BSED | 10.0 mm | 0.30 Torr |

4 μm
Retro

Figure IV-5.
SEM image of TiO$_2$ film cross-section prepared under a solution flow rate of 10 ml/min.

The refractive index n(λ) and the extinction coefficient k(λ) values are calculated using Eqs. IV-5, IV-7. Their dependences on wavelength are presented in Figure IV-6. The values of both n and k decrease with increasing of wavelength then remain practically constant up to 650 nm. This behavior is in agreement with the ones reported in the literature [7,8]. The refractive index decreases from a value of 2.25-1.84 at λ=500 nm reaching a value of 2.21-1.80 at λ=650 nm. These values are smaller than 2.5, which is the characteristic value for the anatase phase [7]. In the case of sprayed solution flow rate of 10 ml/min, the refractive index value is higher than those obtained at 2, 7 and 15 ml/min which indicates the crystallinity improvement of the layer [9] as shown in XRD spectra (Figure III-1). Extinction coefficient reflects the absorption of electromagnetic waves in the semiconductor due to inelastic scattering events. Their values do not exceed 0.25 in the transparency region, which is a low value indicating the surface smoothness and homogeneity of the films. The variation of the optical constants, n and k, can be related to the film thickness change [9].

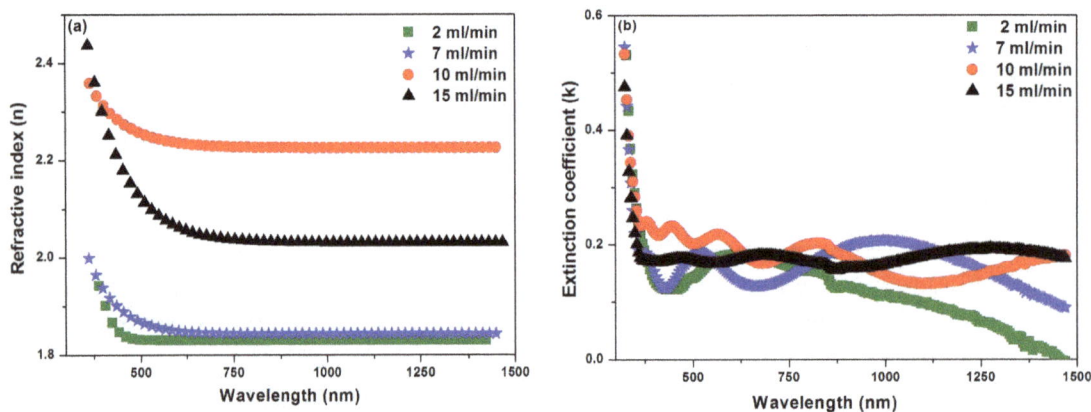

Figure IV-6. Variation of the refractive index (n) and the extinction coefficient (k) with wavelength for TiO$_2$ thin films obtained at different flow rates

The real (εr) and imaginary (εi) parts of the dielectric constant ε are related to the n and k values. They are calculated using the Eqs. IV-8, IV-9. Figure IV-7 shows their dependence on wavelength. Thus, at shorter wavelength, the values of the real and imaginary parts of the dielectric constant are very high. However, at longer wavelength, both of them decrease and show almost a wavelength independent behavior. Furthermore, it is worth noting that εr is a measure of dispersion of photons in a medium and εi measures the rate of dissipation in the medium which explains that they follow the same behavior of the refractive index and the extinction coefficient, respectively. Moreover, we can remark that the values of εi are lower than εr which indicates that there is a low light energy loss through the deposited films [10].

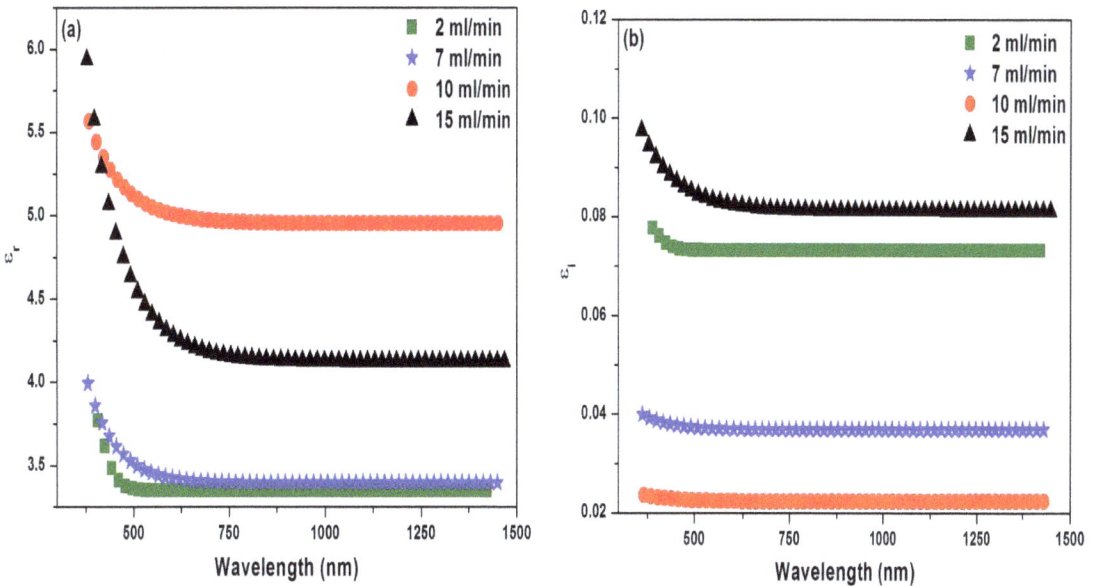

Figure IV-7. Variation of real and imaginary parts (εr and εi) of dielectric constant with wavelength for TiO_2 thin films obtained at different flow rates.

4.1.1. Effect of sprayed solution volume

Transmission and reflection measurements are performed over a large spectral range (250 to 2500 nm) in order to study the effect of sprayed solution volume on the optical performances of TiO_2 films. The obtained transmission and reflection spectra are displayed in Figure IV-8.

It is seen from this Figure that all the as-grown films show good transmission in the visible region ranging from 63 % to 90 %. Moreover, we can remark that for volume value higher that 60 ml, the optical transmission of TiO_2 thin films decreases with further increase of sprayed solution volume. This reduction may be caused by the fact that with increasing thickness, photon scattering increases too by the increment of surface roughness as shown in AFM results (Figure III-9). The broad cut off is attributed to intrinsic inter-band absorption in the TiO_2 compound. It shifts towards longer wavelengths (red shift) with increasing thickness, which indicates a smaller band gap. This behavior can be ascribed to an increase of the defects with increasing thickness [11]. In fact, as the sprayed volume increases, both crystallite size and film thickness increase too. However, defects inside the deposited films occur during the growth process which would substantially reduce the band gap [12]. This result is consistent with that obtained from X-ray diffraction (Figure III-4 and Figure III-7). A sinusoidal behavior of transmission curves is observed due to the interference phenomena (air and substrate interfaces). This reveals a good surface quality, homogeneity of the film and reduces scattering losses. The highest transmission in the transparency domain is obtained for sprayed solution volume equals to 60 ml. This result may be related to the good crystalline quality of the deposited film.

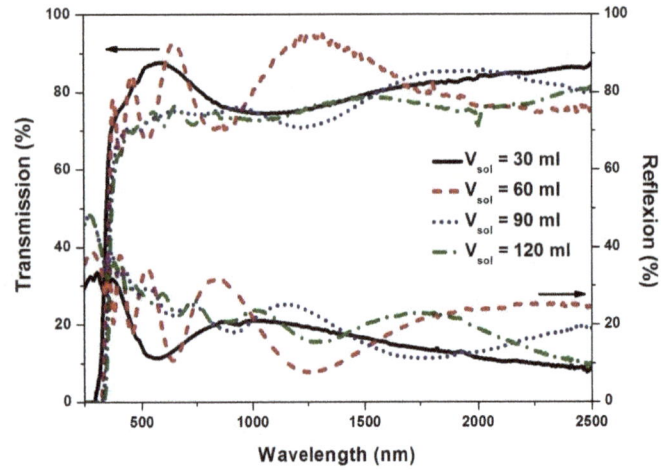

Figure IV-8.
Reflection and transmission spectra of TiO$_2$ thin films deposited at various solution volumes.

The extrapolation of linearly part of $(\alpha h\nu)(1/2)$ versus $(h\nu)$ gives a value of the indirect band energy of anatase TiO$_2$. The variation of the optical energy band gap Eg values with sprayed volumes is summarized in Table IV-2.

Table IV-2. Variation of optical gap energy for anatase TiO$_2$ thin films prepared with different solution volumes.

Vsol (ml)	30	60	90	120
Eg (eV)	3.57	3.46	3.41	3.39

It is found to decrease from 3.57 eV to 3.39 eV with increasing thickness. In fact, increasing film thickness associated with the increased defects (Figure III-7), would be expected to shift the absorption edge to higher wavelengths, so to reduce the band gap [12]. This induces the occurrence of a red shift in transmission spectra as seen in Figure IV-8. In the case of anatase TiO$_2$ thin films grown at sprayed solution volume of about 60 ml, the indirect band gap is found to be equal to 3.46 eV.

Film thickness (eth) is calculated using the envelope method [5]. Their dependence on the sprayed solution volume is depicted in Table IV-3. It is found that eth increases from 140 to 590 nm with solution volume increment.

Table IV-3. Theoretical values of thickness of TiO$_2$ thin films obtained for different solution volumes.

Vsol (ml)	30	60	90	120
Thickness estimated from envelope method eth (nm)	140 ± 5	300 ± 12	520 ± 20	590 ± 23

The dependence of refractive index and extinction coefficient versus wavelength for different sprayed solution volumes is presented in Figure IV-9. It is clear that the highest refractive index is obtained in the case of sprayed solution volume equals to 60 ml, it is about 2.22 above 700 nm. The variation in refractive index versus sprayed solution volumes can be attributed to the change of layers density since it allows us to have information about voids present in the films [9]. The extinction coefficient is about 0.27 in the transparency region, this low value indicates the surface homogeneity and smoothness of the grown layers. A similar behavior of optical parameters was observed in other reports [7,8].

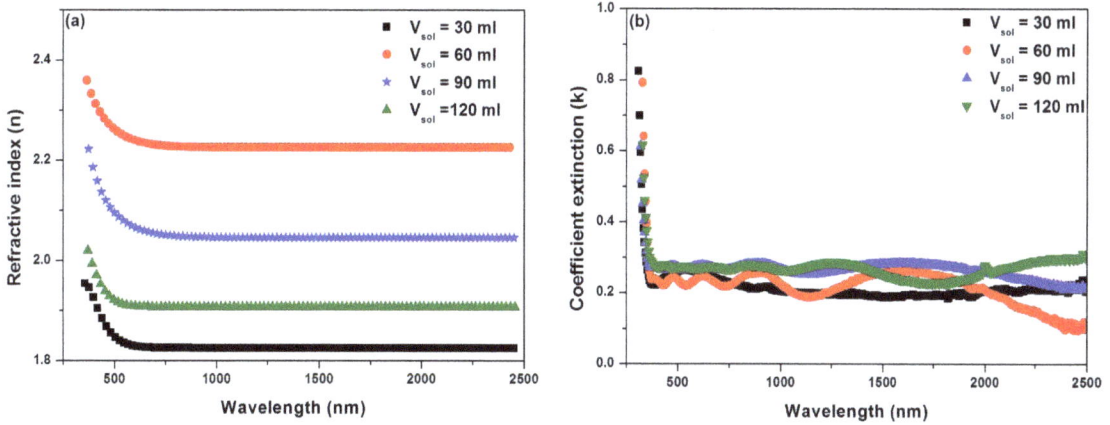

Figure IV-9. The variations of refractive index (n) and extinction coefficient (k) of TiO_2 films grown at different sprayed solution volumes.

The dependence of the real εr and imaginary εi parts of dielectric constant (ε) on wavelengths are shown in Figure IV-10. We can remark that εr decreases with increasing wavelengths and then remains practically constant. The same behavior is observed for the imaginary part (εi) of the dielectric constant.

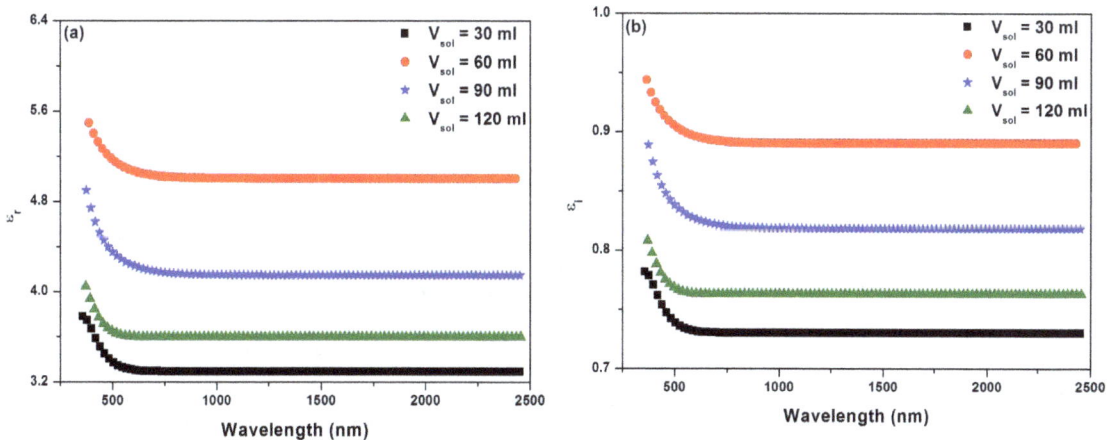

Figure IV-10. The real (εr) and imaginary (εi) parts of the dielectric constant ε of TiO_2 thin layers grown at different sprayed solution volumes.

The dispersion of refractive index plays an important role in the determination of optical properties of diverse materials. It can be described by a single effective oscillatory of the Wemple-Di Domenico model given by the following relation [13]:

$$n^2(h\nu) = 1 + \frac{E_d E_0}{E_0^2 - (h\nu)^2}$$

(IV-10)

Where hν is the photon energy, E0 is the oscillator energy between conduction and valence bands of the semiconductor and Ed is the dispersion energy inside the bulk. The values of E0 and Ed are estimated from the variation of (n2-1)-1 versus (hν)2 shown in Figure IV-11. The refractive index n(0) at zero photon energy can be deduced from the dispersion relationship.

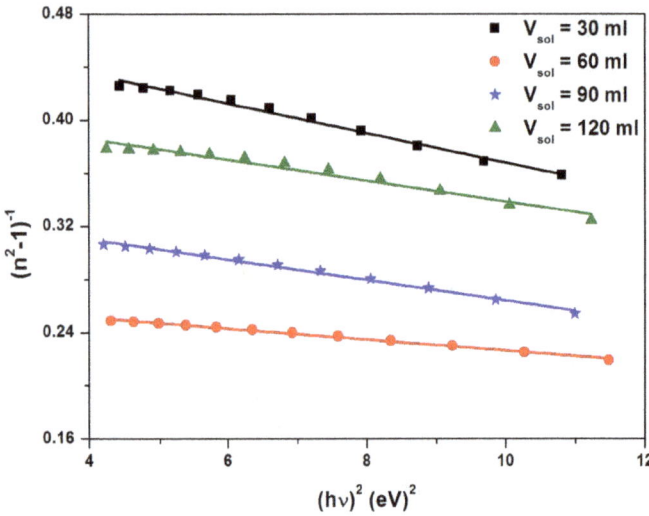

Figure IV-11.
Plot of $(n_2-1)^{-1}$ versus $(h\nu)^2$ for TiO$_2$ thin films deposited with different sprayed volumes.

For further optical analysis, the dependence of the free carrier electric susceptibility χe on the real dielectric constant is discussed according to the Spitzer–Fan model [14]:

$$\varepsilon_r = n^2 - k^2 = \varepsilon_\infty - \left(\frac{e^2}{\pi c^2}\right)\left(\frac{N}{m^*}\right)\lambda^2$$

(IV-11)

$$\left(\frac{e^2}{\pi c^2}\right)\left(\frac{N}{m^*}\right)\lambda^2 = -4\pi\chi_e$$

(IV-12)

where ε_∞ is the high-frequency dielectric constant in the absence of any contribution from free carrier, $\frac{N}{m^*}$ is the carrier concentration to the effective mass ratio, c is the light velocity and e is the elementary charge of an electron.

Figure IV-12 shows the plots of εr versus λ2 for films deposited from different sprayed solution volumes. The values of ε_∞ and $\frac{N}{m^*}$ are obtained from the intercept and the slope resulting from the extrapolation of these curves.

We note that χe increases as the wavelengths increase and then it becomes sufficiently large, in the near-infrared region, indicating the reduction of the refractive index as well as the dielectric constant. All the obtained results are summarized in Table IV-4.

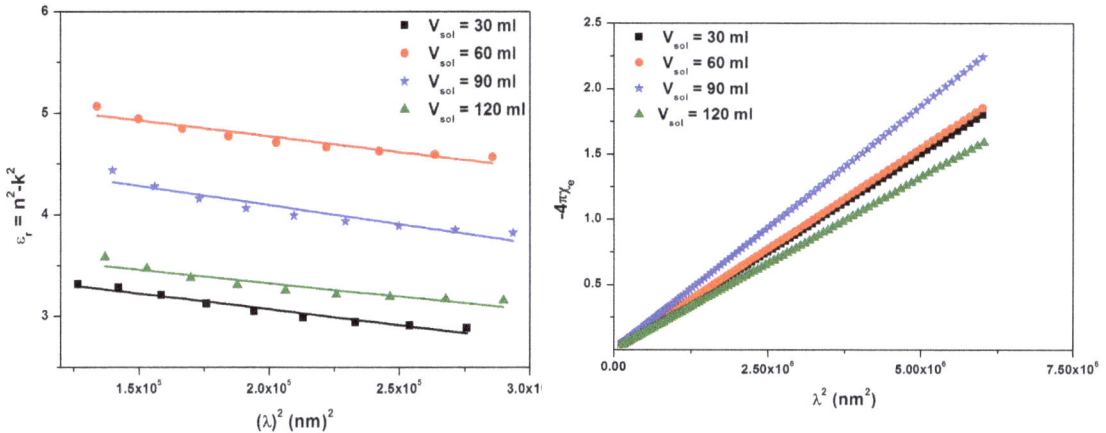

Figure IV-12. A plot of the optical dielectric constant $\varepsilon_r = n^2 - k^2$ and $(-4\pi\chi e)$ versus λ^2 of TiO_2 thin films deposited with different solution volumes.

Table IV-4. Estimated values of optical parameters for TiO_2 thin films deposited at different sprayed solution volumes.

Vsol (ml)	E0 (eV)	Ed (eV)	$\dfrac{E_0}{E_g}$	n(0)	$\dfrac{N}{m^*}$ (kg-1m-3)	ε_∞^W	ε_∞^S	$-\chi e \ (10^{-2})$
30	6.56	13.68	1.83	1.75	3.45*1049	3.08	3.69	0.5-14
60	7.35	26.45	2.12	2.14	3.48*1049	4.59	5.19	0.5-15
90	6.67	19.58	1.95	1.98	4.12*1049	3.93	4.84	0.5-17
120	6.60	15.80	1.94	1.84	3.89*1049	3.39	3.85	0.5-12

From this table, it is clear that all the optical parameters depend on sprayed solution volume. In fact, the dispersion energy Ed increases from 13.68 to 26.45 eV, where the highest value is obtained in the case of solution volume of about 60 ml. E0 is an 'average' energy gap which can be related to the optical band gap Eg by an empirical formula: $E_0 \approx 2E_g$ [15] and it is noticed that the obtained values are in agreement. The refractive

$$n(0) = \sqrt{1 + {E_d}/{E_0}}$$

index n (0) ( ) varies from 1.75 to 2.14. The ratio $\dfrac{N}{m^*}$ is in the order of $4*1049$ (kg-1m-3) for all the samples. The comparison of ε_∞^W values obtained from the Wemple–Di Domenico model (Figure IV-11) with that obtained from the Spitzer-Fan model (Figure IV-12) show satisfactory agreement. The free carrier electric susceptibility χe does not exceed $17*10-2$ for all the synthesized layers.

4.1.2. Effect of substrate temperature

Figure IV-14 displays the optical transmission (T) and reflection (R) spectra in the wave range of 250 to 2500 nm of TiO_2 thin films deposited at different substrate temperatures. It is observed that the entire spectra exhibits very pronounced interference fringes in the transparency region ranging from 400 to 2500 nm, with sharp fall of transmission at 400 nm corresponding to the absorption edge of anatase TiO$_2$ material. The transmission values are over 60 % in the transparency region for all the films. Moreover, it is noted that transmission spectra do not change obviously with increasing substrate temperature. For further optical analysis, the variation of optical gap energy for TiO$_2$ thin films deposited at various substrate temperatures is depicted in table IV-5. It varies from 3.43 to 3.53 eV.

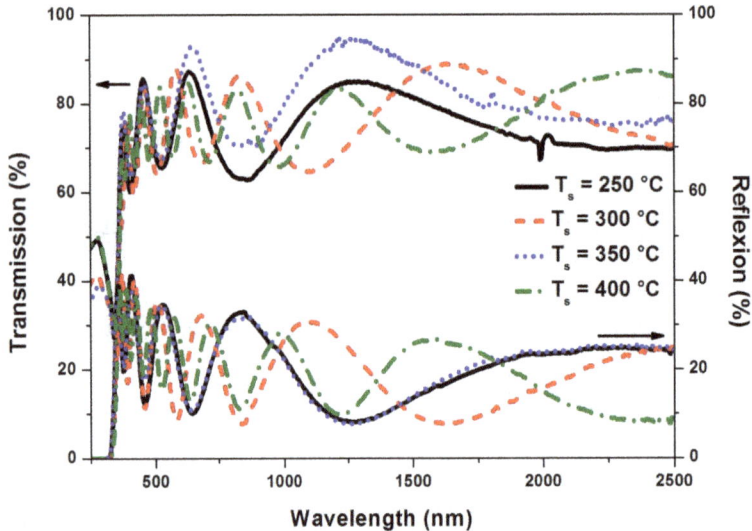

Figure IV-14. Reflection and transmission spectra of TiO$_2$ thin films deposited at different substrate temperatures.

Table IV-5. Variation of optical gap energy (Eg) for TiO$_2$ thin films deposited at various substrate temperatures (Ts).

Ts (°C)	250	300	350	400
Eg (eV)	3.53	3.43	3.46	3.49

4.1.3. Effect of air annealing

The effect of heat treatment on the optical properties of TiO$_2$/quartz thin films has been investigated basing on the optical transmission T(λ) and reflection R(λ) as displayed in Figure IV-15. It is noted from this figure that all the spectra exhibit interference fringes reflecting the homogeneity of the films. A good optical transmission is revealed ranging from 60 to 90 %, in the transparent region. However, a slight decrease of transmission is noticed with annealing temperature increase. This reduction may be explained by the scattering improvement caused by surface roughness increment (Table III-3).

Table IV-6 depicts the variation of optical gap energy for TiO$_2$/quartz thin films annealed at various annealing temperatures. We note that Eg decreases from 3.45 to 3.37 eV when annealing temperature increases from 500 to 900 °C. This effect may be related to grain size increment as shown in Figure III-15. For annealing temperature higher than 900°C, a continuous decrease of Eg is revealed. This reduction may be caused by the formation of rutile phase characterized with a lower optical band gap compared to the anatase one.

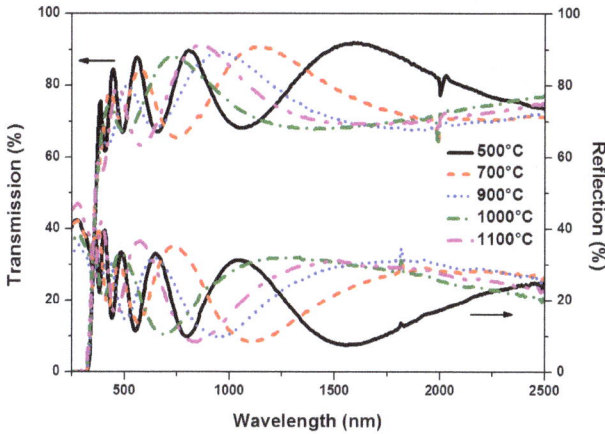

Figure IV-15.
Reflection and transmission spectra of TiO$_2$/quartz thin films annealed at different temperatures.

Table IV-6. Variation of optical gap energy (Eg) for TiO$_2$/quartz thin films annealed at various annealing temperatures (Ta).

Ta (°C)	500	700	900	1000	1100
Eg (eV)	3.45	3.44	3.37	3.35	3.33

4.1.4. Effect of Samarium doping

Transmission and reflection measurements of samarium doped TiO$_2$thin films are performed over a large spectral range (250 - 2500 nm). The corresponding spectra are illustrated in Figure IV-16. We remark the presence of interference fringes in the visible and near infrared regions which reveals a good surface quality, thickness uniformity and surface homogeneity for all the deposited films. Furthermore, a good optical transmission in the range of 60 - 90 % is noticed for all of them, even if its value decreases slightly when y (y=[Sm^{3+}/Ti^{4+}] at.%) increases. This reduction may be explained by the fact that samarium doping creates defects in the crystal which leads to an increment of photon scattering.

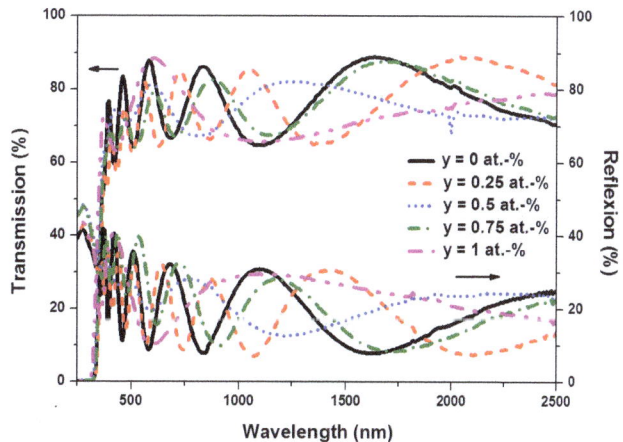

Figure IV-16.
Reflection and transmission spectra of samarium doped TiO$_2$ thin films at different ratios (y=[Sm^{3+}/Ti^{4+}] at.%).

The indirect band energy of Sm doped TiO$_2$ thin films is obtained using the extrapolation of the linear part of the (αhv)(1/2) versus (hv) curve. Its variation versus the nominal Sm concentration is summarized in Table IV-7. It ranges from 3.33 to 3.51 eV, where the minimum value is obtained for the film corresponding to a nominal Sm content of

0.25 at.%, which exhibits the largest average crystal size over all the doped film series. This behavior may be due to the increase of grain size after Sm doping at 0.25 at.% [16] as displayed in XRD analysis (Figure III-17 and III-19).

Table IV-7. Variation of optical gap energy for anatase TiO₂:Sm thin films as a function of Sm doping ratio (y=[Sm^{3+}/Ti^{4+}] at.%).

y (at.%)	0	0.25	0.5	0.75	1
Eg (eV)	3.43	3.33	3.45	3.40	3.51

Envelope method is employed in order to determine film thickness and some optical constants [5] for all the studied samples. Table IV-8 displays the theoretical values of thickness (eth) of TiO₂:Sm thin films versus their nominal Sm doping ratio. It varies from 180 to 285 nm, the thickest layer being obtained for the poorest Sm doped series, namely that corresponding to a nominal Sm content of 0.25 at.%.

Table IV-8. Theoretical values of thickness of TiO₂:Sm thin films versus Sm doping ratio (y=[Sm^{3+}/Ti^{4+}] at.%).

y (at.%)	0	0.25	0.5	0.75	1
eth (nm)	250 ± 10	285 ± 11	260 ± 10	220 ± 8	180 ± 8

The variation of the refractive index (n) and that of the extinction coefficient (k) as a function of the wavelength of all the produced films are presented in Figure IV-17. Above 750 nm, the refractive index ranges from 1.77 to 1.91, the highest values being obtained once again for the poorest Sm doped sample in the studied series (0.25 at.%). This result agrees fairly with the good microstructural properties of this film specifically compared to its richly doped parents (high density, large crystal size, best roughness, large thickness…). The extinction coefficient does not exceed 0.1 in the transparency region which reveals a surface homogeneity of the deposited films. A practically similar behavior of the refractive index and the extinction coefficient was observed in other investigations [7,8].

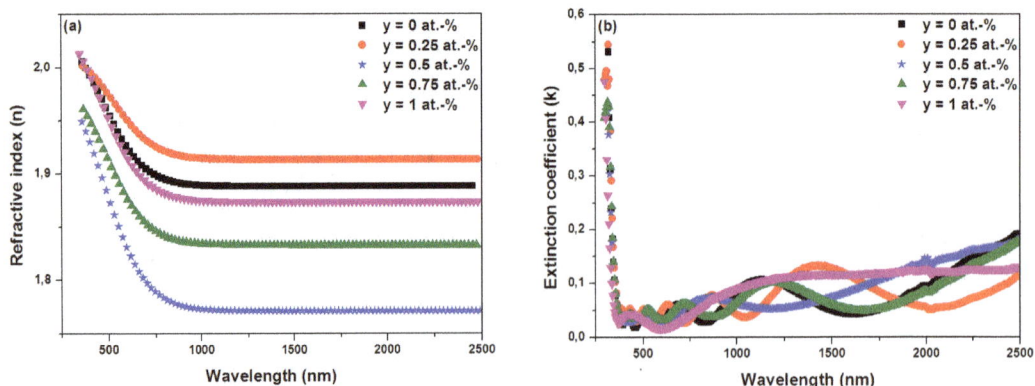

Figure IV-17. Variation of the refractive index (n) and the extinction coefficient (k) versus wavelength of samarium doped TiO₂ thin films for different ratios (y=[Sm^{3+}/Ti^{4+}] at.%).

Figure IV-18 depicts the variation of εr and εi of the sprayed TiO_2:Sm thin film as a function of the wavelength. We note that both of them decrease when the wavelength increases and tend to remain practically constant at wavelength values higher than 750 nm.

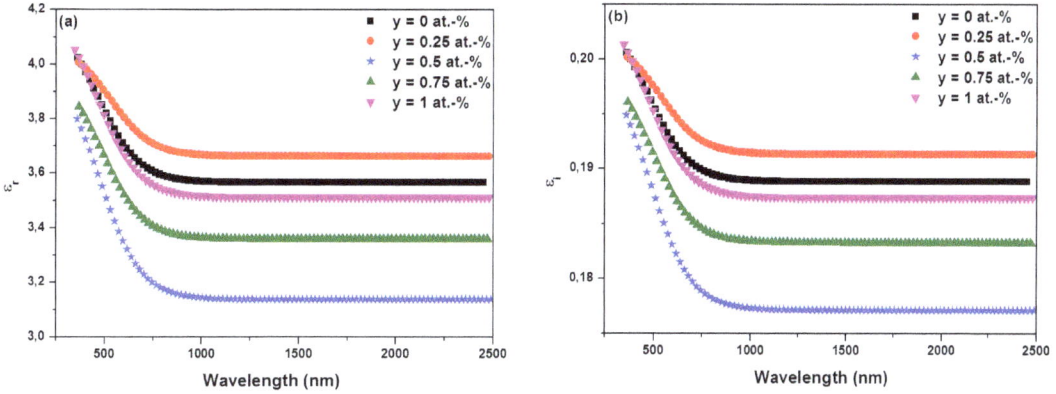

Figure IV-18. Variation of real and imaginary parts (εr and εi) of dielectric constant of samarium doped TiO_2 thin films for different ratios (y=[Sm^{3+}/Ti^{4+}] at.%).

Furthermore, some of dispersion parameters, namely E0, Ed, n(0) and λ0, are achieved within the single effective oscillatory theory using the Wemple-Di Domenico model [13] as described previously (Eq. IV-10).

λ0 is the oscillator wavelength usually expressed as follows:

$$\lambda_0 = \frac{hc}{E_0}$$

(IV-13)

The refractive index n(0) at zero photon energy is calculated:

$$n(0) = \sqrt{1 + \frac{E_d}{E_0}}$$

(IV-14)

Both E0 and Ed values are estimated from the variation of (n2-1)-1 versus (hv)2 displayed in Figure IV-19. The obtained results are summarized in Table IV-9. It is found that all the dispersion parameters depend on the samarium doping ratio. Indeed, E0 values obey the empirical formula: $E_0 \sim 2E_g$ [15] since it is an 'average' energy gap. Ed increases from 12.46 to 22.35 eV as a function of samarium doping ratio. n(0) values ranges between 1.74 and 1.90 whereas the oscillator wavelength does not exceed 205.63 nm.

Figure IV-19.
Plot of (n2-1)-1 versus (hv)2 of samarium doped TiO_2 thin films for different ratios (y=[Sm^{3+}/Ti^{4+}] at.%).

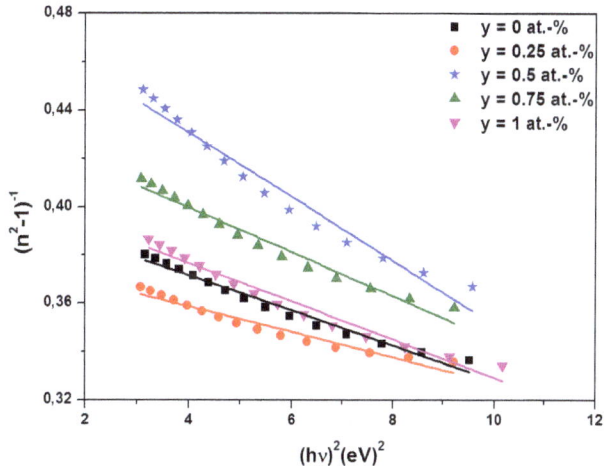

Table IV-9. Calculating dispersion and oscillator energies using Wemple model for samarium doped TiO$_2$ thin films for different ratios ((y=[Sm^{3+}/Ti^{4+}] at.%).

y (at.%)	E0 (eV)	$\dfrac{E_0}{E_g}$	Ed (eV)	n(0)	λ0 (nm)
0	6.03	1.80	12.46	1.74	205.63
0.25	6.89	2.06	15.79	1.81	179.97
0.5	7.17	2.07	17.57	1.85	172.94
0.75	7.41	2.17	18.51	1.86	167.34
1	8.48	2.41	22.35	1.90	146.22

4.1.5. Effect of Manganese doping

Optical transmission and reflection of undoped and Mn doped TiO$_2$ thin films are shown in Figure IV-20. We note that all the films are transparent in the visible range with the presence of very pronounced interference fringes reflecting the smoothness and homogeneity of the grown layers. Below 400 nm, a sharp fall of the transmission is observed corresponding to the fundamental absorption of anatase TiO$_2$ compound. After manganese doping, a slight decrease of the optical transmission is revealed, which may be resulted from more scattering by more defects in the lattice.

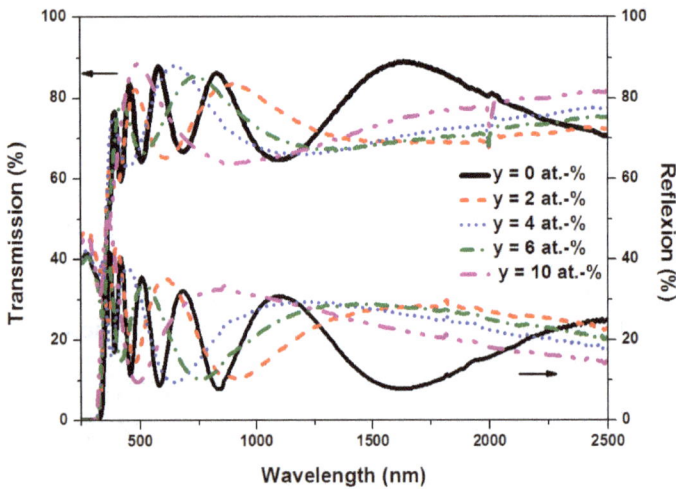

Figure IV-20. Reflection and transmission spectra of manganese doped TiO$_2$ thin films for different ratios (y=[Mn^{2+}/Ti^{4+}] at.%).

Basing on reflection spectra, the theoretical thickness (eth) of TiO$_2$:Mn thin films can be estimated using envelope method [5]. The refractive index is given by [17]:

$$n = \sqrt{n_s \frac{1+\sqrt{R_{max}}}{1-\sqrt{R_{max}}}}$$

(IV-15)

With ns is the substrate index ($n_s \approx 1.52$) and Rmax is the reflection maxima. Table IV-10 depicts the estimated thickness of TiO$_2$:Mn thin films versus Mn doping concentration. It is found to vary from 200 to 272 nm where the thickest layer is obtained at

2 at.% Mn. Thus, even a small amount of manganese induces important changes in the layer growth process. The optical energy band gap (Eg) is obtained by extrapolating the linear part of the plot $(\alpha h\nu)(1/2)$ versus (hv). The results are summarized in Table IV-10. We can remark that Eg values decrease from 3.45 to 3.38 eV with Mn doping concentration increment which may be due to the increase of defects within the forbidden band.

The Urbach energy, namely the width of the band tail of the localized states in the band gap, measures the degree of structural disorder within the investigated compound. It is related to the absorption coefficient (α) by the empirical Urbach law [18]:

$$\alpha = \alpha_0 exp\left(\frac{h\nu}{E_u}\right)$$

(IV-16)

where hv denotes the energy of the incident photon, $\alpha 0$ is a constant and Eu the Urbach energy. The values of Eu are calculated from the inverse of the slope of Ln(α) versus hv plots [19] presented in Figure IV-21.

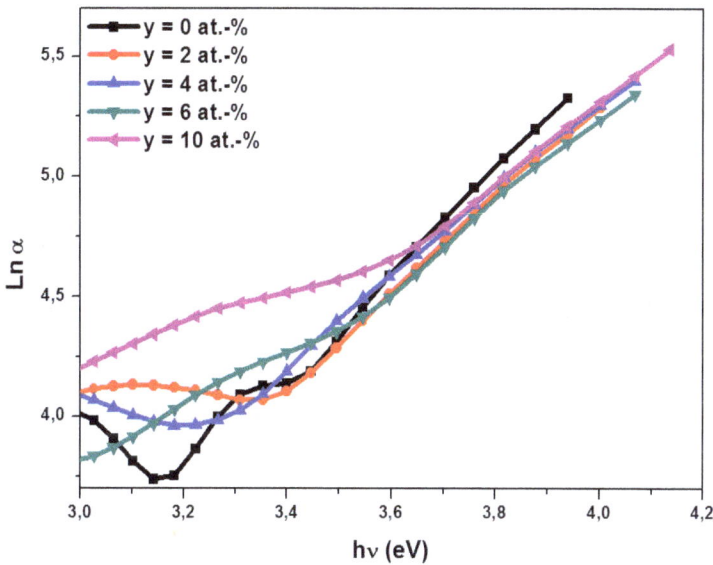

Figure IV-21. Ln(α) versus energy hv plots of TiO_2:Mn.

The obtained values are given in Table IV-10, and are found to rise after Mn doping indicating an increase of the structural disorder and the number of defects within the crystal lattice, as proved in XRD analysis. Therefore, a considerable introduction of tail states at the band edges is noticed. Thus, transition between band to tail and tail to tail takes place, leading to simultaneous widening of Urbach energy and shrinkage of band gap [20].

Table IV-10. Estimated thickness (eth), optical band gap (Eg) and Urbach energy (Eu) of manganese doped TiO_2 thin films for different ratios (y=[Mn^{2+}/Ti^{4+}] at.%).

y (at.%)	0	2	4	6	10
eth(nm)	250 ± 10	272 ± 10	264 ± 10	235 ± 9	200 ± 8
Eg (eV)	3.45	3.44	3.41	3.40	3.38
Eu (eV)	0.48	0.58	0.55	0.65	0.81

4.2. Photoluminescence

4.2.1. Effect of sprayed solution flow rate

Photoluminescence emission (PL) results from recombination of free carriers and can be used to determine the efficiency of charge trapping. The emission spectra of anatase TiO_2 thin films are shown in Figure IV-22 for various sprayed solution flow rates.

These spectra are recorded at room temperature with an excitation wavelength of 320 nm, corresponding to photon energy of 3.88 eV, larger than the band gap of investigated films, as reported by other works [21, 22]. PL spectra are featured by emissions in both UV and visible regions. As shown in Figure IV-22, a strong PL band is detected at ~361 nm (3.43 eV) which can be attributed to the intrinsic emission of the anatase TiO_2 thin films. Three other peaks located at 400 nm (3.1 eV), 486 nm (2.55 eV) and 530 nm (2.33 eV) corresponding to blue and green are also observed. These visible emissions are caused by intrinsic defect states within the crystal. In fact, defect levels deep inside the band gap are trap levels for both electrons and holes. These trap levels contribute to emission in visible region [22]. The intensity of PL emissions in both UV and visible region is maxim for the sample grown with solution flow rate at 10 ml/min. This behavior can be attributed to the improved crystallization and increased crystal size of the film [23] which is supported by the obtained XRD pattern (Figure III-1).

4.2.2. Effect of sprayed solution volume

The effect of sprayed solution volume on the optical properties is investigated and photoluminescence (PL) spectra of anatase TiO_2 samples grown on glass substrates excited at 320 nm are shown in Figure IV-23. A strong PL band at 361 nm is observed which can be attributed to the intrinsic emission of anatase TiO_2 thin films. Three broad visible emissions appear in the spectra at 400 (blue emission), 486 (UV/blue emission) and 529 nm (blue/green emission) corresponding to trap levels inside the band gap of the investigated crystal. These peaks are shifted towards longer wavelengths (lower energy). This shift is related to reduction in optical band gap as the defects increase with film thickness increment. A higher PL intensity is obtained for sample synthesized at sprayed solution volume equals to 60 ml since it possesses the best crystallinity (Figure III-4) as well as the reduced defects (Figure III-7).

4.2.3. Effect of substrate temperature

Figure IV-24 depicts the room temperature PL spectra of TiO_2 thin films deposited at various substrate temperatures. All spectra are investigated by exciting the 320 nm wavelength light of Xenon Lamp. As it can be seen from this Figure, the PL spectra show a strong PL emission in UV region located at ~ 360 (3.44 eV, blue). This intense blue emission may be ascribed to the electron transition from the bottom of the conduction band to the top of the valence band. Moreover, we note that PL intensity increases with substrate temperature increment up to 350°C, and then it decreases for substrate temperature of about 400°C, which may be related to the crystalline quality of the deposited layers.

4.2.4. Effect of air annealing

PL spectra of titanium dioxide thin films deposited on quartz substrates and heated at different annealing temperatures are displayed in Figure IV-25. We note that there is a prominent PL emission in the UV region located around 365 nm which may be attributed to the edge emission of titanium dioxide material. Some other peaks occur in the visible region (at ~453, 500 and 540 nm) corresponding to the electron transfer from the conduction band to trap levels inside the forbidden one. A red shifting is observed in the PL spectra with increasing annealing temperature which may be related to the decrease of the band gap after formation of the rutile phase, as discussed earlier (Table IV-6). Furthermore, both UV and visible emission intensities initially increase with annealing temperature increment up to 900°C and subsequently decrease with further increase of annealing temperature. This behavior may be ascribed to film crystalline quality (Figure III-14).

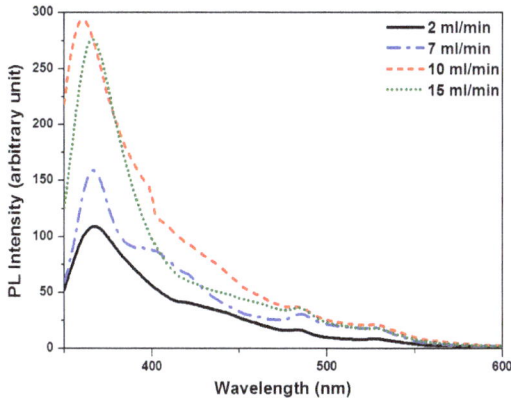

Figure IV-22. Photoluminescence (PL) spectra of TiO_2 thin films obtained at different solution flow rates

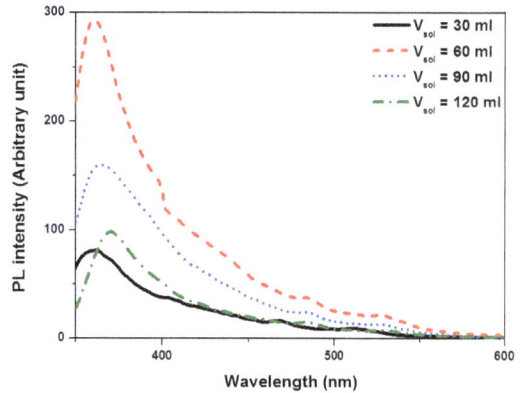

Figure IV-23. Photoluminescence spectra of TiO_2 thin layers grown at different solution volumes.

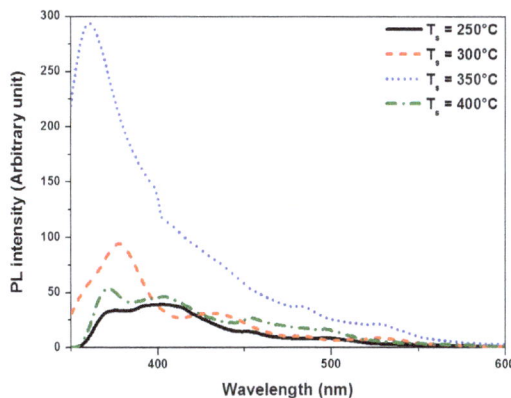

Figure IV-24. Photoluminescence spectra of TiO_2 thin layers deposited at various substrate temperatures.

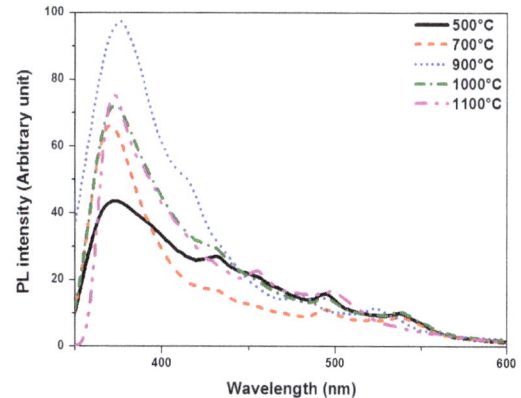

Figure IV-25. Photoluminescence spectra of TiO_2 thin layers annealed at various annealing temperatures.

4.2.5. Effect of Samarium doping

The photoluminescence spectra of TiO_2:Sm thin films excited at 320 nm, according to other works [21] are recorded at room temperature. They are displayed in Figure IV-26. All spectra include a strong PL peak centered at 363 nm which may be ascribed to the intrinsic emission of anatase TiO$_2$ films. Three other emission peaks are detected at around 440, 480 and 530 nm corresponding to the contribution of oxygen vacancies [24]. Besides, we note that samarium doping gives rise to new PL phenomena. Indeed, after doping, three emission peaks centered at 581, 610 and 662 nm appear. They are assigned to the electron transitions of Sm^{3+} ions from their excited 4G5/2 atomic state to the 6H5/2, 6H7/2, 6H9/2 ones, respectively [25]. Furthermore, a higher PL intensity is obtained for TiO$_2$ film deposited with a nominal Sm doping ratio of about 0.25 at.% since it possesses the best structural and microstructural properties as shown previously (Figure III-17).

4.2.6. Effect of Manganese doping

PL spectra of undoped and Mn doped TiO$_2$ thin films are displayed in Figure IV-27. The used excitation wavelength is about 320 nm [26]. As it can be seen from this Figure, two main peaks can be observed. These peaks are situated at 370 and 440 nm which may be attributed to the indirect band to band transition of TiO$_2$ thin films and oxygen vacancies [24], respectively. Moreover, we can remark that no Mn emission peak is present in all PL spectra. However, a red shift toward higher wavelength (lower energy) can be observed after Mn doping which is consistent with reduction in optical band gap (Table IV-10). A higher PL intensity is obtained for 2 at-% Mn doping ratio which deals clearly with XRD results (Figure III-22).

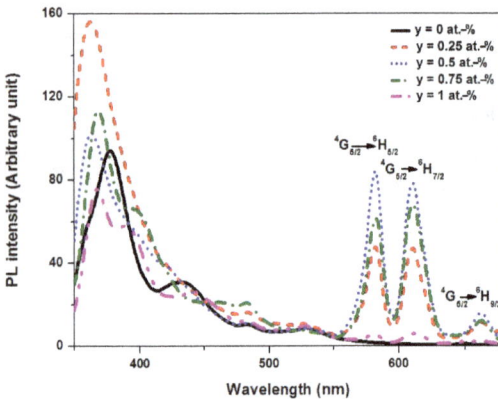

Figure IV-26. Room temperature photoluminescence spectra of samarium doped TiO$_2$ thin films excited at 320 nm.

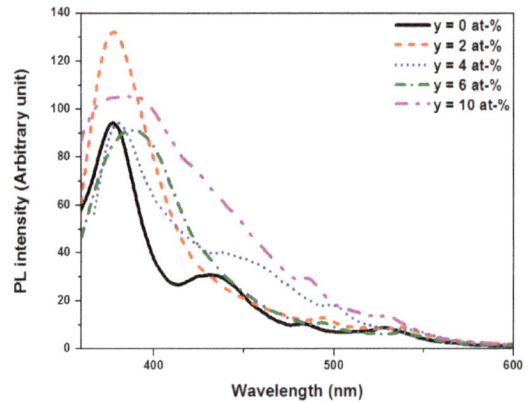

Figure IV-27. Photoluminescence spectra of manganese doped TiO$_2$ thin films.

Conclusions

This section focused on the effect of deposition parameters, heat treatment and doping on the optical properties of chemically sprayed titanium dioxide thin films. It was found that all transmission and reflection spectra exhibited a high optical transmission of 80% in the transparency region with an indirect band gap around 3.40 eV. Moreover, a sinusoidal behavior was observed indicating a good surface quality and homogeneity of the deposited films. Envelope method was employed in order to estimate the film thickness and some optical constants. Obtained low value of the extinction coefficient in the transparency domain is a good indication of film surface smoothness and homogeneity. The estimation of certain optical constants, such as: dispersion energy Ed, oscillator energy E0, the high-frequency dielectric constant ε_∞ and free carrier electric susceptibility χe, was achieved from the Wemple–Di Domenico and Spitzer–Fan models. Photoluminescence spectra were featured by several emissions in both UV and visible regions corresponding to intrinsic emission of the anatase TiO_2 thin films and defect states within the crystal, respectively. However, doping with samarium ion gave rise to new emission peaks corresponding to electron transitions from the excited 4G5/2 state to the 6H5/2, 6H7/2, 6H9/2 states. This investigation may be useful guidelines to the synthesis of some highly effective luminescence devices based on anatase TiO_2 material.

References

[1] S. Zhang, Y. F. Zhu, D. E. Brodie, Thin Solid Films 213, 265 (1992).

[2] D.Y.Lee,J.T.Kim,J.H.Park,Y.H.Kim,I.K.Lee,M.H.Lee, Curr.Appl.Phys. 13 (2013) 1301

[3] J.I.Pankove, Optical Processes in Semiconductors (Prentice-Hall, Inc., New Jersey, 1971), p. 93.

[4] J.Tauc, Amorphous and Liquid Semiconductors (Plenum Press, NewYork, 1974), p 159.

[5] J. C. Manifacier, J. Gassiot, J. P. Fillard, J. Phys. E: Sci. Instrum. 9, 1002 (1976).

[6] S. Belgacem, R. Bennaceur, Rev. Phys. Appl. 25, 1245 (1990).

[7] A.V.Manole,M.Dobromir,M.Gîrtan,R.Mallet,G.Rusu,D.Luca, Ceram.Int. 39(2013)4771

[8] B. Houng, C.C. Liu, M.T. Hung, Ceram. Int. 39, 3669 (2013).

[9] N. Revathi, P. Prathap, K.T. Ramakrishna Reddy, Solid State Sci. 11, 1288 (2009).

[10] A. Barhoumi, G. Leroy, B. Duponchel , J. Gest, L. Yang, N. Waldhoff, S. Guermazi, Superlattices Microstruct. 82, 483 (2015).

[11] S.H. Mohamed, J. Phys. Chem. Solids 69, 2378 (2008).

[12] R. Rajalakshmi, S. Angappane, J. Alloys Comp. 615, 355 (2014).

[13] S.H. Wemple, M. Di Domenico, Phys. Rev. B3 4, 1338 (1971).

[14] W.G. Spitzer, H.Y. Fan, Phys. Rev. 106, 882 (1957).

[15] A.F. Qasrawi, Semicond. Sci. Technol. 20, 765 (2005).

[16] S. Benramache, B. Benhaoua, O. Belahssen, Optik 125, 5864 (2014).

[17] S.K. Bahl, K.L. Chopra, J. Appl. Phys. 40, 4940 (1959).

[18] G.D. Cody, Semiconductors and Semimetals, 1984, ed J.I. Pankove Academic, 21B, Ch.2

[19] F. Yakuphanoglu, A. Cukurovalib, I. Yilmaz, Opt. Mater. 27, 1363-1368 (2005).

[20] K.Punitha, R.Sivakumar, C.Sanjeeviraja, V.Ganesan, Appl. Surf. Sci. 344, 89 (2015).

[21] P. B. Nair, V.B. Justinvictor, G. P. Daniel, K. Joy, V. Ramakrishnan, D. D. Kumar, P.V. Thomas, Thin Solid Films 550, 121 (2014).

[22] P. B. Nair, V.B. Justinvictor, G. P. Daniel, K. Joy, V. Ramakrishnan, P.V. Thomas, Appl. Surf. Sci. 257, 10869 (2011).

[23] K. Murali Krishna, M. Mosaddcq-ur-Rahman, T. Miki, T. Soga, K. Igarashi, S. Tanemura, M. Umeno, Appl. Surf. Sci. 113, 149 (1997).

[24] Q. Xiao, Z. Si, Z. Yu, G. Qiu, Mater. Sci. Eng. B 137, 189 (2007).

[25] G. An, S. Jin, C. Yang, Y. Zhou, X. Zhao, Opt. Mater. 35, 45 (2012).

[26] B. Choudhury, A. Choudhury, Curr. Appl. Phys. 13, 1025 (2013).

V. General conclusions

Titanium dioxide thin films have been successfully elaborated using spray pyrolysis technique. Two main objectives were pursued in this investigation: the theoretical study of various physical properties of anatase TiO_2 material as well as the optimization of processing condition and parameters for better performance in optoelectronic devices.

Thus, structural, electronic, thermal and transport properties of anatase TiO_2 compound had been investigated using the relativistic full-potential linearized augmented plane wave method (FPLAPW) within density functional theory (DFT). The obtained lattice parameters were consistent with available experimental data. Anatase TiO_2 compound was characterized by an indirect band gap of about 3.05 eV, as revealed by MBJ calculated band structure. The investigation of some optical parameters such as dielectric constant, refractive index, extinction coefficient, reflectivity and electron energy loss function, was also achieved. The application of Debye quasi-harmonic approximation revealed that both temperature and pressure had an important effect on thermal properties of anatase TiO_2 material. The transport properties, achieved using BoltzTraP code as embodied in the WIEN2k package, showed that calculated Seebeck coefficient is higher at room temperature. However, a slightly decrease of electrical conductivity with temperature increment could be noticed, contrary to thermal conductivity which gave higher values when temperature increased from 300 to 900K. Furthermore, the calculated power factor average was appreciable indicating the performance of anatase TiO_2 material in thermoelectric devices.

Experimental study showed that both deposition parameters and doping process had an important effect on the physical properties of chemically sprayed TiO_2 thin films. It was particularly observed that a flow rate of about 10 ml/min improved the crystalline structure of TiO_2 thin films. Film thickness was estimated of about 300 nm from envelope method which was correlated to the experimental value obtained by SEM cross section. Obtained low value of the extinction coefficient in the transparency domain was a good indication of film surface smoothness and homogeneity. This reflected a good optical quality which permitted to use this binary material as optical window layer in photovoltaic device. PL spectra of the samples were featured by emissions in both UV and visible regions.

Furthermore, the study of the variation of solution volumes proved that the deposited films revealed a good crystallinity at solution volume equals to 60 ml. The obtained TiO_2 thin films were dense, compact and uniform as revealed by AFM. The optical transmission was about 85% in the transparency region for film grown at solution volume equals to 60 ml. The estimation of certain optical constants, such as: dispersion energy Ed, oscillator energy E0, the high-frequency dielectric constant ε_∞ and free carrier electric susceptibility χe, was achieved from the Wemple–Di Domenico and Spitzer–Fan models. It was found that all of the optical constants depend on the sprayed solution volume.

The effect of substrate temperature was also investigated. It is worth mentioning that increasing substrate temperature not only improved the crystallinity of TiO_2 thin films but also significantly enhanced the RMS roughness and decreases the band gap energy up to 3.43 eV.

A phase transition from the anatase to the rutile phase had been observed after heat treatment in air at 500°C for 2 hours. A granular morphology started to appear at annealing temperature > 700°C consisting of spherical shaped grains distributed homogenously, as revealed by AFM topographies. Optical studies proved that annealed TiO_2/quartz thin films exhibited a good optical transmission ranging from 60 to 90%, in the transparent region. However, this average decreased slightly with annealing temperature increment due to surface roughness increasing.

Doping with 0, 0.25, 0.5, 0.75 and 1% of samarium ions (Sm) was also performed. Well crystallized TiO_2 thin films were obtained for a nominal atomic Sm content of 0.25 at.% Sm. For this film specifically the largest crystal size as well as smallest RMS roughness were measured (39 nm and 4.3 nm, respectively). It also exhibited a high optical transmission of 85% in the transparency region with an indirect band gap of 3.33 eV. Doping with samarium ion was found to strongly enhance the photoluminescence properties of titanium dioxide thin films. Furthermore, it gave rise to new emission peaks corresponding to electron transitions from the excited 4G5/2 state to the 6H5/2, 6H7/2, 6H9/2 states.

TiO_2:Mn thin films were chemically sprayed on ordinary glass substrates. It was particularly observed that well crystallized film was obtained at y = 2 at.% Mn. The surface roughness (RMS) value decreased, slightly, from 5.45 nm for undoped TiO_2 to 5.30 nm for TiO_2:Mn (y = 2 at.%). Optical transmission showed a slightly decrease after manganese doping. The increase of Mn doping concentration in the sprayed solution was related to the decrease of the optical band gap. No Mn emission peak was detected for all photoluminescence spectra.

Finally, the optimization of deposition parameters, heat treatment and doping process investigation may be useful guidelines to the synthesis of some optoelectronic devices based on TiO_2 material. As suggestions for further work, there are several areas of research involving TiO_2 that would appear to be promising. Indeed, other dopants will be envisaged in order to get further improvement of photocatalytic activity. Moreover, we may include titanium dioxide thin films as an optical window in the solar cell Au/CZTS/$In_{2-x}Al_xS_3$/TiO_2/SnO_2:F/substrate where Au and SnO_2:F will be used as an ohmic contact, CZTS and $In_{2-x}Al_xS_3$ as absorber and buffer layer, respectively. The optimized flow rate, sprayed volume as well as substrate temperature used for the deposition of TiO_2 thin films are 10 ml/min, 60 ml and 300°C, respectively. The growth by spray pyrolysis and CBD techniques of SnO_2:F, CZTS and $In_{2-x}Al_xS_3$ thin films is well controlled in our laboratory [1-3].

References

[1] N. Jebbari, B. Ouertani, M. Ramonda, C. Guasch, N. Kamoun Turki and R. Bennaceur, Energy Procedia 2 (2010) 79.
[2] Z. Seboui, A. Gassoumi, N. K. Turki, Mater. Sci. Semicond. Process. 26, 360 (2014).
[3] M. Kilani, B. Yahmadi, N.K. Turki, M. Castagné, Mater. Sci. 46, 6293 (2011).

Synthesis and Physical Study of Codoped ZnO Ultrathin Films along with Photocatalytic and Wettability Applications

Refka Mimouni and Mosbah Amlouk

Unité de physique des dispositifs à semi-conducteurs, Faculté des sciences
de Tunis, Université de Tunis, El Manar, 2092 Tunis, TUNISIA.

Abstract: ZnO thin films have attracted tremendous research interest in the last three decades. The photosensitivity devices using codoped metal transitions ZnO as nanocomposites and as thin films largely depend on the structural and the optical properties of such films. In the beginning of this chapter, the physical properties, the fabrication processes as well as the applications of ZnO thin films were reviewed. Then, a detailed characterization of the codoped ZnO thin films by means of various techniques such as XRD, SEM, AFM, optical as well as photoluminescence have been performed, to investigate the use of these films as photocatalysts and/or corrosion protective coating systems prepared through the economic spray pyrolysis technique. Finally, the possibility of their application as photocatalysts under sunlight irradiation against dyes have been carried out.

Table of Contents

Introduction

Environmental contaminants such as pesticides, heavy metals and dyes in water need treatment as well as possible removal to make the water suitable for human consumption. A lot of research works deal with water purification from its contaminants especially the organic ones which constitute a major portion in nature. In this context, harvesting solar energy and degradation of harmful and organic pollutants in water by solar light irradiation is a highly promising strategy. Nowadays, semiconducting materials are increasingly used for the photocatalytic degradation and the decomposing of organic pollutants, contaminants, harmful dyes and industrial wastewater from water. Among various semiconductors, TiO_2 is the primary candidate in photocatalysis based semiconductor. Also, it has been widely applied for environmental interest problems. By contrast, it has been proved that ZnO has a higher photocatalytic efficiency compared to TiO_2 because it absorbs a larger fraction and quanta of the light spectrum than TiO_2 [1].

In fact, the high recombination rate of photogenerated (electron-hole) pairs in the semiconductor material tends to reduce their photocatalytic efficiency[2]. Indeed, all the previous studies have reported that ZnO has a wide band-gap which easily achieves photodegradation through UV irradiation[3] while this type of sources consumes an important energy amount and it constitutes only 4% of the solar spectrum, whereas the photocatalysis research goal is to exploit the total solar energy (especially visible light domain). Currently, the big challenge in the photocatalysis field is to develop the photocatalysts that enhance the charge separation and to exploit sunlight instead of other expensive sources of irradiation. Whence, it is needful and important to modify the band gap of ZnO by using codoping process which leads to the creation of intermediate energy levels in the band gap and enhances the charge separation in order to exploit the total solar energy and enhance the photocatalytic degradation. Moreover, most works have previously utilized ZnO in its nano-powder form; it was dispersed within industrial water and harmful dyes for the photo-degradation of organic pollutants. Nevertheless, there is an additional necessity of this process to re-filter such water, which requires additional constraints of obtaining pure water due to the difficulty of re-filtering process of the nano-sized components[4, 5].

Otherwise, extending the time life of materials and the improvement of resistance to humidity, corrosion and deterioration by applying a protective thin film are main challenges in surface engineering. Also, there is a need to develop environmentally friendly, non-toxic, corrosion protection systems with a low cost. The fabrication of hydrophobic surfaces and coatings (water repellent surfaces) is a novel method to control corrosion and humidity since water is a main part of corrosion mechanism. They tend not to adsorb water or be wetted by it and are a new type of coating which is highly desirable for a broad range of applications such as microfluidic devices, optical devices, self-cleaning, anti-icing and anti-corrosion coatings, textiles. etc. [6-8]. In the last decade, there has been a huge interest in studying the wetting behaviors of various transition-metal oxides, such as ZnO [9], V_2O_5 [10], TiO_2 [11], Ag_2O[12]and so on. Among these oxides, zinc oxide has received much attention due to its fascinating properties and applications. This binary material has a direct wide band gap energy of 3.37 eV, a large exciton binding energy (60 meV), high optical gain (320 cm^{-1}) at room temperature [13, 14], high optical transparency in the visible region, low electrical resistivity as well as high electrochemical stability, non-toxicity and abundance in nature [15]. Besides, it is highly resistant to chemical attack even under plasma of hydrogen. Furthermore, the adding of hydrophobic properties into ZnO surfaces may expand the domain of its classic applications such as electrical and optoelectronic devices.

1. Physico-chemical properties of used elements

1.1. Physico-chemical properties of ZnO basic elements: Zinc (Zn) and Oxygen (O)

Zinc is a chemical element with the symbol (**Zn**) and atomic number **30**. It is a transition blue-gray metal, medium reactive, which combines with oxygen and other non-metals. Zinc crystallizes in hexagonal prism, these crystals can be obtained by the method used to cool slowly from the molten zinc, and to cleanse the part remaining liquid, after partial solidification of the metal on the walls of the vase. Above its boiling temperature (1180 K) is zinc distillation, but it emits vapors well below this temperature.

Oxygen is a gaseous chemical element of non-metal group, with the symbol (**O**) and atomic number **8**. It forms easily compounds including oxides with any other chemical element. It possesses a great importance because it is essential in the respiratory processes of the living cells. It is odorless, colorless and slightly magnetic. Also, it crystallizes in a cubic structure. The different physico-chemical properties of Zinc (Zn) and oxygen (O) chemical elements have been listed in Table 1. Among different doping elements of such oxide, cobalt, chromium and indium are chosen to be used as codopants in this work.

Table 1. Physico-chemical properties of zinc (Zn) and oxygen (O) elements

Physico-chemical properties	Zinc (Zn)	Oxygen (O)
Electronic configuration	[Ar] $3d^{10}4s^2$	[He] $2s^2\,2p^4$
Atomic number	30	8
Atomic mass	65.39 g.mol^{-1}	15.99 g.mol^{-1}
Density	7.13 g.cm^3	1.43 g.m^{-3}
Atomic radius	142 pm	65 pm
Ionic radius	0.074 nm (+2)	0.14 nm (-2)
Boiling Point	907°C	-182, 95°C
Electronegativity	1.65	3.44
Oxidation state	2	-1, 2

1.2. Physico-chemical properties of used codopants: Cobalt (Co), Chromium (Cr) and Indium (In) elements.

2. 2. 1. Physico-chemical properties of Cobalt (Co)

Cobalt is a chemical element with symbol **(Co)** and atomic number 27. Cobalt is hard, brittle, silver-grey transition metal that resembles iron and nickel in appearance. It is a moderately reactive element. The metal and the most common ions are ferromagnetic. It crystallizes in hexagonal close-packed. It combines slowly with oxygen in the air and it reacts with most acids to produce hydrogen gas. Cobalt is primarily used as a metal, in the preparation of magnetic, wear-resistant and high-strength alloys.

1. 2. 2. Physico-chemical properties of Chromium (Cr)

Chromium is a chemical element with symbol **(Cr)** and atomic number **24**. It is hard, brittle and a steely-gray transition metal. It has a bright, shiny surface and high melting points. It crystallizes in a face-centered cubic structure. It is the only solid elemental which shows antiferromagnetic ordering at room temperature (and below). Above 311K, it transforms into a paramagnetic state. Chromium is a fairly active metal. To form chromium oxide (Cr_2O_3) it combines with oxygen at room temperature.

1. 2. 3. Physico-chemical properties of Indium (In)

Indium is a chemical element with **(In)** symbol and atomic number 49. It is a silvery-white, shiny metal and highly ductile. It crystallizes in a tetragonal structure. It is one of the softest metals known. Indium metal dissolves in acids, but does not react with oxygen at room temperature. At higher temperatures, it combines with oxygen to form indium oxide (In_2O_3). It is of considerable industrial importance, most notably in the transparent conductive coatings production. The principle physico-chemical properties of these elements are listed in the Table 2.

Table 2. Physico-chemical properties of cobalt, chromium and indium elements.

Physico-chemical properties	Cobalt (Co)	Chromium (Cr)	Indium (In)
Electronic configuration	[Ar] $3d^7$ $4s^2$	[Ar] $3d^5$ $4s^1$	[Kr] $4d^{10}$ $5s^2$ $5p^1$
Atomic number	27	24	49
Atomic mass	58.93	52 g.mol^{-1}	114.82 g.mol^{-1}
Density	8.90 g.m^{-3}	7.19 g cm^{-3}	7.31 g.cm^3
Atomic radius	135 pm	140 pm	167 pm
Ionic radius	0.070 (+2) 0.060 (+3)	0.062 nm (+3)	0.080 nm (+3)
Boiling Point	2927 °C	2671 °C	2072 °C
Electronegativity	1.88	1.66	1.79
Oxidation state	+1, +2, +3, +4	+2, +3, +6	+3

1.3. Bibliographical review on ZnO Properties

ZnO is a II-VI compound semiconductor. It can adapt to three different crystallographic phases (Fig. 1): phase (a) cubic (Rocksalt), phase (b) Blende and phase (c) hexagonal (wurtzite). At ambient conditions, the thermodynamically stable phase is hexagonal wurtzite. The blende structure can be stabilized only by growth on cubic substrates, and the rocksalt structure (NaCl) may be obtained at relatively high pressures (10-15 GPa).

Zinc oxide hexagonal wurtzite structure has a polar hexagonal c-axis which is chosen to be parallel to z. Wurtzite structure of this binary belongs to space group of $P6_3mc$ (C_{6v}^4). Each cation (zinc ion) is surrounded by four anions (oxygen ions) at the corners of a tetrahedron, and vice versa. This arrangement gives rise to a polar symmetry along the c-axis. Also, this tetrahedral coordination is characteristic of sp^3 covalent bonding nature. The primitive unit cell contains two units of ZnO. Also, this binary can be described by a hexagonal Bravais lattice with constants **a=b** =3. 2496Å and **c** =5. 2042Å; $\alpha = \beta = 90°$ and γ =120°[16]. In this compact hexagonal lattice, 6 atoms can be seen in the same compact plan, 3 in the above plan and 3 in the below plan which makes 12 neighboring atoms.

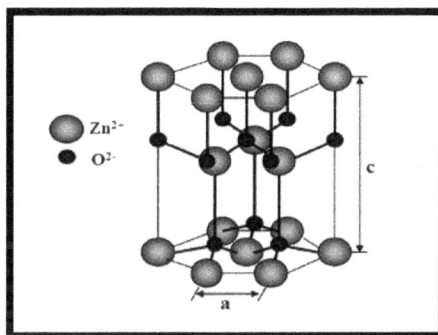

Figure 1.
The unit cell wurtzite structure ZnO.

Table 3. Physical properties of zinc oxide [17].

Lattice parameter "a_0" at 300K	0.32495 nm
Lattice parameter "c_0"at 300K	0.52069 nm
c_0/a_0	1.602
Density	5.606 g cm^{-3}
Coordination (Z)	2
Stable phase at 300K	Wurtzite
Space group	$P6_3mc$ (C_{6v})
Melting point	1975°C
Band gap energy	3.3 eV (Direct)
Exciton binding energy	60 meV
Transmittance	> 90%

The intrinsic and the extrinsic defects present in the crystal structures as well as the efficient radiative recombination have made ZnO so attractive in optoelectronics applications. The ZnO band structure has donor and accepter energy levels which are present at below and above conduction band (CB) and valance band (VB). The observed UV peak is due to the transition recombination phenomena of free excitons in the near band-edge of ZnO. Moreover, the emission of a whole visible region is due to the presence of different deep energy levels within the band gap. Several research groups have reported different origins for these deep level defects as described in references [16, 18]. It is generally accepted that the visible range in ZnO is attributed to the presence of various intrinsic defects in the ZnO lattice such as oxygen vacancies (V_O), interstitial oxygen (O_i), antisite oxygen (O_{Zn}), zinc vacancies (V_{Zn}), interstitial zinc (Zn_i), and antisite zinc (Zn_O) [19].

ZnO is a semiconductor intrinsically doped n-type. This behavior has been reported in the literature and it is assigned to the presence of native defects in its crystal structure like oxygen vacancies and zinc interstitials [20]. The electron mobility values in un-doped ZnO nanostructures are estimated varying from 120 to 440 cm^2Vs^{-1} at room temperature, also it depends on the fabrication methods [16]. Moreover, it is possible to modify the zinc oxide properties by doping with metal ions. In addition, it has been reported that after ZnO doping, the highest carrier concentration for holes and electrons $\sim 10^{19}\,cm^{-3}$ and $10^{20}\,cm^{-3}$ respectively were obtained [21].

1.4. Strategies for ZnO performance: Co-doping effect

The codoping consists in doping a material with a mixture of two types of dopants (elements). Generally, it is an effective way for improving the physical properties of the materials such as crystallinity, electrical properties, optical properties and specific surface area. This changing technique has gained remarkable attention in recent years.

Various chemical methods have been adopted to engineer oxide semiconductors, TiO_2, SnO_2 and ZnO such as anion-anion codoping, anion-cation codoping, and cation-cation cooping [22]. The cation-cation codoping approach has been pursued due to the easy experimental conditions for their fabrication [22]. On the one hand, Moharrami *et al.* [23] have worked on S and Al codoped SnO_2 semiconductor thin films prepared by spray pyrolysis technique.

On the other hand, Chen *et al.* [24] prepared successfully spin coated cobalt and vanadium co-doped TiO_2 thin films to enhance the photocatalytic performance, as well as Jabbari *et al.* [25] who synthesized by a photochemical reduction technique (In, V) codoped TiO_2 catalysts. Furthermore, it is well known that codoping has been suggested to be an effective method to achieve p-type conductivity of ZnO. Indeed, Ga (donor) and N (acceptor) as dopants improve this type of conductivity for epitaxial ZnO as reported previously [26]. Nevertheless, the codoping inducing n type ZnO thin films is used to reinforce the gas sensors sensitivity [27] and to enhance the photocatalysis.

1.5. Experimental procedure

The spray pyrolysis process is a useful technique in material elaboration which has been widely applied in the past few years. Among the aerosol processes, spray is the most suitable one, with its numerous advantages such as: simple equipment, short running time, high homogeneity, uniform distribution, purity of products, large scale, non-toxic, low cost method and so on. Moreover, the application of this technique in industry is very promising.

Its principle consists in making liquid drops from material solution and producing fine particles by solvent evaporation, drying and pyrolysis reaction of liquid drops at high temperature. The size and the distribution of particles depend on liquid drops, the evaporation process of solvent and the property of starting material. Each droplet has similar composition, and multicomponent particles are facile to form by controlling the precursors[28]. The spray pyrolysis deposition system is represented schematically in Figure 2. It contains essentially a round spray nozzle, a pump and a heater plate.

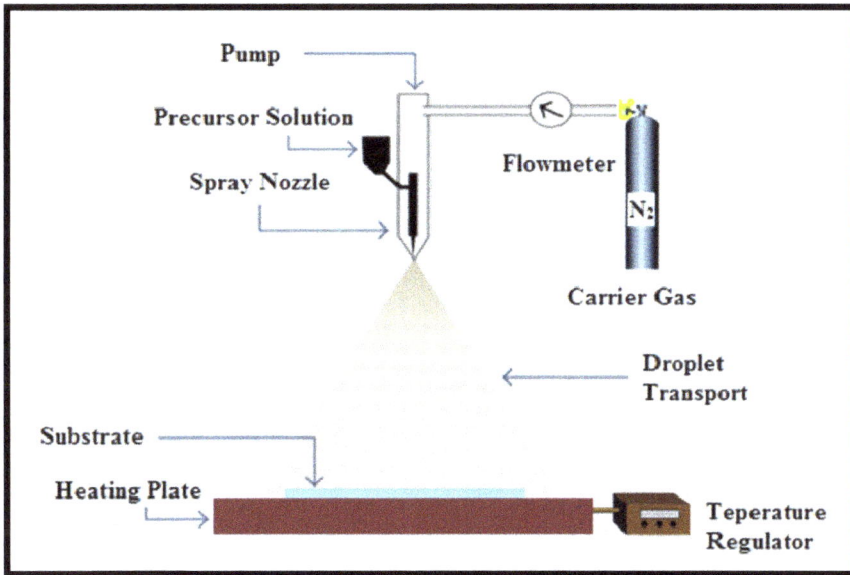

Figure 2. Spray pyrolysis deposition process schema.

Un-doped ZnO and (Cr, Co), (Co, In) and (Cr, In) codoped ZnO thin filmswere deposited on glass substrates at 460°C, using the chemical spray technique [28]. The aqueous solution of zinc acetate dihydrate (Zn $(CH_3COO)H_2$. $2H_2O$). $10^{-2}M$ contained a mixture of distilled water and propanol with fraction volumes of 50 cm^3 and 150 cm^3 respectively. It was acidified with acetic acid (pH=5) [7, 8]. Cobalt (II) chloride hexahydrate ($CoCl_26H_2O$), Potassium chromate (K_2CrO_4) and Indium chloride ($InCl_3$) were used as a source of Co, Cr and In respectively. Co, Cr and In to Zinc molar ratios (Co/Zn), (Cr/Zn) and (In/Zn) were 0%, 1% and 2%. Nitrogen was used as the gas carrier (pressure at 0. 35 bar) through a 0. 5 mm diameter nozzle. The distance between the nozzle and the substrate plane was fixed at the optimal value of 27/cm[28]. During the deposition process, the solution flow rate was taken at 4mL/min throughout the thin films deposition.

2. Structural characterization results

2.1. Structural characterization results on (Cr, Co) codoped ZnO nanofilms

Figure 3. shows X-ray diffraction patterns of (Cr, Co) codoped ZnO sprayed nanofilms. Pure ZnO film shows diffraction peaks at2□□values of 34. 5 and 36. 26

attributed to (002) and (101) planes of hexagonal wurtzite structure (JCPDS N°. 36-1451), respectively. (002) peak is the most intense and the narrowest one, revealing the highly crystalline character of pure ZnO thin film. By (Cr, Co) codoping, all nanofilms are predominantly oriented along(002) orientation which is characteristic of the würtzite structure. The intensity of the privileged peak increases which reflects an improvement in the crystallinity of the nanofilms, particularly for ZnO:Cr1%:Co2% nanofilm. It is also worth noting that neither peak related to chromium oxide phase (Cr_xO_y) nor that to cobalt oxide phase (Co_xO_y) has been found in these spectra, which is due to the low Cr and Co contents. This indicates that little amount of codopants does not change wurtzite structure and incorporates successfully in the ZnO lattice.

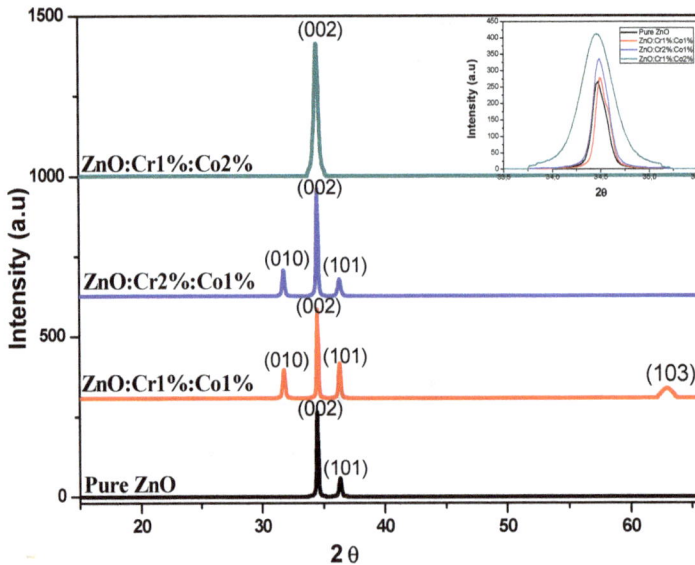

Figure 3. X-ray diffraction spectra of (Cr, Co) codoped ZnO thin films.

As shown in Table 4, the highly textured peak centered at 34. 49° characterizing the hexagonal wurtzite structure, was shifted toward smaller angles indicating an increase in the interplanar spacing (d_{hkl}) from 2. 5979 to 2. 6022 Å with the increase in Cr and Co doping content. This variation can be related to the microstrain, structural disorder and rearrangement of the lattice due to codoping. As a result of this phenomenon, we have a variation in lattice parameters due to the incorporation of the dopants in the ZnO matrix. This incorporation may possibly be a substitution of Zn^{2+} by Cr^{3+} and Co^{2+}ions. Indeed, (002) peak position of ZnO:Cr:Co nanofilms moves to a slightly weak angles as compared to that of undoped ZnO because Cr^{3+} and Co^{2+}ions have smaller ionic radius than Zn^{2+} ones. This indicates that most Cr and Co atoms replace Zn and the films are under compressive stress in the film plane. This is a consequence of the microstrain, the structural disorder and the rearrangement of the lattice due to codoping. Asa resultof this phenomenon, we have a change in the lattice parameters, as indicated in Table 4 possibly due to the incorporation of both doping agents in ZnO matrix.

Table 4. Different interplanar spacing dhkl and lattice parameters of (Cr, Co) codoped ZnO sprayed thin films.

Sample	Pure ZnO	ZnO:Cr1% :Co1%	ZnO:Cr2%: Co1%	ZnO:Cr1%: Co2%
Position of (002) peak in terms of 2Θ (∘)	34.49	34.50	34.48	34.44
Position of (101) peak in terms of 2 Θ (∘)	36.26	36.29	36.28	–
d 002 (Å)	2.59	2.59	2.60	2.60
d 101 (Å)	2.48	2.47	2.47	–
Lattice Parameters (Å)	a =3.25 c= 5.19	a = 3.25 c= 5.19	a = 3.25 c = 5.20	a = - c = 5.20
c/a ratio	1.59	1.60	1.60	–

From Table 5, we can note: On the one hand, the Crystallites size is ranging from 22.64 to 62.05 nm with increasing both Cr and Co concentrations. On the other hand, the microstrain varies from $18.8 10^{-4}$ to $51.7 10^{-4}$ with (Cr, Co) codoping. These results denote that the crystallites sizes and the microstrain are sensitive to the codoping effect.

Table 5. Grain size and microstrain values of (Cr, Co) codoped ZnO sprayed thin films.

Samples	Pure ZnO	ZnO:Cr1%: Co1%	ZnO:Cr2%: Co1%	ZnO:Cr1%: Co2%
Full width at half maximum β(002) (°)	0.18	0.16	0.13	0.37
Crystallites size (nm)	45	52	62	23
Microstrain (10-4)	25.86	22.49	18.84	51.70

Figure 4 shows that all the nanofilms surfaces are compact without any defect zones or cracks and with uniform grains distribution. Also, the grain growth followed the c-axis which is perpendicular to the film plane and it is found as a polar axis[29]. By (Cr, Co) codoping, the morphology and the geometry of obtained nanofilms seems unaffected by Cr and Co concentration, it is always polygonal for all nanofilms, only the grains size varies. For ZnO:Cr2%:Co1% nanofilm there is an expansion of the grains size, likewise, it exhibits a rough surface compared to other codoped nanofilms. Whereas for ZnO:Cr1%:Co1% and ZnO:Cr1%:Co2% nanofilms, the grains size decreases. The average diameter size measured from SEM data is found ranging from 50 to 110 nm while, the calculated crystallites size from the XRD investigations is lower than these values. This might be due to the agglomeration of the smaller grains for clustered formation occurring during endothermic reaction between precursors.

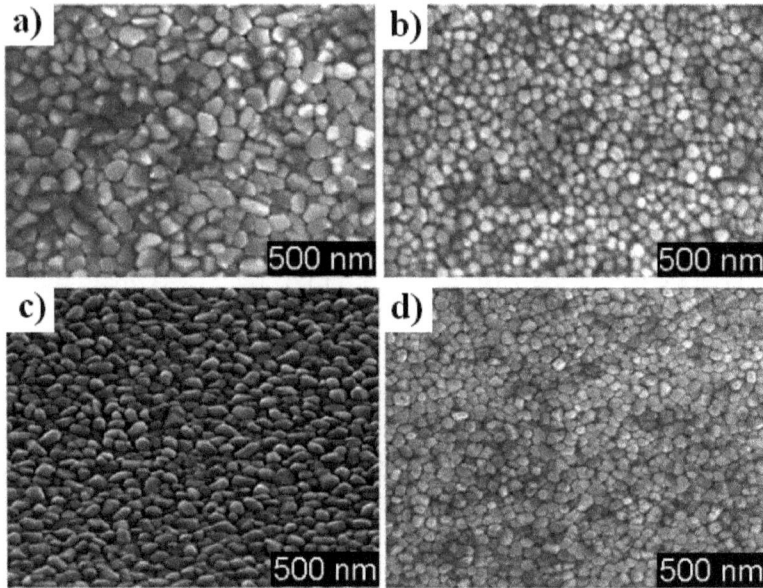

Figure 4. SEM micrographs of: Pure ZnO(a), ZnO:Cr1%:Co1% (b) ZnO:Cr2%:Co1% (c) and ZnO:Cr1%:Co2% (d) thin films.

On another side, to obtain well cross- sectional micrographs and evaluate the thickness values of codoped ZnO nanofilms, the (AZ5214E) resin was spin-coated on the top of ZnO thin films. Then, the hetero-structure (AZ5214E/ZnO/glass) was heated at 125°C for 1 min. After that, we cleaved the samples. Next, the AZ5214E was dissolved by acetone. Then, we put the samples on the cleaved part of the (XHR) SEM, then, we took micrographs. Figure 5 exhibits the cross sectional Photographs of (Cr, Co) codoped ZnO thin films. It is clear that the grains are equiaxed, compact and closely spaced. The thickness values are lying in 40-80 nm domains, Table 6. These films thickness variation is due to different morphologies and sizes of the grains which are dependent on the codopants concentration.

Table 6. Thickness values of (Cr, Co) codopedZnOnanofilms

sample	Pure ZnO	ZnO:Cr1%: Co1%	ZnO:Cr2%: Co1%	ZnO:Cr1%: Co2%
Thickness (nm)	72	60	78	45

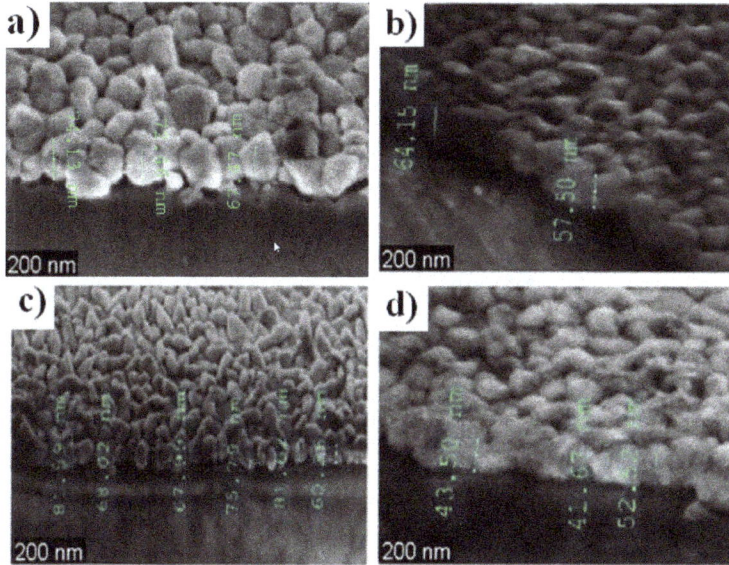

Figure 5. Cross-section SEM micrographs of: Pure ZnO (a),
ZnO:Cr1%:Co1%; (b) ZnO:Cr2%:Co1%; (c) and ZnO:Cr1%:Co2%; (d) thin films.

Figure 6. presents the two-dimensional (2D) images of AFM micrographs. As seen, all thin films are homogeneous and uniform; they are distinguished by rounded clusters limited by grain boundaries. It is also clear that no segregation of dopants has been observed. The (Cr, Co) codoped nanofilms show perturbed surfaces with obtained grains that are oriented along c axis.

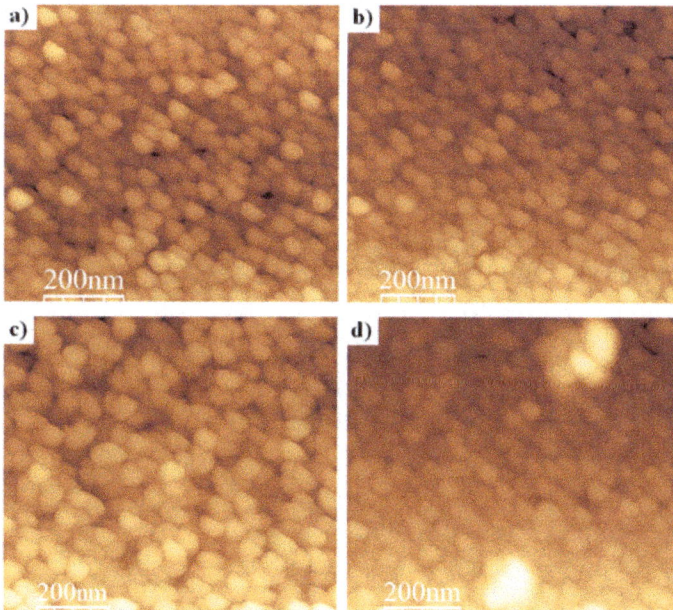

Figure 6. 2D AFM topography of (a) Pure ZnO, (b) ZnO:Cr1%:Co1%,
(c) ZnO:Cr2%:Co1% and (d) ZnO:Cr1%:Co2% thin films.

From roughness values listed in Table 7, it is found that the surface roughness decreases with increasing dopants concentration especially for ZnO:Cr1%:Co2% nanofilm which present a relatively smooth surface, this result is coherent with the thickness values found by Cross-section SEM investigations when the thickness decreases, the root mean square roughness (Rq) decreases and vice versa. Hence, it is more suitable for optoelectronic applications. On the other side, the surface roughness of ZnO:Cr2%:Co1% nanofilm increases to reach 11 nm. This increase in roughness may be attributed to the growth of surface crystallites with a cluster formation. However, this surface is quite favorable for the use of such component type in the sensitivity applications (gas sensors, photocatalysis, etc.), because they depend strongly on the quality of the surface and the roughness. A similar effect has also been observed previously in ZnO:Cr[30] thin films.

Table 7. Roughness parameter extracted from the surface analysis of AFM topography.

Sample	Pure ZnO	ZnO:Cr1%: Co1%	ZnO:Cr2%: Co1%	ZnO:Cr1%: Co2%
Roughness (nm)	8	5	11	3

2. 2. Structural characterization results on (Co, In) codoped ZnO nanofilms

Figure 7 depicts X-ray diffraction patterns of (Co, In) codoped ZnO sprayed nanofilms. Pure ZnO film shows diffraction peaks at2θvalues of 34. 5 and 36. 26 attributed to (002) and (101) planes of hexagonal würtzite structure (JCPDS N°. 36-1451), respectively. The (002) peak is the most intense and the narrowest one, revealing the highly crystalline character of pure ZnO thin film. By (Co, In) codoping, all nanofilms are predominantly oriented along (002) orientation which is characteristic of the würtzite structure. The intensity of the privileged peak increases which reflects an improvement in the crystallinity of the nanofilms, particularly for ZnO:Co1%:In2% nanofilm. Also, we note that neither peak related to chromium oxide phase (Cr_xO_y) nor that to cobalt oxide phase (Co_xO_y) has been found in these spectra, which is due to the low Cr and Co contents. This indicates that little amount of codopants does not change wurtzite structure and incorporates successfully in the ZnO lattice. Obviously, all the prepared (Co, In) codoped ZnO samples show neither cobalt oxides (Co_xO_y)nor Indium oxide (In_2O_3) related phases due to the low Co and In concentrations.

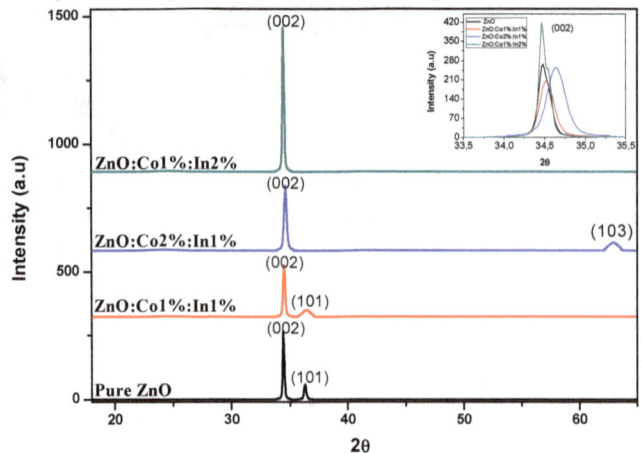

Figure 7.
X-ray diffraction spectra of
(Co, In) codoped ZnO thin films.

As shown in Table 8, the privileged peak centered at 34. 49°, was shifted toward bigger angles showing a decrease in the interplanar spacing (d_{hkl}) from 2. 60 to 2. 57 Å with codoping. This fluctuation may be attributed to the microstrain, the structural disorder and the rearrangement of the lattice due to codoping. As a result of this phenomenon, we have a change in the lattice parameters, as indicated in Table 8, possibly due to the incorporation of both doping agents in ZnO matrix.

From Table 9, we can note that the grain sizeis ranging from 35. 52 to 67. 06 nm with increasing both In and Co concentrations. On the other hand, the microstrain varies from 17. 4 10^{-4} to 32. 8 10^{-4} with (Co, In) codoping. These results denote that the crystallites size and the microstrain are sensitive to the codoping effect.

Table 8. Different interplanar spacing dhkl and lattice parameters of (In, Co) codoped ZnO sprayed thin films.

Sample	Pure ZnO	ZnO:Co1%: In1%	ZnO:Co2%: In1%	ZnO:Co1% :In2%
Position of (002) peak in terms of 2Θ (°)	34. 49	34. 50	34. 62	34. 50
Position of (101) peak in terms of 2Θ (°)	36. 26	36. 41	–	–
d 002 (Å)	2. 59	2. 57	2. 59	2. 60
d 101 (Å)	2. 48	2. 47	–	–
Lattice Parameters (Å)	a =3. 25 c= 5. 19	a = 3. 25 c= 5. 14	a = - c = 5. 18	a = _ c = 5. 20
c/a ratio	1. 59	1. 59	–	–

Table 9. Grain size as well as microstrain values of (In, Co) codoped ZnO sprayed thin films.

Sample	Pure ZnO	ZnO:Co1%: In1%	ZnO:Co2%: In1%	ZnO:Co1%: In2%
Full width at half maximum β(002) (°)	0.18	0.17	0.23	0.22
Crystallites size (nm)	45	50	36	67
Microstrain (10-4)	25.86	23.51	32.80	17.40

Figure 8 shows that all films have rather homogeneous grains distribution. Besides, all films top surfaces were parallel to the substrate plane, showing that the grain growth followed c-axis that proved to be a polar axis[29]. The doping with Co and In elements seems preserving ZnO microstructure characterized by the presence of grains emerging perpendicularly to the film surface.

One can observe a slight influence of the codoping on increasing grains size which is in good agreement with the variation of crystallite size observed along XRD data. The crystallites of the deposited thin films have an average size ranging from 80 to 120 nm. It can be seen that there is an expansion of grains size for ZnO:Co1%:In1% and ZnO:Co1%:In2% nanofilms. However, the size measured from (XHRSEM) images was found always higher than the calculated size from the XRD reflections. This might be considered due to agglomeration towards clustered formation occurring during endothermic reaction between precursors. All nanofilms are uniform without any remarkable defect in the structure. After codoping, the morphology of obtained films changes slightly from spherical for pure ZnO nanofilm to polygonal for ZnO:Co1%:In1% and ZnO:Co1%:In2% ones. It is clear that these grains are densely packed with each other but still there are voids which made them perturbed. For ZnO:Co2%:In1% sample, the grains are stacked on each other in such a special and particular manner that made a mixture of polygonal and sheet-like morphology observed, Figure 8. Also, it is noted that the roughness of such film is important, the grains are densely packed and there are no voids among them which made them highly glued.

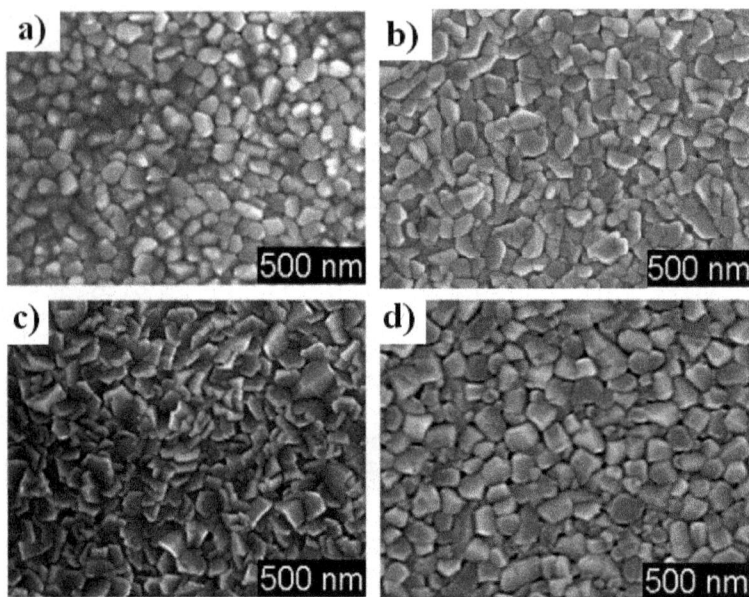

Figure 8. SEM micrographs of: Pure ZnO(a), ZnO:Co1%:In1% (b), ZnO:Co2%:In1% (c) and ZnO:Co1%:In2% (d) thin films.

Figure 9 exhibits the cross sectional Photographs of (Co, In) codoped ZnO thin films. It is clear that the grains are equiaxed, compact and closely spaced. The thickness values are lying in 70-100 nm domain, Table 10. These films thickness variation is due to different morphologies and sizes of the grains which are dependent on the codopants concentration.

Figure 10 illustrates the two-dimensional (2D) images of AFM micrographs. All the thin films are homogeneous and uniform. Also, they are distinguished by rounded clusters limited by grain boundaries. Besides, it is clear that no segregation of dopants has been observed. All the (Co, In) codoped nanofilms depict perturbed surfaces with elongated grains and hollow zones showing a rough surface. The grains are oriented along c-axis which leads to the roughness improvement.

From roughness values listed in Table 11, it is found that the surface roughness increases with(Co, In) codopingespecially for ZnO:Co1%:In1% and ZnO:Co2%:In1% nanofilms. This may be attributed to the growth of crystallites surface with a cluster formation. A similar effect has also been observed previously for Indium doped ZnO thin films[31]. Hence, these surfaces are favorable for the use of such components type in the sensitivity applications (gas sensors, photocatalysis, etc.), because they strongly depend on the roughness and the quality of the surface. The roughness values are in agreement with the thickness values found by Cross-section SEM investigations; when the thickness increases, the root mean square roughness (Rq) increases and vice versa.

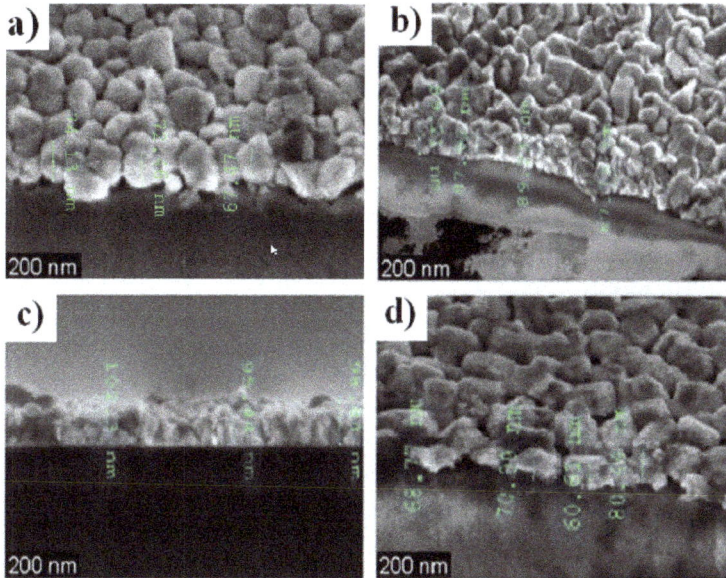

Figure 9. Cross-section SEM micrographs of: (a) Pure ZnO, (b) ZnO:Co1%:In1%, (c) ZnO:Co2%:In1% and (d) ZnO:Co1%:In2% thin films.

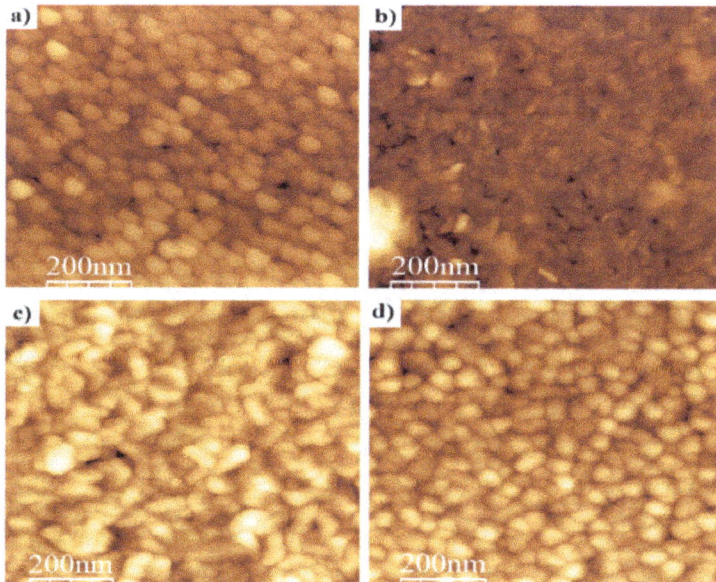

Figure 10. 2D AFM topography of of ZnO:Co:In thin films (a) Pure ZnO, (b) ZnO:Co1%:In1%, (c) ZnO:Co2%:In1% and (d) ZnO:Co1%:In2% respectively.

Table 10. Thickness values of (Co, In) codopedZnOnanofilms.

Sample	Pure ZnO	ZnO:Co1%:In1%	ZnO:Co2%:In1%	ZnO:Co1%:In2%
Thickness (nm)	72	89	99	71

Table 11. Roughness parameter extracted from the surface analysis of AFM topography.

Sample	Pure ZnO	ZnO:Co1%:In1%	ZnO:Co2%:In1%	ZnO:Co1%:In2%
Roughness (nm)	8	10	12	7

2.3. Structural characterization results on(Cr, In) codoped ZnO nanofilms

Figure 11. shows X-ray diffraction of (Cr, In) codoped ZnO sprayed nanofilms. Pure ZnO film shows diffraction peaks at2θvalues of 34. 5 and 36. 26 assigned to (002) and (101) planes of hexagonal würtzite structure (JCPDS N°. 36-1451), respectively. For ZnO:Cr1%:In1% and ZnO:Cr1%:In1% nanofilms, no other peaks of ZnO are observed except the most intense (002) one which is characteristic of the würtzite structure. This indicates that these codoped nanofilms have high privileged c-axis orientation and the crystallinity had been improved by codoping, especially for ZnO:Cr1%:In1% nanofilm. This is consistent with the results found in the literature for (Cu, Co) codoped ZnO thin films prepared using RF magnetron sputtering method[32]. Besides, all the prepared (Cr, In) codoped ZnO samples show neither chromium oxide (Cr_xO_y) nor Indium oxide (In_2O_3) related phases due to the low Cr and In amount. Whence, a little amount of codopants incorporates successfully in the ZnO lattice and does not modify wurtzite structure.

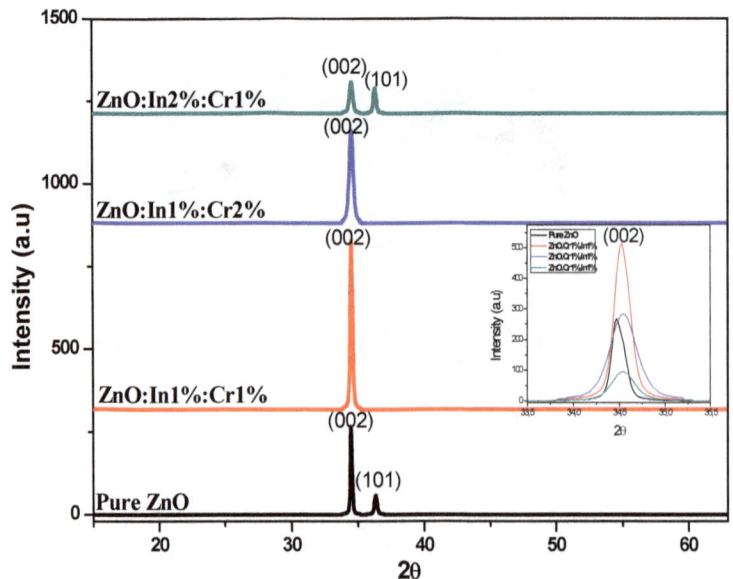

Figure 11. X-ray diffraction spectra of (Cr, In) codoped ZnO thin films.

As shown in Table 12, the privileged peak centered at 34. 49° shifts toward bigger angles revealing a decrease in the interplanar spacing (d_{hkl}) values from 2. 60 to 2. 5963Å with increasing the amount of codoping.

This variation may be assigned to the structural disorder, the microstrain and the rearrangement of the lattice as a result of codoping. Hence, we obtain a change in the lattice parameters, as mentioned in Table 12, possibly attributed to the incorporation of both Cr and In elements in ZnO matrix.

Table 12. Different interplanar spacing dhkl and lattice parameters of (In, Cr) codoped ZnO sprayed thin films.

Sample	Pure ZnO	ZnO:Cr1%: In1%	ZnO:Cr2%: In1%	ZnO:Cr1%: In2%
Position of (002) peak in terms of 2Θ (○)	34. 49	34. 51	34. 52	34. 51
Position of (101) peak in terms of 2Θ (○)	36. 26	−	−	36. 29
d 002 (Å)	2. 59	2. 60	2. 60	2. 60
d 101 (Å)	2. 48	−	−	2. 47
Lattice Parameters (Å)	a =3. 25 c= 5. 19	a = - c= 5. 20	a = - c = 5. 19	a = 3. 25 c = 5. 19
c/a ratio	1. 59	−	−	1. 60

From Table 13, we can note that the crystallites size values decrease from 45. 19 to 25. 48 nm by (Cr, In) codoping. On the other hand, the microstrain values increase from 25. 9 10^{-4} to 45. 8$10^{-4}$ when both In and Cr ratio increases. So, the crystallite sizes and the microstrain are sensitive to the codoping effect.

Table 13. Grain size as well as microstrain values of (Cr, In)codoped ZnO sprayed thin films.

Sample	Pure ZnO	ZnO:Cr1%: In1%	ZnO:Cr2%: In1%	ZnO:Cr1%: In2%
Full width at half maximum β(002) (°)	0. 18	0. 21	0. 33	0. 27
Crystallites size (nm)	45	41	25	31
Microstrain (10-4)	25. 86	28. 80	45. 84	37. 59

Figure 12 illustrates that all the nanofilms surfaces are compact without any defect zones or visible voids and with almost homogeneous and uniform grains distribution. The grains are densely packed with each other and highly glued. Also, they perfectly adhere to the substrate. In addition, the grain growth followed the c-axis emerging perpendicularly to the film surface and it is revealed to be a polar axis[29]. Due to the codoping effect, it is clear that the surface morphology of the nanofilms is influenced by the introduction of chromium and indium atoms. A remarkable change and perturbation in the grain morphologies are observed. For undoped ZnO nanofilm, the film surface is composed of spherical and polygonal grains. After codoping, the obtained nanofilms have polygonal and prismatic geometries. Some white spots are seen on the surface of ZnO:Cr2%:In1% nanofilm. They may be attributed to the discharge effect. [33, 34]. Also, a variation of crystallites size is seen. For ZnO:Cr1%:In2% nanofilm there is an expansion of the grains size, while for ZnO:Cr1%:In1% and ZnO:Cr2%:In1% nanofilms there is a decrease in the grains size. These results are coherent with the XRD data. The average crystallites size measured from SEM investigations is found ranging from ~ 50 to 100 nm; whereas, the calculated grain size from the XRD data are lower than these values. This might be assigned to the agglomeration of the smaller grains for clustered formation occurring during endothermic reaction between precursors

Besides, these images exhibit rough surfaces which suggest that the codoped nanofilms are potential candidates for sensitivity applications.

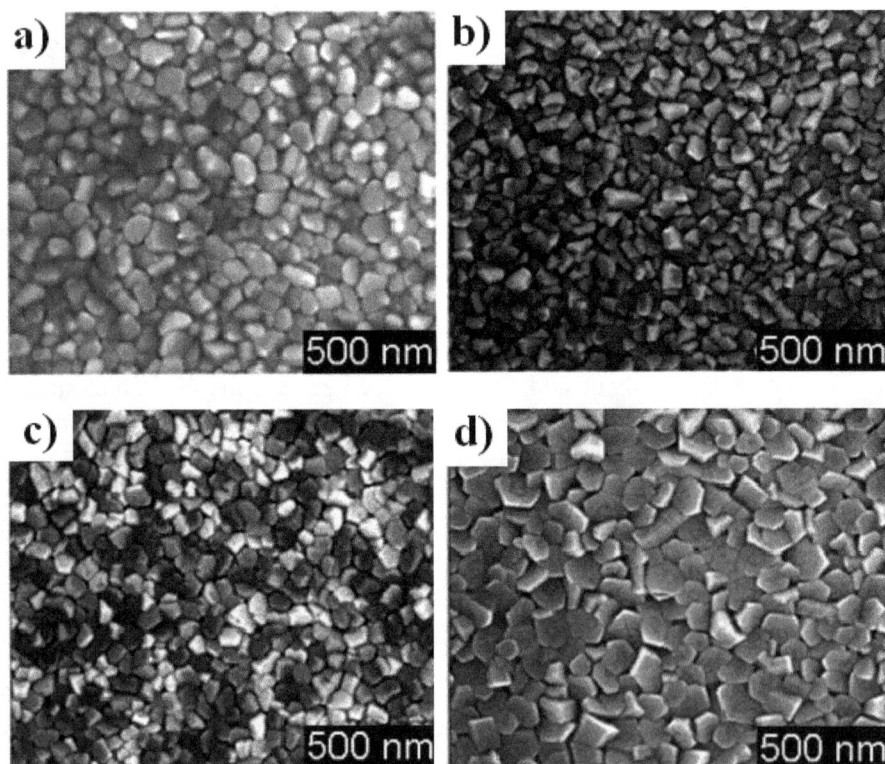

Figure 12. (XHR) SEM images of: (a) Pure ZnO, (b) ZnO:Cr1%:In1%, (c) ZnO:Cr2%:In1% and (d) ZnO:Cr1%:In2% thin films.

Figure 13. exhibits the cross sectional Photographs of (Cr, In) codopedZnO thin films. It is clear that the grains are equiaxed, compact and closely spaced. The thickness values are lying in 60-100 nm domain, Table 14. These films thickness variation is due to different morphologies and sizes of the grains which are dependent on the codopants concentration.

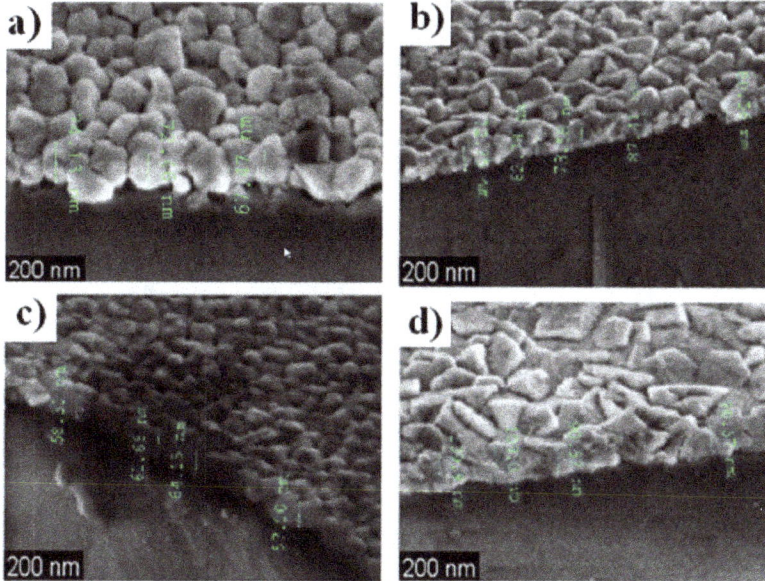

Figure 13. Cross-section SEM micrographs of: (a) Pure ZnO, (b) ZnO:Cr1%:In1%, (c) ZnO:Cr2%:In1% and (d) ZnO:Cr1%:In2% thin films.

Table 14. Thickness values of (In, Cr) codoped ZnO nanofilms.

Sample	Pure ZnO	ZnO:Cr1%: In1%	ZnO:Cr2%: In1%	ZnO:Cr1%: In2%
Thickness (nm)	72	80	60	100

Figure 14 illustrates the two-dimensional (2D) images of AFM micrographs. It is obvious that all the thin films are homogeneous and uniform; they are distinguished by rounded clusters limited by grain boundaries. Besides, it is clear that no segregation of dopants has been observed. All the (Cr, In) codoped nanofilms depict perturbed surfaces with elongated grains and hollow zones showing a rough surface, the grains are oriented along c-ais which leads to the roughness improvement.

From roughness values listed in Table 15, it is found that the surface roughness increases with (Cr, In) codopingespecially for ZnO:Cr1%:In1% and ZnO:Cr1%:In2% nanofilms. This may be attributed to the growth of crystallites surface with a cluster formation. A similar effect has also been observed previously for Cr doped ZnO thin films [30] and for Indium doped ZnO thin films [31]. Hence, these nanofilms surfaces are favorable for the use of such components type in the sensitivity applications (gas sensors, photocatalysis, etc.), because theystrongly depend on the roughness and the quality of the surface. The roughness values are in agreement with the thickness values found by Cross-section SEM investigations; when the thickness increases, the root mean square roughness (Rq) increases and vice versa.

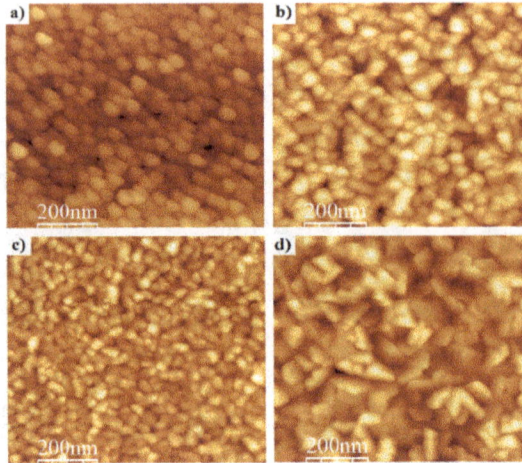

Figure 14. 2D AFM topography of of ZnO:Cr:In thin films (a) Pure ZnO, (b) ZnO:Cr1%:In1%, (c) ZnO:Cr2%:In1% and (d) ZnO:Cr1%:In2% respectively.

Table 15. Roughness parameter extracted from the surface analysis of AFM topography.

Sample	Pure ZnO	ZnO:Cr1%: In1%	ZnO:Cr2%: In1%	ZnO:Cr1%: In2%
Roughness (nm)	8	11	4	13

3. Optical and Photoluminescence results

3.1. Optical and Photoluminescence characterization results of (Cr, Co) codoped ZnO nanofilms

From the optical transmission and reflectivity spectra of (Cr, Co) codoped ZnO nanofilms, Figure 15 and Figure 16 respectively, we note that these films show a high transparency within the visible range with an average transmittance lying between 79% and 90%. Also, the spectra depict a reflectance less than 30%, hence their absorption is almost worthless. This may be due to the thickness of the films and the doping effect. The decrease in optical transmission with higher concentration of Cr can be associated with the scattering phenomenon at grain boundaries. It can also be seen that the interference fringe patterns are absent in all transmittance and reflectance spectra due to weak multiple reflections at the interface. This phenomenon can contribute to more scattering effects and to the crystalline state. Furthermore, within the IR region, the reflectance is on the order of 10%, this indicates that the films seem exempt of free charges, although codoping elements have been used. This may be due to the incorporation of Co and Cr in the ZnO matrix. Hence, the layers are anti-reflective.

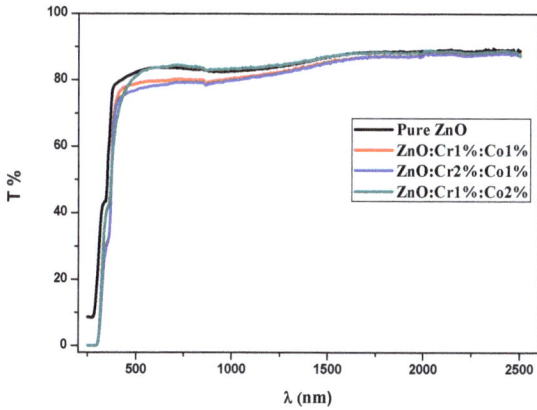

Figure 15. Transmission spectra of (Cr, Co) codoped ZnO thin films.

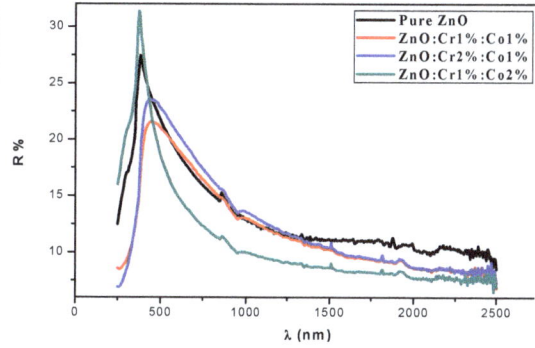

Figure 16. Reflectivity spectra of (Cr, Co) codoped ZnO thin films.

Figure 17 and Figure 18 show the plots of $(\alpha h\nu)^2$ versus the photon energy $h\nu$ and the plots of Ln(α) versus the photon energy $h\nu$ respectively. They lead to the determination of gap energy values as well as the Urbach energy values as a function of Co and Cr ratios which are listed in Table 16.

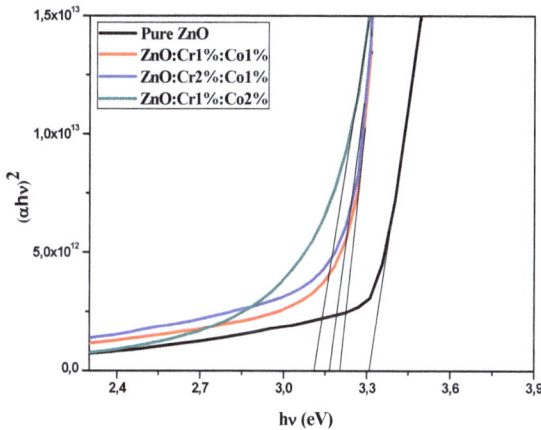

Figure 17. Plots of $(\alpha h\nu)^2$ versus photon energy $h\nu$.

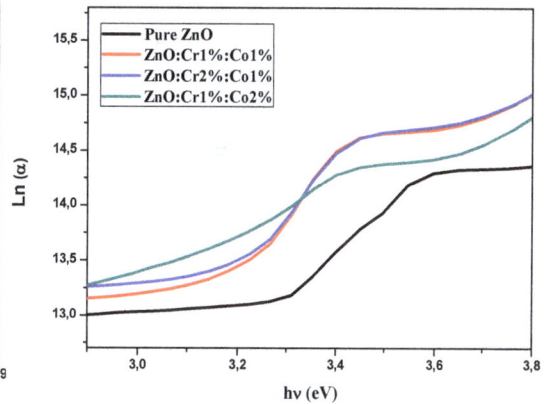

Figure 18. Plots of ln(α) versus $h\nu$.

On the one hand, we observed a decrease of the optical band gap from 3.3 eV to 3.12 eV with Co and Cr content in ZnO thin films. Similar results have also been reported for Cr contents doped ZnO nanostructures [35]. This band gap narrowing of transition metal doped II-IV semiconductor can be best interpreted in terms of sp–d spin exchange interactions between the conduction band electrons and the localized electrons of transition metals ions Co^{2+} and Cr^{3+}. This red shift of band-gap energy into co doped ZnO thin films has already been denoted by Liu *et al.* for (Cu, Co) codoped ZnO thin films prepared by RF magnetron sputtering [32]. On the other hand, E_U increases from 89 to 186 meV with Co and Cr contents. The increase in the amount of Urbach energies indicates the poor crystallinity of the samples. This means that there are a large number of defects in the film structure. Sayed *et al.* indicated similar results [35]. This behavior corresponds primarily to optical transitions between occupied states in the valence band tail to unoccupied states at the conduction band edge due to the density of defects and donor levels due to the interstitial Zn, Co and Cr atoms.

Table 16. Calculated values of Band gap energy Eg and Urbach energy EU.

Sample	Eg (eV)	EU (meV)
Pure ZnO	3. 30	79
ZnO:Cr1%:Co1%	3. 22	107
ZnO:Cr2%:Co1%	3. 17	155
ZnO:Cr1%:Co2%	3. 12	186

The Photoluminescence properties of the samples were studied under 325 nm excitation wavelengths at room temperature. Figure 19. displays two emission peaks; the first is assigned to the near band-edge emission (NBE) at 387 nm which is originated from recombination of free excitons from conduction band (CB) to valence band (VB). It has been emphasized that both free exciton and phonon-assisted emission may contribute to Room Temperature near-band-edge emission (NBE) [36]. Furthermore, this emission decreases with (Cr, Co) codoping and reaches a minimum for ZnO:Cr1%:Co2% film which has the lowest band gap energy value. Thus, the density of free exciton in ZnO nanofilms affects the intensity of ultraviolet emission [37]. The gradual decrease of PL intensities with increasing Cr and Co contents in ZnO implies the trapping of photo-generated electrons by Co^{2+} and Cr^{3+} ions, then an enhancement of charge separation in the ZnO thin films. A similar effect has also been observed previously for (Na, Mg) codoped ZnO thin films prepared by spin coating technique [32]. The second peak corresponds to the visible band emission which is mainly related to the intrinsic defects of ZnO lattice such as oxygen vacancies (V_O), interstitial oxygen (O_i), antisite oxygen (O_{Zn}), zinc vacancies (V_{Zn}), interstitial zinc (Zn_i), and antisite zinc (Zn_O) [19]. Furthermore, it can be generally attributed to electron transfer from the conductor band to defect levels (or impurity levels) in the forbidden band or from defect levels to the valence band or between different defect levels (or impurity levels). This result implies that this material can be effectively excited by the energy below the band gap. With codoping by Cr and Co elements, the visible emission intensity decreases which is assumed to be a result of possible reduction of intrinsic defects. However, ZnO:Cr1%:Co1% film has the highest visible intensity which is probably attributed to high amount of oxygen vacancies V_O [38] as well as extrinsic impurities or structural defaults [39].

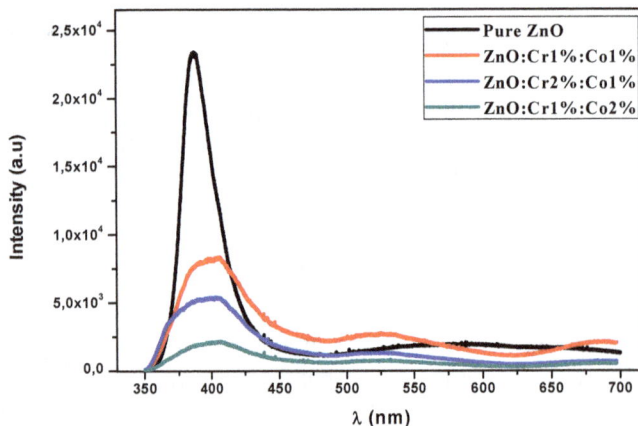

Figure 19.
PL spectra of (Cr, Co) codoped ZnO thin films.

3.2. Optical and Photoluminescence characterization results of (Co, In) codoped ZnO nanofilms

According to the optical transmission and reflectivity spectra of (Co, In) codoped ZnO nanofilms given in Figure 20 and Figure 21 respectively, it is obvious that the codoped nanofilms show a high transparency into the visible range with an average transmittance variable between 75% and 85%. Further, the spectra depict a reflectance less than 28%, therefore their absorption is quasi futile. This behavior may be attributed to the codoping effect and the variation of films thickness (as shown by SEM observations). The decrease in optical transmission with (Co, In) codoping can be associated with the scattering losses phenomenon at grain boundaries. In addition, the observed shoulders in UV region of transmission and reflectivity spectra, suggested probably the presence of defaults and impurities inside the films. Moreover, the reflectance is on the order of 10% into the IR region which indicates that (Co, In) codoped ZnO thin films appear exempt of free charges which may be due to the incorporation of Co and In elements in the ZnO matrix. As a consequence, the codoped thin films are anti-reflective.

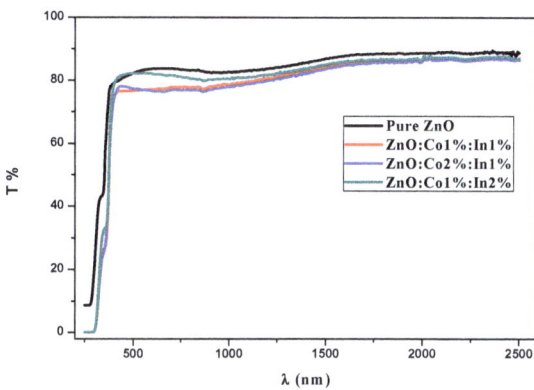

Figure 20. Transmission spectra of (Co, In) codoped ZnO thin films.

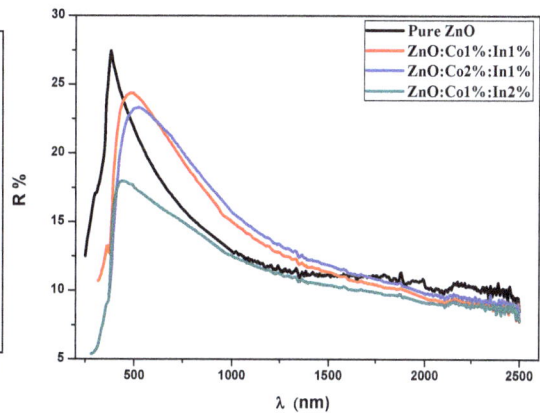

Figure 21. Reflectivity spectra of (Co, In) codoped ZnO thin films.

Figure 22 and Figure 23 show the plots of $(\alpha h\nu)^2$ versus the photon energy and the plots of $Ln(\alpha)$ versus the photon energy respectively. They lead to the determination of gap energy values as well as the Urbach energy values as a function of Co and In doping elements which are listed in Table 17.

On the one hand, the optical band gap decreases from 3.3 eV to 3.12 eV with codoping by Co and In elements. This band gap narrowing may be explained by sp–d spin exchange interactions between electrons in conduction band and the localized d electrons of In^{3+} and Co^{2+} ions as well as the increase in carrier concentration by codoping. The band-gap Energy narrowing of (Cu, Co) codoped ZnO thin films prepared by RF magnetron sputtering has previously been pointed out by Islam *et al.* [40]. On the other hand, the Urbach energy increases from 89 to 186 meV by (Co, In) codoping. This increase of Urbach energy values indicates that there is a large density of defects in (Co, In) codoped thin films attributable to the donor levels due to the interstitial Zn, In and Co atoms as well as the transitions between the occupied states in the valence band tail and the unoccupied states at the conduction band edge.

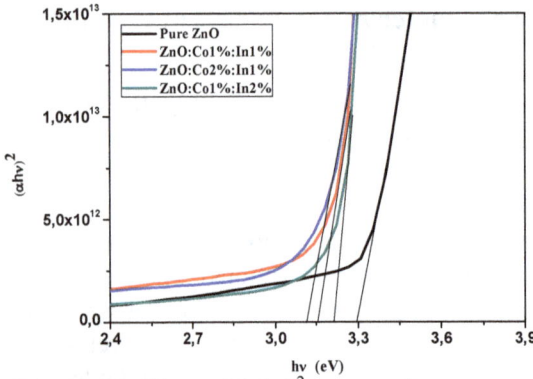

Figure 22. Plots of $(\alpha h\nu)^2$ versus photon energy hν.

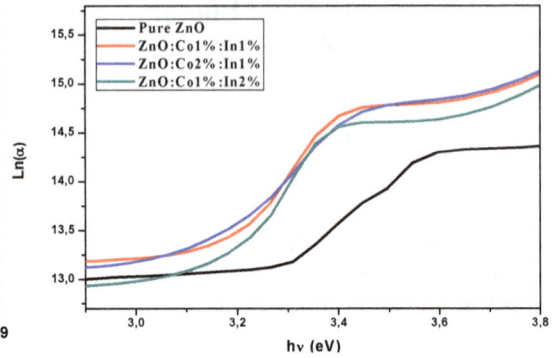

Figure 23. Plots of Ln(α) versus hν.

Table 17. Calculated values of Band gap energy Eg and Urbach energy EU.

Sample	Eg (eV)	EU (meV)
Pure ZnO	3.30	79
ZnO:Co 1%: In 1%	3.16	176
ZnO:Co 2%: In 1%	3.12	186
ZnO:Co 1%: In 2%	3.22	109

Figure 24 displays two emission peaks; the first is assigned to the near band-edge emission (NBE) at 387 nm. The NBE emission originated from recombination of free excitons from conduction band (CB) to valence band (VB). It has been pointed out that both free exciton and phonon-assisted emission may contribute to Room Temperature near-band-edge emission (NBE) [36]. Furthermore, this emission decreases with (Co, In) codoping and reaches a minimum for ZnO:Co2%:In1% film which has the lowest band gap energy value. The intensity of a PL spectrum is useful for estimating the rate of the electron-hole recombination, *i.e.* a lower rate of radiative electron-hole recombination leads to a weak intense spectrum [37]. The gradual decrease of PL intensities with increasing Co and In contents in ZnO implies the trapping of photo-generated electrons by Co^{2+} and In^{3+} ions, then an enhancement of charge separation in the ZnO thin films. A similar effect has also been observed previously for (Na, Mg) codoped ZnO thin films [32]. The second peak corresponds to the visible band emission which is mainly related to the intrinsic defects of ZnO lattice such as oxygen vacancies (V_O), interstitial oxygen (O_i), antisite oxygen (O_{Zn}), zinc vacancies (V_{Zn}), interstitial zinc (Zn_i), and antisite zinc (Zn_O) [19]. With codoping by Co

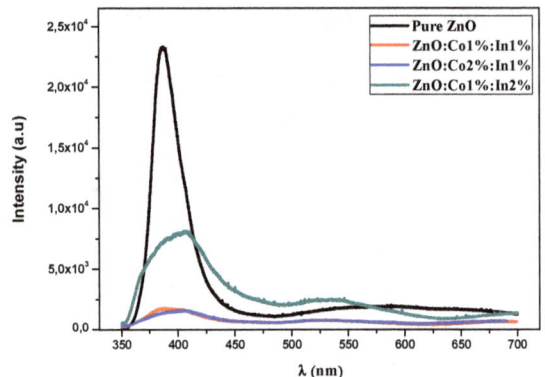

Figure 24. PL spectra of (Co, In) codoped ZnO thin films.

and In elements, the visible emission intensity decreases which is assumed to be a result of possible reduction of intrinsic defects. However, ZnO:Co1%:In2% film has the highest visible intensity which is probably attributed to high amount of oxygen vacancies V_O [38] as well as extrinsic impurities or structural defaults [39].

3.3. Optical and Photoluminescence characterization results of (Cr, In) ZnO nanofilms

From the optical transmission and reflectivity spectra of (Cr, In) ZnO nanofilms presented in Figure 25 and Figure 26 respectively, it is clear that the codoped nanofilms exhibit a high transparency into the visible range with an average transmittance variable between 75% and 87%. Similar results were previously reported by Hadri *et al.* [41]. Moreover, they depict a reflectance less than 33%, thus their absorption is insignificant. This may be due to the codoping effect as well as the films thickness variation (as shown by SEM observations). With (Cr, In) codoping, the optical transmission decreases slightly which can be attributed to the scattering losses phenomenon at grain boundaries. Within the IR region, the reflectance is on the order of 10%. This points out that even by (Cr, In) codoping, the nanofilms are exempt of free charges which can be assigned to the incorporation of Cr and In elements in the ZnO host matrix. Whence, the studied nanofilms are anti-reflective.

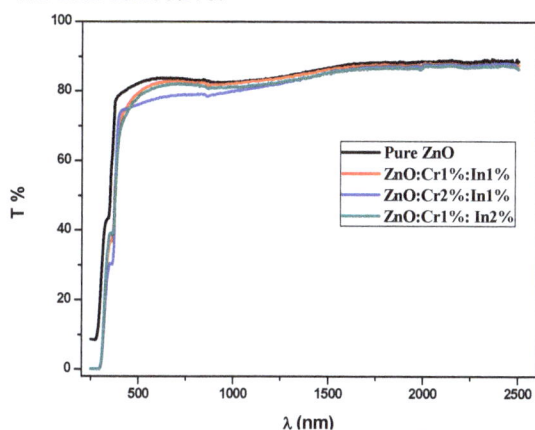

Figure 25. Transmission spectra of (Cr, In) codoped ZnO thin films.

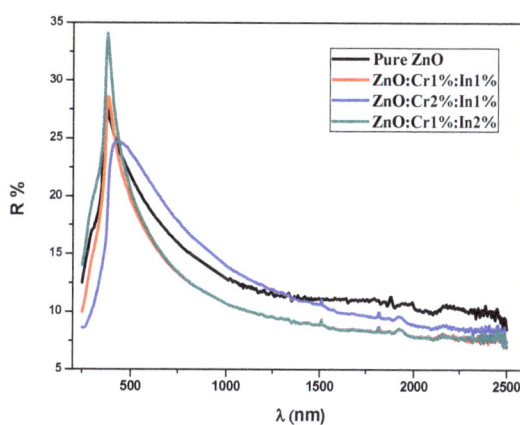

Figure 26. Reflectivity spectra of (Cr, In) codoped ZnO thin films.

Figure 27 and Figure 28 show the plots of $(\alpha h v)^2$ versus the photon energy and the plots of $Ln(\alpha)$ versus the photon energy respectively. They lead to the determination of gap energy values as well as the Urbach energy values which are listed in Table 18.

On the one hand, with codoping by Cr and In elements, the optical band gap energy values decrease from 3.3 eV to 3.10 eV. The band gap narrowing is probably explained in terms of the increase in carrier concentration by codoping as well as sp–d spin exchange interactions between the localized d electrons of In^{3+} and Cr^{3+} ions and electrons in conduction band. The band-gap red shift of codoped ZnO nanofilms prepared by RF magnetron sputtering has previously been reported by Islam *et al.* [40]. On the other hand, by (Cr, In) codoping, the Urbach energy values increase from 89 to 210 meV. This broadening of Urbach energy values denotes that there is a large density of defects in (Cr,

In) codoped nanofilms assigned to the donor levels (interstitial Zn, In and Cr atoms) and the transitions between the occupied states in the valence band tail and the unoccupied states at the conduction band edge.

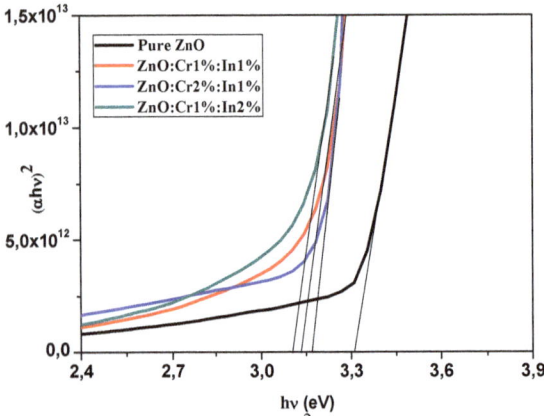

Figure 27. Plots of $(\alpha h\nu)^2$ versus photon energy $h\nu$.

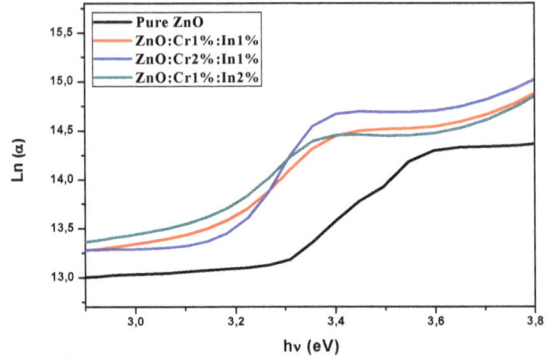

Figure 28. Plots of $\ln(\alpha)$ versus $h\nu$.

Table 18. Calculated values of Band gap energy Eg and Urbach energy EU.

Sample	Eg (eV)	EU (meV)
Pure ZnO	3.30	79
ZnO:Cr1%: In1%	3.13	195
ZnO:Cr2%:In1%	3.17	155
ZnO:Cr1%:In2%	3.10	210

Figure 29 exhibits two emission peaks; the first one is attributed to the near band-edge emission (NBE) towards 387 nm which is generated by recombination of free excitons from conduction band (CB) to valence band (VB). It has been affirmed that both free exciton and phonon-assisted emission may contribute to Room Temperature near-band-edge emission (NBE) [36]. Moreover, the gradual decrease of the (NBE) intensity with Cr and In ratio in ZnO involves the trapping of photo-generated electrons by In^{3+} and Cr^{3+} ions, whence, an enhancement of charge separation in the ZnO nanofilms especially for ZnO:Cr1%:In2% nanofilm which has the minimum emission intensity and the lowest band gap energy value. A similar effect has also been observed previously for (Na, Mg) codoped ZnO thin films [32]. Thus, the density of free exciton in ZnO nanofilms affects the intensity of ultraviolet emission [37]. The decrease and decline of the PL emission is assigned to defects which act as electron traps created by Cr and In dopants having low energy levels to absorb emissions generated from higher energy levels. Whence, the recombination of photo-generated electrons and holes is indeed suppressed by codoping which is one of the reasons to expect the photocatalytic activity enhancement of the (Cr, In) codoped ZnO samples.

The second visible band is assigned to deep-level emission which is mainly related to the intrinsic defects of ZnO lattice such as oxygen vacancies (V_O), interstitial oxygen (O_i), antisite oxygen (O_{Zn}), zinc vacancies (V_{Zn}), interstitial zinc (Zn_i), and antisite zinc (Zn_O) [19]. By (Cr, In) codoping, a decrease in the visible emission intensity is observed which is supposed to be a consequence of prospective diminution of intrinsic defects. Nevertheless, ZnO:Cr1%:In1% nanofilm has the uppermost visible intensity which is assigned to extrinsic impurities or structural defaults [39]such as high amount of oxygen vacancies V_O [38].

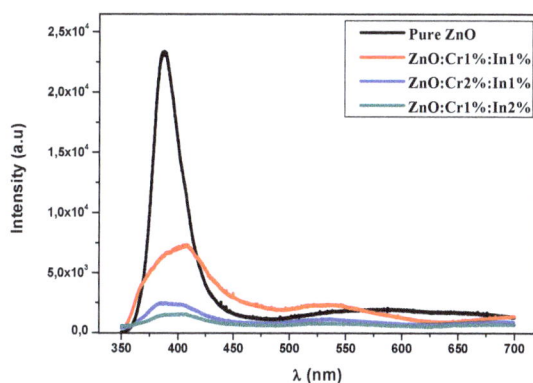

Figure 29. PL spectra of (Cr, In) codoped ZnO thin films.

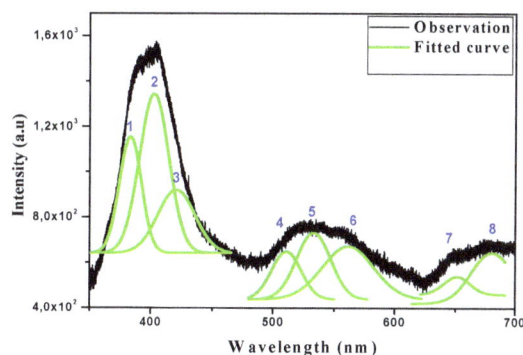

Figure 30. Gaussian deconvolution of the PL spectrum of the ZnO:Cr2%:In1% nanofilm.

Furthermore, the PL spectra can be Gaussian deconvoluted to the subpeaks according to their origination [19]. The deconvoluted PL spectra are shown in Figure 30 together with the subpeak contents, their corresponding positions and energies are summarized in Table 19. The UV emission peaks at382 nm (3.25 eV) nm results from the free excitons recombination (donor-acceptor pair combination). The first intense visible peak corresponds to the violet emission located at 405 nm (3.06 eV) deemed to be connected to the electron transition from the bottom of the conduction band to the zinc vacancies (V_{Zn}) level. While, is believed that the blue-violet emission toward 425 nm (2.92eV) originates from the energy interval from the donor level of Zni to the top of the valence band [42, 43]. The green luminescence bands at 510-565 nm (2.43-2.2eV) can be attributed to the electron transition from deep oxygen vacancies level to the top of valance band [44]. Whereas, the red emissions at 650-680 nm (1.90-1.83eV) are related to the energy intervals between the top of valence band and Oi levels [45].

Table 19. Positions of emission peaks of the ZnO:Cr1%:In2% nanofilm

Peaks	1	2	3	4	5	6	7	8
Position (nm)	382	405	425	510	533	565	650	680
Energy (eV)	3. 25	3. 06	2. 92	2. 43	2. 33	2.2	1. 91	1. 83

4. Photocatalytic activity measurement

The codoped ZnO nanofilms with different codopant contents and with a dimension of 2 cm^2 were placed in a quartz beaker containing 20 mL of 3 mmolL^{-1} of Methylene Blue. This dye is used as a model of organic pollutant which is known as a typical pollutant in textile industry and a probe for studying the interfacial electron transfer in semiconductor systems [46]. The quartz cell was covered by an extremely thin transparent plastic sheet in order to avoid the evaporation of dye solution under sunlight irradiation. Prior to photocatalytic reaction, the mixture of the catalyst and the solution were kept in darkness for 30 min until reaching adsorption/desorption equilibrium between the dye molecule and the catalyst on the surface of the nanofilm. The experiments were carried out under sunlight irradiation at a temperature of about 25°C. The photocatalytic oxidation process is based on degradation and destraction of organic pollutants in the presence of semiconductor photocatalysts (ZnO), and energetic light source (sunlight). The interaction of the semiconductor with light results in the generation of electron-hole pairs in the semiconductor [47]. The incoming photons having energies larger than the band gap one of the photocatalyst are absorbed. The electrons situated in the valence band are excited toward the conduction band of the semiconductor. They leave behind an equal number of holes. The photogenerated electron-hole pairs are separated from each other. Then, they migrate to the active sites at the interface between the semiconductor and the liquid. Also, they react with the adsorbed species (Figure 31). Moreover, the transfer of the photo-induced electrons to adsorbed organic or inorganic species is achieved through migration of holes and electrons to the semiconductor surface. When electrons from the semiconductor are given to reduce acceptor electrons, the holes migrate to the surface. There, a donor electron is able to combine with the surface hole oxidizing the donor species.

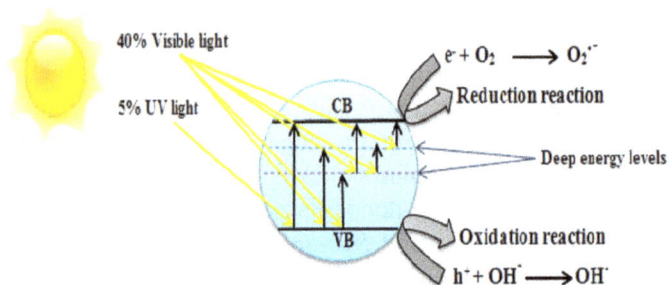

Figure 31. Schematic diagram illustrating the photocatalysis principle.

4.1. Photocatalytic activity Evaluation of(Cr, Co) codoped ZnO nanofilms

The photocatalytic performance was evaluated by the Methylene Blue photodegradation by the synthesized (Cr, Co) codoped ZnO nanofilms, under sunlight irradiation. Figure 32. exhibits the change of absorbance spectra of MB aqueous solutions in the absence and in the presence of different photocatalysts under sunlight irradiation for different durations. The visible absorbance spectrum of MB aqueous solution displays two characteristic peaks at 610 and 665 nm. These peaks are attributed to the absorbance of n π^* transitions[48]. The photodegradation was monitored by the decay of the MB absorption as a function of irradiation time. It is clear that the MB was stable under sunlight irradiation in the absence of the catalyst and its photodegradation alone is insignificant while, with the assistance of (Cr, Co) codoped ZnO nanofilms under sunlight the MB characteristics absorption decreases speedily with the exposure time extension and it disappears almost completely after 180 min for ZnO:Cr1%:Co2% nanofilm.

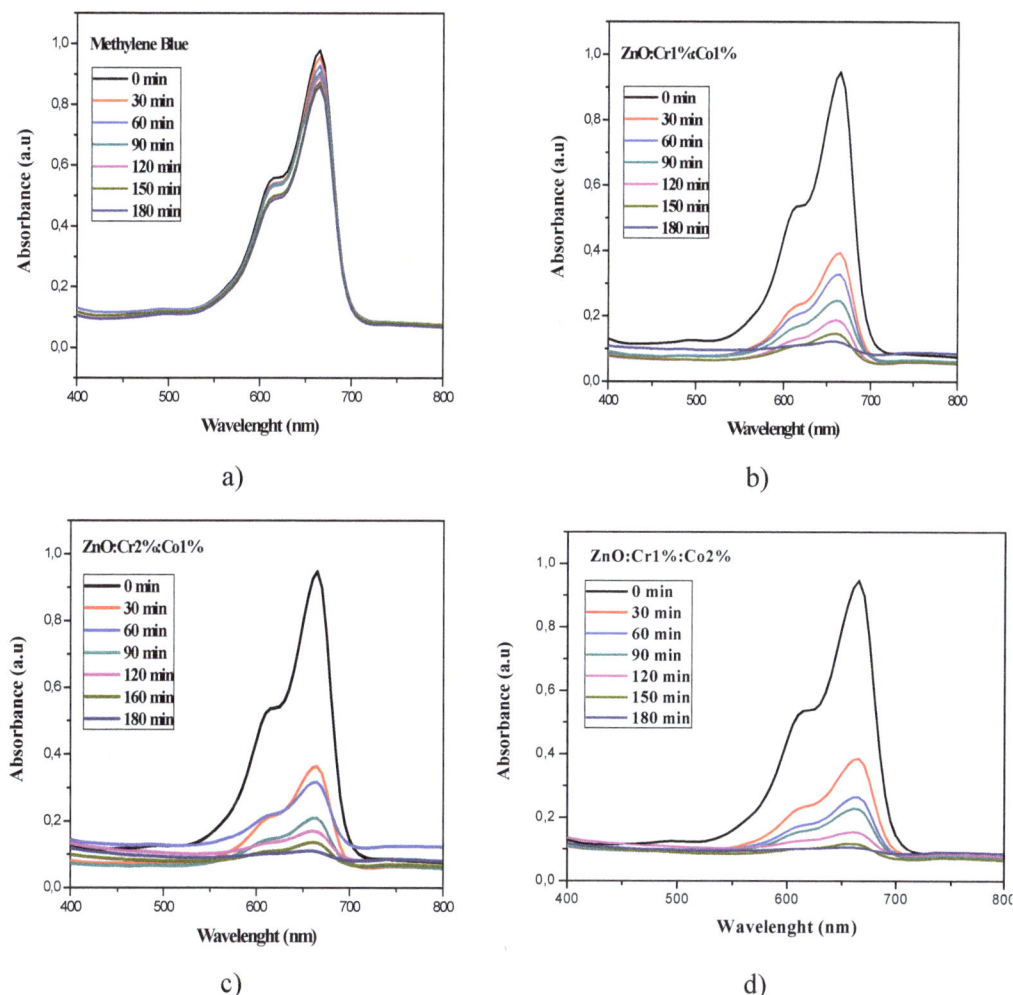

Figure 32. Time-dependent absorption spectra of MB solution under sunlight irradiation without and in the presence of (Cr, Co) codoped ZnO nanofilms.

From, Figure 33 it is noted that the MB decolorization occurs in the presence of the codoped ZnO nanofilms after 3h. Hence, the (Cr, Co) codoping induces an improvement in the photodegradation efficiency.

The logarithmic plots (Figure 34) show that the first-order reaction is taking place during the photodegradation of MB molecules. The kinetic rate constants were determined from the slope of fitted curves, and their calculated values are listed in Table 20.

The first-order degradation rate constants for undoped ZnO and (Cr, Co) codoped ZnO are found to be $k_{ZnO} = 6. 5×0^{-3}min^{-1}$ and $k_{ZnO:Cr1\%:Co2\%} = 11. 5×10^{-3}min^{-1}$, respectively. So, the degradation rate of MB by (Cr, Co) codoped samples is faster than by undoped ZnO sample.

From Figure 35, it is obvious that the degradation of MB molecules was about 90% for (Cr, Co) codoped ZnO nanofilms after irradiation for 180 min, while the MB photodegradation rate by pure ZnO nanofilm was 17% less for the same time of irradiation. Whence, the (Cr, Co) codoping improves the photocatalytic activity of the nanofilms because it increases their specific surface area. Likewise, it leads to a narrow optical band gap. Further, the large amount of impurities (dopants: Cr and Co) and defects as well as oxygen vacancies in ZnO lattice act as trapping centers for the photogenerated electrons, and inhibit the recombination with holes [49], thus, fostering the interfacial charge transfer.

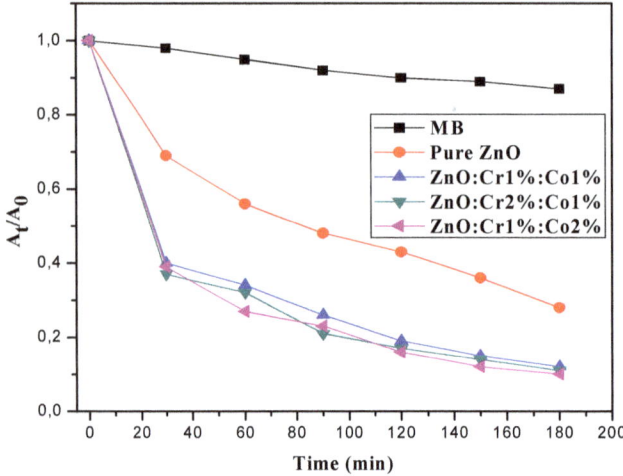

Figure 33. Photocatalytic degradation of MB dye for (Cr, Co) codoped ZnO nanofilms.

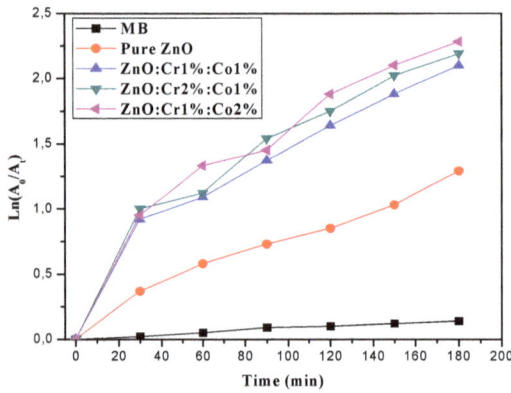

Figure 34. Photocatalytic degradation kinetics of MB dye for (Cr, Co) codoped ZnO nanofilms.

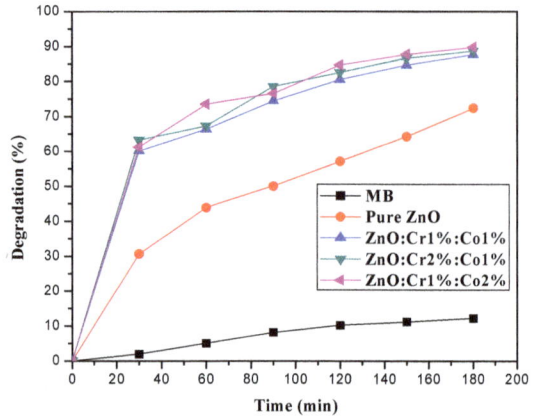

Figure 35. Photocatalytic degradation efficiency of MB dye for (Cr, Co) codoped ZnO nanofilms.

Table 20. Kinetic parameter of (Cr, Co) codoped ZnO nanofilms for MB dye.

Sample	Rate constant k ($10^{-3}.min^{-1}$)
Methylene blue	0.8
Pure ZnO	6.5
ZnO:Cr1%:Co1%	10.4
ZnO:Cr2%:Co1%	11
ZnO:Cr1%:Co2%	11.5

4.2. Photocatalytic activity Evaluation of (Co, In) codoped ZnO nanofilms

Figure 36 shows the change of absorbance spectra of MB aqueous solutions in the absence and in the presence of different synthesized (Co, In) codoped ZnO nanofilms as photocatalysts under sunlight irradiation for different durations. The photodegradation was controlled by the decay of the MB absorption as a function of irradiation time. It is clear that the MB was stable under sunlight irradiation in the absence of the catalyst and its photodegradation alone is insignificant. Although, with the assistance of (Co, In) codoped ZnO nanofilms under sunlight, the MB characteristics absorption decreases speedily with the exposure time extension and it disappears completely after 180 min for ZnO:Co2%:In1% and ZnO:Co1%:In2% nanofilms.

The relative change of the MB concentration controls the photocatalytic activity. From, Figure 37, it is noted that the MB decolorization occurs in the presence of the codoped ZnO nanofilms after 3h.

The logarithmic plots (Figure 38) point out that the first-order reaction is taking place during the photodegradation of MB molecules. The kinetic rate constants were determined from the slope of fitted curves, and their calculated values are listed in Table 21. The first-order degradation rate constants for undoped ZnO and (Co, In) codoped ZnO are found to be $k_{ZnO} = 6.5 \times 10^{-3} min^{-1}$ and $k_{ZnO:Co2\%:In1\%} = 14.2 \times 10^{-3} min^{-1}$, respectively. So, the degradation rate of MB by (Co, In) codoped samples is very fast compared to undoped ZnO nanofilm.

From Figure 39, it is obvious that the degradation of MB molecules was almost entire (94%) for (Co, In) codoped ZnO nanofilms after irradiation for 180 min, while the MB photodegradation rate by pure ZnO nanofilm was 22% less for the same time of irradiation. Whence, the (Co, In) codoping improves the photocatalytic activity of the nanofilms because it increases their specific surface area and their roughness. Likewise, it leads to a narrow optical band gap. Further, the large amount of impurities (dopants: Co and In) and defects as well as oxygen vacancies in ZnO lattice act as trapping centers for the photogenerated electrons, and inhibit the recombination with holes [49], thus, fostering the interfacial charge transfer.

Table 21. Kinetic parameter of (Co, In) codoped ZnO nanofilms for MB dye.

Sample	Rate constant k ($10^{-3}min^{-1}$)
Methylene blue	0.8
Pure ZnO	6.5
ZnO:Co1%:In1%	12.1
ZnO:Co2%:In1%	14.2
ZnO:Co1%:In2%	11.9

a)

b)

c)

d)

e)

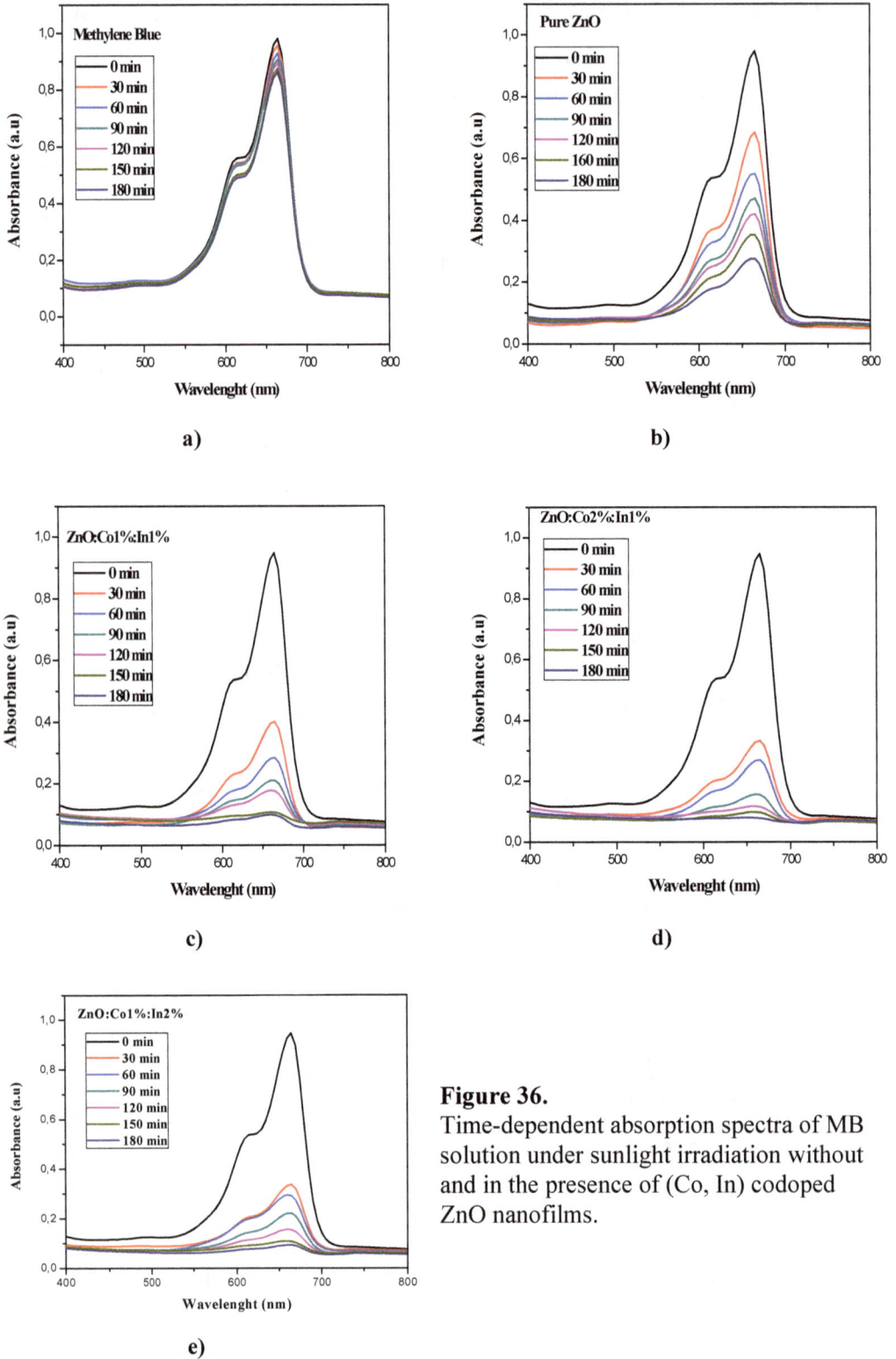

Figure 36.
Time-dependent absorption spectra of MB solution under sunlight irradiation without and in the presence of (Co, In) codoped ZnO nanofilms.

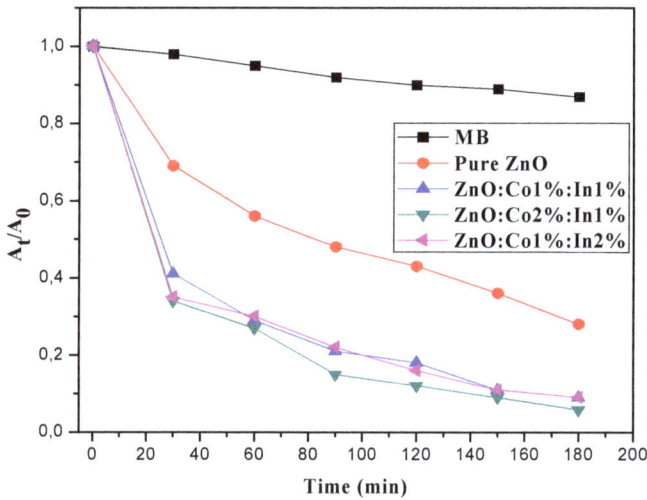

Figure 37. Photocatalytic degradation of MB dye for (Co, In) codoped ZnO nanofilms.

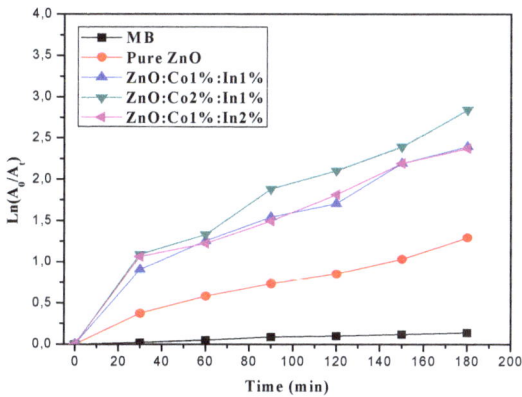

Figure 38. Photocatalytic degradation kinetics of MB dye for (Co, In) codoped ZnO nanofilms.

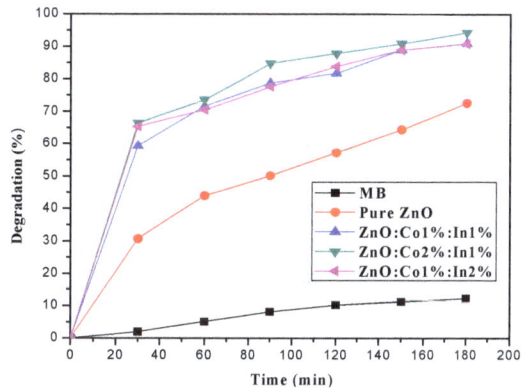

Figure 39. Photocatalytic degradation efficiency of MB dye for (Co, In) codoped ZnO nanofilms.

4.3 Photocatalytic activity Evaluation of (Cr, In) codoped ZnO nanofilms

The photocatalytic performance was evaluated by the Methylene Blue photodegradation by the synthesized (Cr, In) codoped ZnO nanofilms, under sunlight irradiation. Figure 40 denotes the change of absorbance spectra of MB aqueous solutions in the absence and in the presence of different photocatalysts under sunlight irradiation for different durations. The photodegradation was controlled by the decay of the MB absorption as a function of irradiation time. When the MB solution is irradiated for 180 min, its degradation is very small (8%). So, it is pointed out that the organic dye was difficult to decompose in the absence of the catalyst. Whereas, the (Cr, In) codoped ZnO nanofilms are more photoactive than the undoped ZnO samples for MB photodegradation and its characteristics absorption decrease speedily with the exposure time extension and disappear completely after 180 min for ZnO:Cr1%:In2% nanofilm.

a)

b)

c)

d)

e)

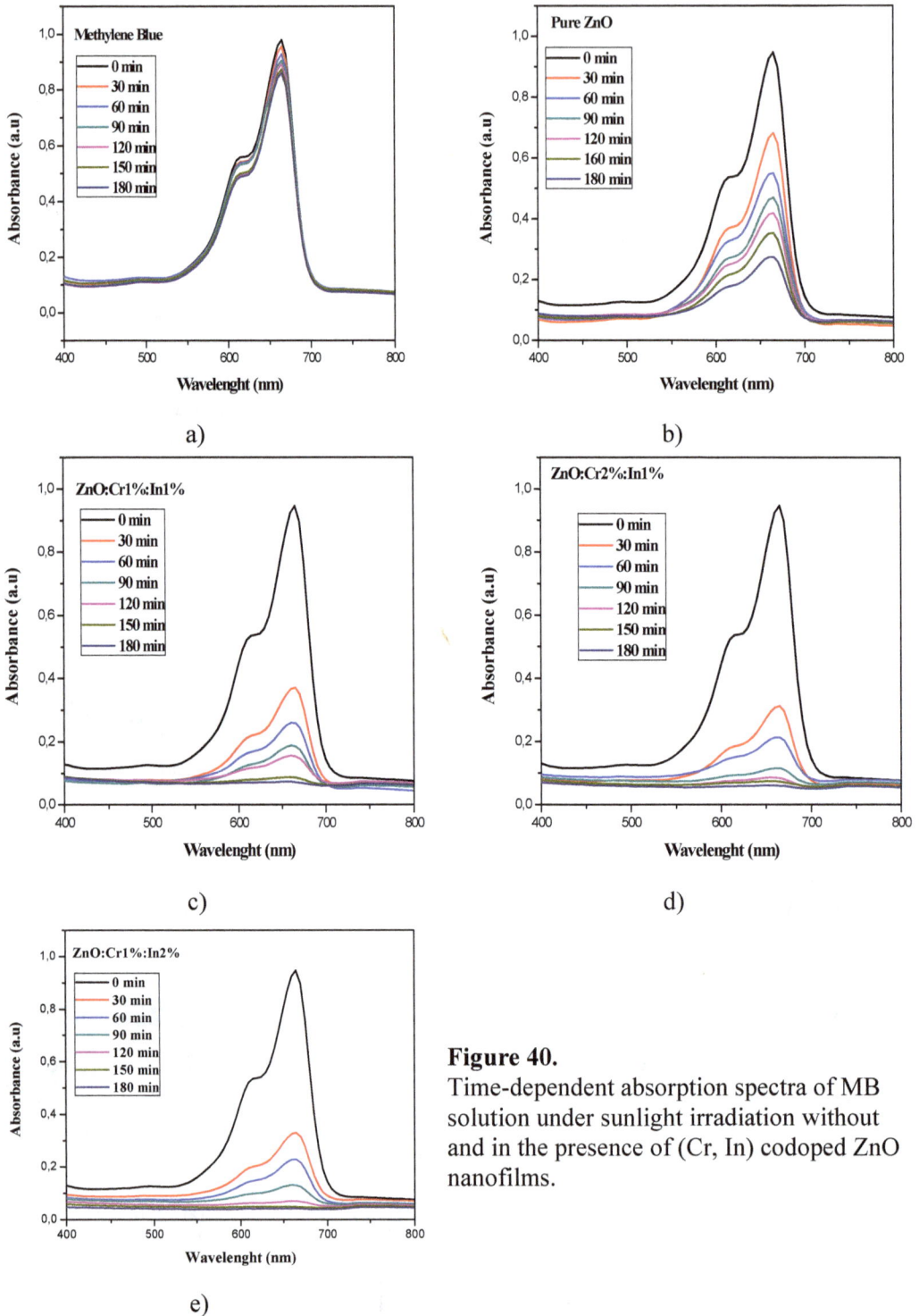

Figure 40.
Time-dependent absorption spectra of MB solution under sunlight irradiation without and in the presence of (Cr, In) codoped ZnO nanofilms.

From Figure 41, it is noted that the MB de-colorization occurs in the presence of the (Cr, In) codoped ZnO nanofilms after 3h.

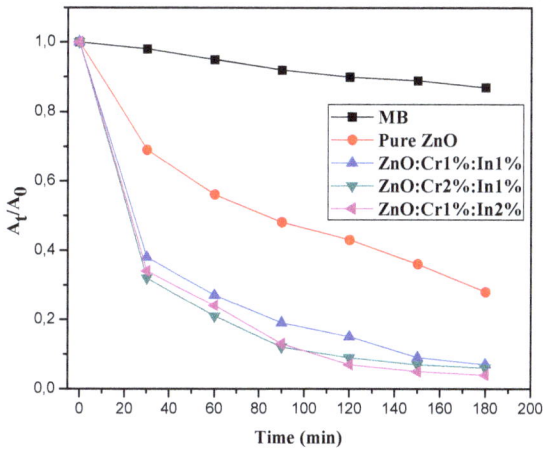

Figure 41.
Photocatalytic degradation of MB dye for (Cr, In) codoped ZnO nanofilms.

The logarithmic plots (Figure 42) indicate that the first-order reaction is taking place during the photodegradation of MB molecules. The kinetic rate constants were determined from the slope of fitted curves, and their calculated values are listed in Table 22. The first-order degradation rate constants for undoped ZnO and (Cr, In) codoped ZnO are found to be $k_{ZnO} = 6.5 \times 10^{-3} min^{-1}$ and $k_{ZnO:Cr1\%:In2\%} = 17.4 \times 10^{-3} min^{-1}$, respectively. So, the degradation rate of MB by (Cr, In) codoped samples is faster than by undoped ZnO samples which presents slow kinetics under sunlight irradiation.

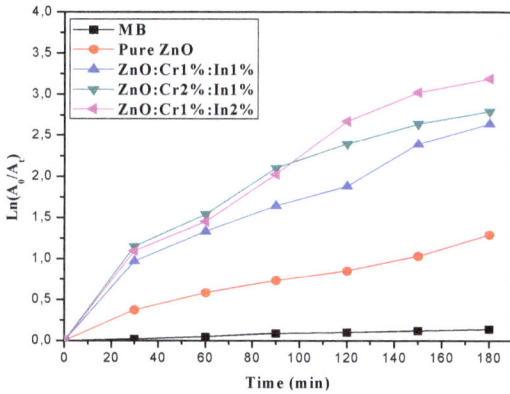

Figure 42. Photocatalytic degradation kinetics of MB dye for (Cr, In) codoped ZnO nanofilms.

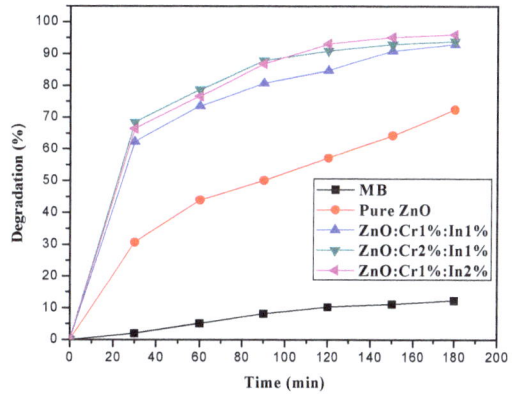

Figure 43. Photocatalytic degradation efficiency of MB dye for (Cr, In) codoped ZnO nanofilms.

Table 22. Kinetic parameter of (Cr, In) codoped ZnO nanofilms for MB dye.

Sample	Rate constant k ($10^{-3} min^{-1}$)
Methylene blue	0.8
Pure ZnO	6.5
ZnO:Cr1%:In1%	13.5
ZnO:Cr2%:In1%	14.5
ZnO:Cr1%:In2%	17.4

From Figure 43, it is obvious that the degradation of MB molecules was almost complete (around 95%) for (Cr, In) codoped ZnO nanofilms after irradiation for 180 min as compared to pure ZnO nanofilm. Whence, the enhancement in photocatalytic activity for (Cr, In) codoped ZnO samples is essentially assigned to the narrowing of its optical band gap. Further, the large amount of impurities (dopants: Cr and In) and defects as well as oxygen vacancies in ZnO lattice act as trapping centers for the photogenerated electrons, and inhibit the recombination with holes [49], thus, fostering the interfacial charge transfer. Likewise, the increase of the specific surface area and the roughness of the nanofilms improve its photocatalytic activity. Moreover, Dong *et al.* have been determined high photodegradation rate of MB using ZnO:Fe array films after 4h under sunlight irradiation [50].

Furthermore, the cycling runs for the MB photodegradation with (Cr, In) codoped ZnO as photocatalyst were carried out to assess its photocatalytic recyclability, durability and stability, Figure 44. After reusing for three cycles, the photocatalytic activity is still approximately the same. This result suggests that the (Cr, In) codoped ZnO catalysts were stable during the degradation process of the MB organic dye. Thereby, the photocatalytic activity amelioration and the stability of the (Cr, In) codoped ZnO samples improve the MB degradation and the removing process of the organic dye existing in the aqueous solution.

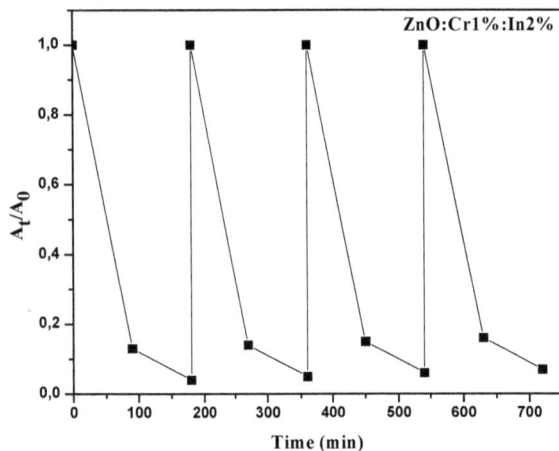

Figure 44.
Cycling runs of the photocatalytic degradation of MB for ZnO:Cr1%:In2% nanofilm under sunlight irradiation.

4.4. Possible Photocatalytic Mechanism

Under sunlight irradiation, there is a generation of electron-hole pairs: the photogenerated electrons are excited from the valence band to the conduction band or from impurity levels to the conduction band or between different defect levels that create holes in the valence band, this phenomena have been obtained by S. Girish Kumar on ZnO and others semiconductor compounds [51]. The MB adsorption occurs through the coulombic interaction since the organic dye has cationic configuration and OH^- ions present on the surface of the catalyst. The generated holes react with OH^- ions and the trapped electrons react with the O_2 absorbed molecule to create the OH^- radicals and the superoxide anion radicals $(O_2^{\cdot-})$ respectively in the aqueous solution [52]. The formed $(O_2^{\cdot-})$ and (OH^-) radicals attack the $C-S^+=C$ functional group of the MB dye which is attached through coulombic interaction on the surface of catalysts [53] and degrade this organic dye into H_2O and CO_2 and other no toxic materials compounds and products such as: NO_3^-, SO_4^{2-} and H^+.

The mechanism of photocatalytic degradation for codoped ZnO nanofilms is given as follows [54, 55]:

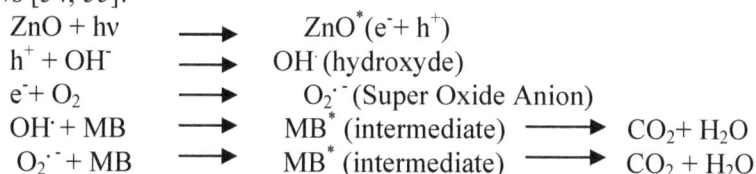

$$ZnO + h\nu \longrightarrow ZnO^*(e^- + h^+)$$
$$h^+ + OH^- \longrightarrow OH^. \text{ (hydroxyde)}$$
$$e^- + O_2 \longrightarrow O_2^{.-} \text{ (Super Oxide Anion)}$$
$$OH^. + MB \longrightarrow MB^* \text{ (intermediate)} \longrightarrow CO_2 + H_2O$$
$$O_2^{.-} + MB \longrightarrow MB^* \text{ (intermediate)} \longrightarrow CO_2 + H_2O$$

where e^- and h^+ are theexcited electrons and the created holes, respectively.

5. Wettability measurement in terms of water contact angle (CA)

Water droplets were deposited on the surface of all studied thin films. The wettability investigations were achieved using Drop Shape Analysis DSA 100 contact angle measuring system at ambient temperature. A water drop with a controlled volume (7µL) was automatically injected through a syringe. Then the droplet was deposited on the surface of the studied nanofilm. During the evaluation, mechanical vibration and air flow were avoided. The altitude of the sample Table was fixed in order to place the water droplet between digital camera and the illumination. High-speed image capture visualizes the shapes of drops which can be analyzed along with numerical integration of the Young–Dupre equation to obtain the best-fit curve. Hence, the contact angle of liquid solid systems is quickly and easily measured.

This contact angle is a consequence of thermodynamic equilibrium of the free energy at the solid-liquid-vapor interphase, it is given by the Young–Dupre relationship [56]:

$$\cos\theta = \frac{\gamma_{SV} - \gamma_{SL}}{\gamma_{LV}}$$

Where γ_{LV}, γ_{SV} are γ_{SL} are assigned to the interfacial tensions with liquid-vapor, solid-vapor and solid-liquid respectively. From Dupre's equation, we can obtain the solid-liquid work of adhesion W_{SL} [57].

$$W_{SL} = \gamma_{SV} + \gamma_{LV} - \gamma_{SL} = \gamma_{LV}(1 + \cos\theta)$$

This indicates that the force of adhesion is directly related to the contact angle between liquid and solid. Similarly, the surface Free Energy byEquation of States model is derived from Dupre's equation[58]:

We have $\quad F(\gamma_S, \gamma_L, \gamma_{SL}) = 0$ Or $\gamma_{SL} = f(\gamma_S, \gamma_L)$

Whence $\quad \gamma_{SL} = \gamma_S + \gamma_L - 2(\gamma_S\gamma_L)^{0.5}$

5.1. Wettability properties of (Cr, Co) codoped ZnO nanofilms

Figure 45 illustrates the shape of water droplet onto different samples, showing the obvious distinction between pure ZnO and codoped ZnO thin films. It is found that the nanofilms present a remarkable hydrophilic character with a CA < 90° except for ZnO:Cr2%:Co1% which shows a hydrophobic character with a CA > 90°. This result indicates that (Cr, Co) codoping can improve the wettability of surfaces which depends on modification of surface morphology and roughness, the shape and the size as well as the chemical composition. The correlation between AFM observations as well as wettability note that when the roughness decreases, the hydrophilic character of the surfaces is enhanced especially for ZnO:Cr1%:Co2% film which presents the lowest contact angle equal to 57°. Also, the variation of adhesion work and surface free energy have effects on surface wettability behavior[59]. Since, the contact angle results from a competition between the cohesion forces responsible for Surface Free Energy and the adhesion force responsible for $W_{adhesion}$. Besides, this surface wettability behavior improved by codoping may be due to the high adhesive force resulting from theVander Waals forces between the water and the solid surfaces[60].

Furthermore, to study the energy aspect of wettability, we have calculated the surface free energy and the adhesion work of the samples. In Table 23, we have summarized the contact angles and the associated calculated values. It is found that the ZnO:Cr1%:Co1% nanofilm has high surface free energy and adhesion work values compared with the pure ZnO nanofilm. Thus, the hydrophilic character depends on the increase in surface free energy and the work of adhesion which can be explained by the fact that the Van der waals interactions at the sample surface are very important.

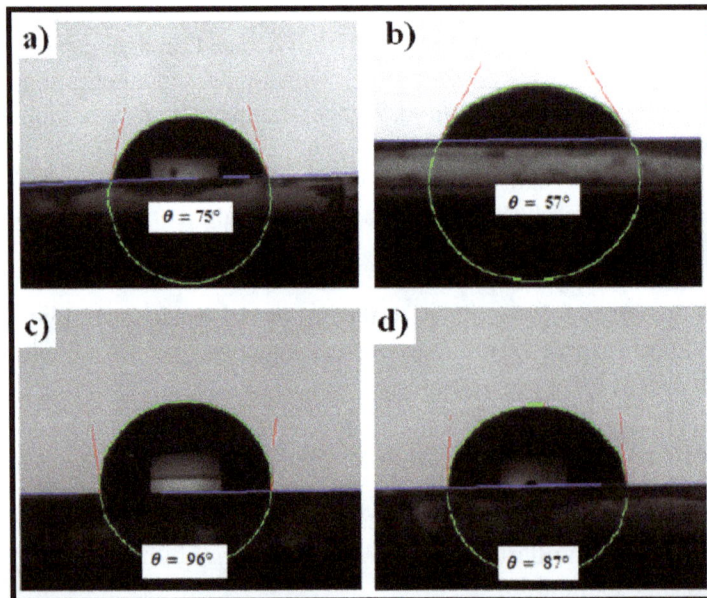

Figure 45. Contact angle measurement of (Cr, Co) codoped ZnO nanofilms:(a) Pure ZnO, (b) ZnO:Cr1%:Co1%, (c) ZnO:Cr2%:Co1% and (d) ZnO:Cr1%:Co2% respectively.

Table 23. Calculated values of Surface Free Energy and the work of adhesion of (Cr, Co) codoped ZnO nanofilms.

Sample	Contact Angle (θ)	Surface Free Energy (mN/m)	$W_{adhesion}$ (mN/m)
Pure ZnO	75	32.3	90.2
ZnO:Cr1%:Co1%	57	45.8	118.4
ZnO:Cr2%:Co1%	96	26.3	65
ZnO:Cr1%:Co 2%	87	31.7	76.4

5.2. Wettability propertiesof (Co, In) codoped ZnO nanofilms

Figure 46 illustrates the shape of water droplet onto different samples, showing the obvious distinction between pure ZnO and (Co, In) codoped ZnO thin films. It is found that codoped ZnO nanofilms present a remarkable hydrophobic character with a CA > 90° compared to pure ZnO film which has a CA = 75°. The same behavior have been obtained by Jamali-Sheini *et al.* for Au–ZnO nanowire films [61]. This result indicates that (Co, In) codoping can improve the hydrophobicity of surfaces which depends on the modification of surface morphology and roughness, the shape and the size as well as the chemical composition. The correlation between AFM observations and hydrophobicity note that when the roughness of the specific surface increases, the hydrophobic character of the surfaces is enhanced especially for ZnO:Co1%:In1% and for ZnO:Co2%:In1% nanofilms.

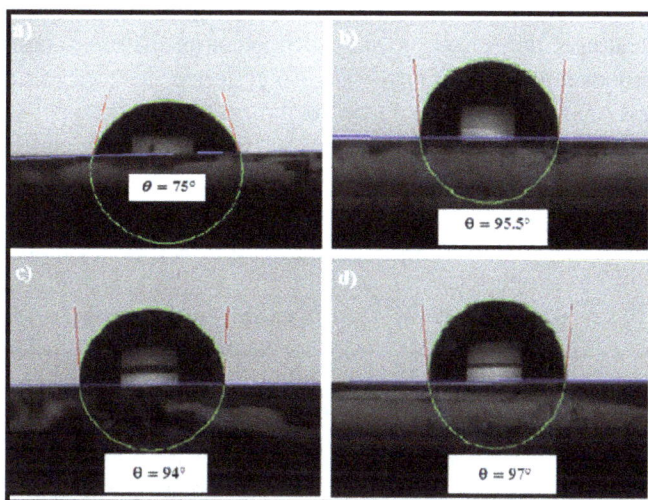

Figure 46. Contact angle measurement of (Co, In)codoped ZnO nanofilms: (a) Pure ZnO, (b) ZnO:Co1%:In1%, (c) ZnO:Co2%:In1% and (d) ZnO:Co1%:In2% respectively.

Also, the variation of adhesion work and surface free energy have effects on surface wettability behavior[59], Since, the contact angle results from a competition between the cohesion forces responsible for Surface Free Energy and the adhesion force responsible for $W_{adhesion}$. Besides, this transformation of the surface nature from hydrophilic to hydrophobic by codoping may be due to the weak adhesive force resulting from theVander Waals forces between the water and the solid surfaces[60].

Furthermore, to study the energy aspect of hydrophobicity, we have calculated the surface free energy and the adhesion work of the samples. In **Table 24**, we have summarized the contact angles and the associated calculated values. It is found that that the codoped ZnO nanofilms have low surface free energy and adhesion work values compared with the pure ZnO nanofilm. Thus, the hydrophobic character depends on the decrease in surface free energy and the work of adhesion which can be explained by the fact that the Van der waals interactions at the sample surface are weak. Hence, from this analysis it is evident that (Co, In) codoped ZnO nanofilms present hydrophobicity, self-cleaning and water repellency properties.

Table 24. Calculated values of Surface Free Energy and the work of adhesion of (Co, In) codoped ZnO nanofilms.

Sample	Contact Angle (θ)	Surface Free Energy (mN/m)	$W_{adhesion}$ (mN/m)
Pure ZnO	75	32.3	90.22
ZnO:Co1%:In1%	95.5	24.8	66.09
ZnO:Co2%:In1%	94	28.6	65.83
ZnO:Co1%:In2%	97	29.6	61.35

5.3. Wettability propertiesof (Cr, In) codoped ZnO nanofilms

Figure 47 illustrates the shape of water droplet onto different samples, showing the obvious distinction between pure ZnO and (Cr, In) codoped ZnO thin films. It is found that codoped ZnO nanofilms present a remarkable hydrophobic character with a CA > 90° compared to pure ZnO film which has a CA = 75°. This result indicates that (Cr, In) codoping can improve the wettability of surfaces which depends on modification of surface morphology and roughness, the shape and the size as well as the chemical composition. Also, the exploitation of AFM observations note that when the roughness of the specific surface increases, the hydrophobic character of the surfaces is enhanced especially for ZnO:Cr1%:In2% film which presents the highest contact angle equal to 110°.

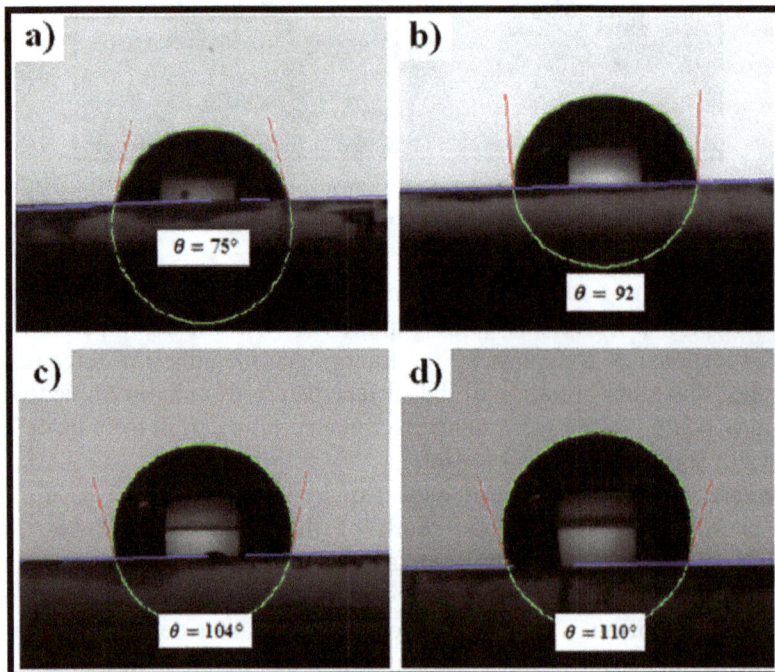

Figure 47. Contact angle measurement of (Cr, In) codoped ZnO nanofilms:
(a) Pure ZnO, (b) ZnO:Cr1%:In1%, (c) ZnO:Cr2%:In1% and
(d) ZnO:Cr1%:In2% respectively.

Also, the variation of adhesion work and surface free energy have effects on surface wettability behavior[59]. Since, the contact angle results from a competition between the cohesion forces responsible for Surface Free Energy and the adhesion force responsible for $W_{adhesion}$. Besides, this surface wettability behavior improved by codoping may be due to the high adhesive force resulting from theVander Waals forces between the water and the solid surfaces[60]. Furthermore, to study the energy aspect of hydrophobicity, we have calculated the surface free energy and the adhesion work of the samples. In **Table 25**, we have summarized the contact angles and the associated calculated values. It is found that the codoped ZnO nanofilms have low surface free energy and adhesion work values compared with the pure ZnO nanofilm. Thus, the hydrophobic character depends on the decrease in surface free energy and work of adhesion which can be explained by the fact that the Van der waals interactions at the sample surface are very important. Hence, from this analysis it is evident that (Cr, In) codoped ZnO nanofilms present hydrophobicity, self-cleaning and water repellency properties.

Furthermore, as shown in Figure 48, the water contact angle on ZnO:Cr1%:In2% surface remains practically constant (is still approximately the same) over time (60 seconds). These results confirm the hydrophobic character of the corresponding surfaces is stable

Table 25. Calculated values of Surface Free Energy and the work of adhesion of (Cr, In) codoped ZnO nanofilms.

Sample	Contact Angle (θ)	Surface Free Energy (mN/m)	Wadhesion (mN/m)
Pure ZnO	75	32.3	90.22
ZnO:Cr1%:In1%	92	29.8	69.81
ZnO:Cr2%:In1%	104	20.09	55.05
ZnO:Cr1%:In2%	110	17.1	47.52

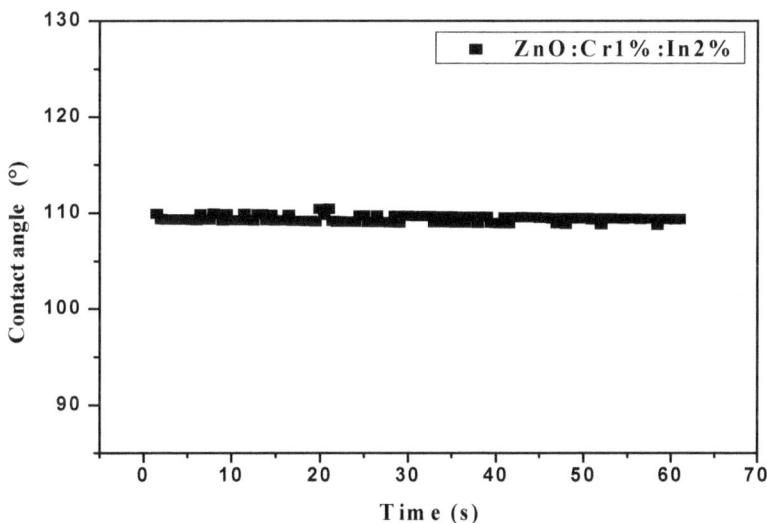

Figure 48. Evolution of water contact angle on ZnO:Cr1%:In2% surface over time.

6. Conclusion and outlook

In summary, we have addressed the issues of the synthesis of the codoped ZnO ultrathin films grown on glass substrates at 460°C by a cost-effective process. Besides the structural and optical investigations, these films (catalysts) showed high photocatalytic activity for the degradation of Methylene Blue dye under sunlight irradiation. Hence, the codoping induces an improvement in the photodegradation efficiency and it is an effective environment remediation for water filtration and purification. Also, hydrophobicity experiments showed that the codoping transforms the surface nature from hydrophilic to hydrophobic. The ZnO:Cr1%:In2% nanofilm presents the highest contact angle equal to110°, hence, it exhibits the best self-cleaning and water repellency, anti-icing and anti-corrosion properties. These results seem so interesting since a low cost-effective method has been used to prepare codoped ZnO nanofilms and pave the way for possible use of such films in various applications such as photocatalysis, protective coating systems.

7. References

[1] S. Rehman, R. Ullah, A. Butt, N. Gohar, Strategies of making TiO 2 and ZnO visible light active, Journal of hazardous materials 170 (2009) 560-569.

[2] M.N.Chong, B.Jin, C.W.Chow, C.Saint, Recent developments in photocatalytic water treatment technology: a review, Water research 44 (2010) 2997-3027.

[3] A. Gupta, J. R. Saurav, S. Bhattacharya, Solar light based degradation of organic pollutants using ZnO nanobrushes for water filtration, RSC Advances 5 (2015) 71472-71481.

[4] D. F. Ollis, Contaminant degradation in water, Environmental science & technology 19 (1985) 480-484.

[5] D.F.Ollis, E.Pelizzetti, N.Serpone, Photocatalyzed destruction of water contaminants, Environmental Science & Technology 25 (1991) 1522-1529.

[6] A.Walther, I.Bjurhager, J.-M.Malho, J.Pere, J. Ruokolainen, L.A.Berglund, O.Ikkala, Large-area, lightweight and thick biomimetic composites with superior material properties via fast, economic, and green pathways, Nano letters 10 (2010) 2742-2748.

[7] P. Guo, Y. Zheng, M. Wen, C. Song, Y. Lin, L. Jiang, Icephobic/Anti-Icing Properties of Micro/Nanostructured Surfaces, Advanced Materials 24 (2012) 2642-2648.

[8] P. Ragesh, V. A. Ganesh, S. V. Nair, A. S. Nair, A review on 'self-cleaning and multifunctional materials', Journal of Materials Chemistry A 2 (2014) 14773-14797.

[9] C.Mondal, M.Ganguly, A.K.Sinha, J.Pal, T.Pal, Fabrication of a ZnO nanocolumnar thin film on a glass slide and its reversible switching from a superhydrophobic to a superhydrophilic state, RSC Advances 3 (2013) 5937-5944.

[10] H. S. Lim, D. Kwak, D. Y. Lee, S. G. Lee, K. Cho, UV-driven reversible switching of a roselike vanadium oxide film between superhydrophobicity and superhydrophilicity, Journal of the American Chemical Society 129 (2007) 4128-4129.

[11] B. Yan, J. Tao, C. Pang, Z. Zheng, Z. Shen, C. H. A. Huan, T. Yu, Reversible UV-light-induced ultrahydrophobic-to-ultrahydrophilic transition in an α-Fe2O3 nanoflakes film, Langmuir 24 (2008) 10569-10571.

[12] A. K. Sinha, M. Basu, M. Pradhan, S. Sarkar, Y. Negishi, T. Pal, Redox-Switchable Superhydrophobic Silver Composite, Langmuir 27 (2011) 11629-11635.

[13] I.Shtepliuk, V.Khomyak, V.Khranovskyy, R.Yakimova, Valence band structure and

optical properties of ZnO 1− x S x ternary alloys, J. Alloys Comp 649 (2015) 878-884.

[14] F. S. Saoud, J. C. Plenet, M. Henini, Band gap and partial density of states for ZnO: Under high pressure, Journal of Alloys and Compounds 619 (2015) 812-819.

[15] B. Ismail, M. Abaab, B. Rezig, Structural and electrical properties of ZnO films prepared by screen printing technique, Thin Solid Films 383 (2001) 92-94.

[16] Ü. Özgür, Y. I. Alivov, C. Liu, A. Teke, M. Reshchikov, S. Doğan, V. Avrutin, S. -J. Cho, H. Morkoc, A comprehensive review of ZnO materials and devices, Journal of applied physics 98 (2005) 041301.

[17] D. P. Norton, Y. Heo, M. Ivill, K. Ip, S. Pearton, M. F. Chisholm, T. Steiner, ZnO: growth, doping & processing, Materials today 7 (2004) 34-40.

[18] K. C. Verma, R. Kotnala, Understanding lattice defects to influence ferromagnetic order of ZnO nanoparticles by Ni, Cu, Ce ions, Journal of Solid State Chemistry 246 (2017) 150-159.

[19] N. Han, X. Wu, L. Chai, H. Liu, Y. Chen, Counterintuitive sensing mechanism of ZnO nanoparticle based gas sensors, Sensors and Actuators B: Chemical 150 (2010) 230-238.

[20] C. G. Van de Walle, Hydrogen as a cause of doping in zinc oxide, Physical Review Letters 85 (2000) 1012.

[21] D.Seghier, H.Gislason, Shallow and deep donors in n-type ZnO characterized by admittance spectroscopy, J. of Materials Science: Materials in Electronics 19 (2008) 687.

[22] V. KUMAR, U. Sitharaman, R. NAGARAJAN, Optical and magnetic properties of (Er, F) co-doped SnO2 nanocrystals, Turkish Journal of Physics 38 (2014) 450-462.

[23] F. Moharrami, M. -M. Bagheri-Mohagheghi, H. Azimi-Juybari, Study of structural, electrical, optical, thermoelectric and photoconductive properties of S and Al co-doped SnO2 semiconductor thin films prepared by spray pyrolysis, Thin solid films 520 (2012) 6503-6509.

[24] W.-F.Chen, P.Koshy, C.C.Sorrell, Effect of intervalence charge transfer on photo-catalytic performance of cobalt-and vanadium-codoped TiO2 thin films, International Journal of Hydrogen Energy 40 (2015) 16215-16229.

[25] V. Jabbari, M. Hamadanian, A. Reisi-Vanani, P. Razi, S. Hoseinifard, D. Villagran, In, V-codoped TiO 2 nanocomposite prepared via a photochemical reduction technique as a novel high efficiency visible-light-driven nanophotocatalyst, RSC Advances 5 (2015) 78128-78135.

[26] M. Kumar, T.-H. Kim, S.-S. Kim, B.-T.Lee, Growth of epitaxial p-type ZnO thin films by codoping of Ga and N, Applied physics letters 89 (2006) 112103-112103.

[27] L.Yu, S. Liu, B. Yang, J. Wei, M. Lei, X. Fan, Sn–Ga co-doped ZnO nanobelts fabricated by thermal evaporation and application to ethanol gas sensors, Materials Letters 141 (2015) 79-82.

[28] K. Boubaker, A. Chaouachi, M. Amlouk, H. Bouzouita, Enhancement of pyrolysis spray disposal performance using thermal time-response to precursor uniform deposition, The European Physical Journal- Applied Physics 37 (2007) 105-109.

[29] R. Laudise, A. Ballman, HYDROTHERMAL SYNTHESIS OF ZINC OXIDE AND ZINC SULFIDE1, The Journal of Physical Chemistry 64 (1960) 688-691.

[30] N. Al-Hardan, M.Abdullah, A.A.Aziz, Performance of Cr-doped ZnO for acetone sensing, Applied Surface Science 270 (2013) 480-485.

[31] S.S.Badadhe, I.Mulla, H 2 S gas sensitive indium-doped ZnO thin films: preparation and characterization, Sensors and Actuators B: Chemical 143 (2009) 164-170.

[32] H.Liu, W.Li, X.Zhang, Y.Sun, J.Song, J.Yang, M.Gao, X.Liu, Comparative study of room temperature ferromagnetism in Cu, Co codoped ZnO film enhanced by hybridization, Ceramics International 41 (2015) 3613-3617.

[33] H. -S. Woo, C. -H. Kwak, I. -D. Kim, J. -H. Lee, Selective, sensitive, and reversible detection of H2S using Mo-doped ZnO nanowire network sensors, Journal of Materials Chemistry A 2 (2014) 6412-6418.

[34] N. Barsan, U. Weimar, Conduction model of metal oxide gas sensors, Journal of Electroceramics 7 (2001) 143-167.

[35] A. El Sayed, S. Taha, G. Said, A. A. Al-Ghamdi, F. Yakuphanoglu, Structural and optical properties of spin coated Zn 1−xCrxO nanostructures, Superlattices and Microstructures 60 (2013) 108-119.

[36] D. Hamby, D. Lucca, M. Klopfstein, G. Cantwell, Temperature dependent exciton photoluminescence of bulk ZnO, Journal of applied physics 93 (2003) 3214-3217.

[37] M. M. Khan, S. A. Ansari, J. Lee, M. H. Cho, Nanoscale 5 (2013) 4427-4435.

[38] N. Han, P. Hu, A. Zuo, D. Zhang, Y. Tian, Y. Chen, Photoluminescence investigation on the gas sensing property of ZnO nanorods prepared by plasma-enhanced CVD method, Sensors and Actuators B: Chemical 145 (2010) 114-119.

[39] N.Ohashi, T.Nakata, T.Sekiguchi, H.Hosono, M.Mizuguchi, T.Tsurumi, J.Tanaka, H.Haneda, Yellow emission from zinc oxide giving an electron spin resonance signal at g= 1. 96, Japanese journal of applied physics 38 (1999) L113.

[40] M. B. Islam, M. M. Rahman, M. Khan, M. Halim, M. Sattar, D. K. Saha, M. Hakim, Spray pyrolized Ag–N co-doped p-type ZnO thin films' preparation and study of their structural, surface morphology and opto-electrical properties, Thin Solid Films 534 (2013) 137-143.

[41] A. Hadri, M. Taibi, C. Nassiri, T. S. Tlemçani, A. Mzerd, Development of transparent conductive indium and fluorine co-doped ZnO thin films: Effect of F concentration and post-annealing temperature, Thin Solid Films (2015).

[42] D. Zhang, Q. Wang, Z. Xue, Photoluminescence of ZnO films excited with light of different wavelength, Applied Surface Science 207 (2003) 20-25.

[43] G. H. Mhlongo, D. E. Motaung, S. S. Nkosi, H. Swart, G. F. Malgas, K. T. Hillie, B. W. Mwakikunga, Temperature-dependence on the structural, optical, and paramagnetic properties of ZnO nanostructures, Applied Surface Science 293 (2014) 62-70.

[44] K. Vanheusden, W. Warren, C. Seager, D. Tallant, J. Voigt, B. Gnade, Mechanisms behind green photoluminescence in ZnO phosphor powders, Journal of Applied Physics 79 (1996) 7983-7990.

[45] P.K.Samanta, P.R.Chaudhuri, Substrate effect on morphology and photoluminescence from ZnO monopods and bipods, Frontiers of Optoelectronics in China 4 (2011) 130-136.

[46] S. Baruah, S. S. Sinha, B. Ghosh, S. K. Pal, A. Raychaudhuri, J. Dutta, Photoreactivity of ZnO nanoparticles in visible light: Effect of surface states on electron transfer reaction, Journal of Applied Physics 105 (2009) 074308.

[47] I. Arslan, I. A. Balcioglu, D. W. Bahnemann, Heterogeneous photocatalytic treatment of simulated dyehouse effluents using novel TiO 2-photocatalysts, Applied Catalysis B: Environmental 26 (2000) 193-206.

[48] D. Heger, J. Jirkovský, P. Klan, Aggregation of methylene blue in frozen aqueous solutions studied by absorption spectroscopy, The Journal of Physical Chemistry A 109 (2005) 6702-6709.

[49] C.Xu, L.Cao, G.Su, W.Liu, X.Qu, Y.Yu, Preparation, characterization and photo-catalytic activity of Co-doped ZnO powders, J Alloys Comp 497 (2010) 373-376.

[50] S. Dong, K. Xu, J. Liu, H. Cui, Photocatalytic performance of ZnO: Fe array films under sunlight irradiation, Physica B: Condensed Matter 406 (2011) 3609-3612.

[51] S. G. Kumar, K. K. Rao, Comparison of modification strategies towards enhanced charge carrier separation and photocatalytic degradation activity of metal oxide semiconductors (TiO 2, WO 3 and ZnO), Applied Surface Science 391 (2017) 124-148.

[52] F. Achouri, S. Corbel, L. Balan, K. Mozet, E. Girot, G. Medjahdi, M. B. Said, A. Ghrabi, R. Schneider, Porous Mn-doped ZnO nanoparticles for enhanced solar and visible light photocatalysis, Materials & Design 101 (2016) 309-316.

[53] A.Houas, H.Lachheb, M.Ksibi, E.Elaloui, C.Guillard, J.-M. Herrmann, Photocatalytic degradation pathway of methylene blue in water, Applied Catalysis B: Environmental 31 (2001) 145-157.

[54] J.-C. Sin, S.-M. Lam, K.-T. Lee, A.R. Mohamed, Preparation of rare earth-doped ZnO hierarchical micro/nanospheres and their enhanced photocatalytic activity under visible light irradiation, Ceramics International 40 (2014) 5431-5440.

[55] S.-M. Lam, J.-C. Sin, A. Z. Abdullah, A. R. Mohamed, Green hydrothermal synthesis of ZnO nanotubes for photocatalytic degradation of methylparaben, Materials Letters 93 (2013) 423-426.

[56] T. Young, An essay on the cohesion of fluids, Philosophical Transactions of the Royal Society of London 95 (1805) 65-87.

[57] D. Packham, Work of adhesion: contact angles and contact mechanics, International journal of adhesion and adhesives 16 (1996) 121-128.

[58] M. Żenkiewicz, Methods for the calculation of surface free energy of solids, Journal of Achievements in Materials and Manufacturing Engineering 24 (2007) 137-145.

[59] L. Jiang, R. Wang, B. Yang, T. Li, D. Tryk, A. Fujishima, K. Hashimoto, D. Zhu, Binary cooperative complementary nanoscale interfacial materials, Pure and applied chemistry 72 (2000) 73-81.

[60] A. K. Geim, S. Dubonos, I. Grigorieva, K. Novoselov, A. Zhukov, S. Y. Shapoval, Microfabricated adhesive mimicking gecko foot-hair, Nature materials 2 (2003) 461-463.

[61] F. Jamali-Sheini, R. Yousefi, K. Patil, Surface characterization of Au–ZnO nanowire films, Ceramics International 38 (2012) 6665-6670.

www.ingramcontent.com/pod-product-compliance
Lightning Source LLC
Chambersburg PA
CBHW061347210326
41598CB00035B/5907